PEARSON ALWAYS LEARNING

Richard P. Waldren

Introductory Crop Science

Seventh Edition

Cover Art: Courtesy of Comstock

Pearson Learning Solutions, 501 Boylston Street, Suite 900, Boston, MA 02116
A Pearson Education Company
www.pearsoned.com

Printed in the United States of America

4 5 6 7 8 9 10 0BRV 17 16 15 14

000200010271732586

CG

PEARSON

ISBN 10: 1-269-61540-8
ISBN 13: 978-1-269-61540-2

Contents

TEXTBOOK

PART III. AGRONOMIC CALCULATIONS

PART IV. PLANT IDENTIFICATION

Preface

This book contains basic information needed by students to learn the principles and practices of crop science. It is both a textbook and a laboratory manual, and is written to give an instructor flexibility in specific course content. It does not go into great detail on the production of individual crops. Instead, it uses specific crops as examples to illustrate the principles being discussed. This allows the instructor to concentrate on those crops that are important in a particular area, and still use the book to its full capacity.

The text provides information important to the study of crop science. The contents of the textbook are divided into three general areas. Part I is introductory and covers the historical development of crop production, the terminology used in studying crops, the relationships of crops to the environment, and cropping systems. Part II discusses the "science" of plants and soils, including an introduction to soil science, and the botany and physiology of crop plants. Part III discusses environmental factors that affect crop plants including climate and weather, crop pests, crop diseases, and weeds.

The laboratory exercises are designed to provide hands-on study of crop science. The exercises are divided into four general areas. Part I studies the botany of crop plants. Part II studies plant growth and development. Part III provides practice in mathematical calculations needed in crop management. Part IV covers identification of important crops, forage and range plants, and weeds.

Each chapter and exercise begins with instructional objectives to aid students in comprehending the material covered. Chapters end with study questions that can be used by students and/or the instructor to further the understanding of the subject.

The author thanks several graduate assistants, including Roger Stockton, Patricia Boehner, and Patrick O'Neill, for providing ideas for improving the laboratory exercises, and Sue Waldren for reading the manuscript for grammatical content.

PART I

INTRODUCTION

CHAPTER ONE

Importance and Development of Agriculture

INSTRUCTIONAL OBJECTIVES

Upon completion of this chapter, you should be able to:

1. Discuss the importance of plants to people and animals. List at least six uses of plants.
2. Discuss the development of agriculture from arable to commercial.
3. Define the following types of agricultural systems: arable, pastoral, shifting, sedentary, subsistence, and commercial. Give an example of each.
4. Define intensive and extensive cropping. Discuss the differences between them regarding level and type of inputs and production.
5. Discuss the distribution of crop land in the world and the United States.
6. List the three major crops grown in the world. List at least three regions or countries where each is grown extensively.
7. List the three major crops grown in the U.S.
8. List at least four other crops. Give an example where each is commonly grown.
9. Discuss ways that increased demand for crops can be met in the future.

INTRODUCTION

The green plant is essential to all forms of life on earth and is especially beneficial to human life. Plants are the source of our food, fiber, livestock feed, extracts and derivatives, many of our construction materials, and a high percentage of our energy needs. Although not always as obvious as the items mentioned above, plants serve important functions for people in the conservation of natural resources and pollution control, and are used extensively for landscaping, beautification, and recreation.

Plants, through the process of **photosynthesis,** convert the radiant light energy of the sun into chemical energy that is stored in plant roots, stems, leaves, fruits, and grains. We use this chemical energy when we eat plant products directly or indirectly as meat, eggs, poultry, and dairy products. Even most fish products are indirectly derived from aquatic plants.

The importance of plants, both directly and indirectly, as a food source assumes a very serious note when we realize that one half of the world's people are hungry, two thirds have a poorly balanced diet, and approximately 25,000 persons die of starvation daily. The improvement, production, distribution, and processing of plant products can alleviate this misery and suffering in the world.

USES OF PLANTS

Cotton and flax still are the major fibers used in clothing; and even with the use of synthetic fibers, most items of clothing are a combination of natural and synthetic fibers. Hemp and jute are used for cordage and burlap while wood and other fibers are used in the manufacture of paper. Hardwoods, softwoods, and bamboo are used as construction materials throughout the world, and in many nations wood is an important fuel source for cooking and heating.

Livestock feed, both forages and concentrates, are derived almost exclusively from plants. In the United States, more than 80% of the corn, sorghum, oats, and barley are fed directly or through processed feed to cattle, hogs, sheep, and poultry. Forages, either as range, pasture, hay, or silage, are used extensively in the diets of ruminant animals. Certain land areas, due to steep topography, shallow soil, or climatic conditions, are suitable only for grass production and this grass can be harvested only by animals.

Oils from soybean, canola, cottonseed, and corn are extracted from seeds and used as cooking oils, for the manufacture of margarine, and in the baking industry. Peanut and sesame oils are important cooking oils in some countries. Plant oils from castor, olive palm, tung, and flax seeds are used industrially in paints, lacquers, plastics, soaps, detergents, and lubricants. The leaves of tea, and seeds of coffee and cacao are used for beverages. Spices, flavorings, and perfumes from plants are used to add zest to our food and bodies. Many medicines, both old and new, are derived from plants.

Plants are used to prevent or control wind and water erosion, to increase the soil organic matter content, and to improve the water intake and absorption capacity of soils. Plants provide food and cover for all forms of wildlife through all seasons of the year. Through a recycling process, plants help maintain the soil nutrient supply so that nutrients are not leached out of the root zone. Plant cover protects the soil and soil organisms from the extremes of temperature and other adverse effects of weather. The establishment of some plant growth is the first step in reclaiming land destroyed by erosion, desertification, or mining operations.

Plants play a subtle but important role in the control and reversal of pollution. Microscopic soil plants, such as fungi and bacteria, decompose waste products as well as insecticides and herbicides which are applied to the soil. Barriers of trees, shrubs, and ornamentals greatly reduce noise levels in urban areas, around airports, and along highways. Plants also provide some protection from air pollution by filtering out particulate matter (solids) from the atmosphere. In the previous paragraph, the effect of plants in erosion control was mentioned. Plants anchor the soil and reduce the amount of sediment in waterways and dust in the air.

Plants provide artistic beauty to our homes, parks, recreation areas, and the landscape in general. Flowers, shrubs, trees, and grass give a variety of colors, shapes, and tones that make our surroundings more pleasant and livable. The earth is dominated with plants that provide a variety of functions and purposes for people. However, in this book we will confine our interests to crop plants. These are plants that are managed in some manner or degree, and are produced primarily for their economic return.

DEVELOPMENT OF AGRICULTURE

Crop production is said to have begun when people first learned about seed collection. They discovered, probably by accident, that seeds placed in the soil grew plants and produced more seeds like the ones planted. Before this time people were **hunters and gatherers** and depended entirely on the bounty of nature. If weather conditions were favorable, game and grains were plentiful and food was abundant. Conversely, if weather conditions were not conducive for the production of feed for wild animals, and seeds, nuts, and fruits for people, the standard of living declined for a year or two and in some extreme cases led to periods of famine.

As people began to domesticate both plants and animals, their level of subsistence changed from a hunter and gatherer to a more formal type of agriculture, arable and pastoral. **Arable agriculture** is the domestication and cultivation of crops; and **pastoral agriculture** is the domestication and husbandry of animals. With arable agriculture, people began to select and propagate the best seeds and cultivate them. The first cultivated food grain crops are believed to have been wheat and barley in the Eastern Mediterranean region about 10,000 years ago. Teosinte, the precursor of corn, was domesticated in Central America about 8,000 years ago. Other crops, vegetables, and fruits were added over the years as expertise in cultivation and propagation increased. Pastoral farming involved the grazing of animals to provide food and easier transportation. In all probability, early agriculture involved a combination of arable and pastoral farming, but pastoral tended to move to land areas unsuitable for crop production, such as those that were too dry, too steep, infertile, or too far away for easy access for crop tending.

Land was cultivated or grazed until it was no longer productive and then the farmers moved to new areas. The period of productivity depended on the inherent level of soil productivity (fertility) and the intensity of cultivation or grazing. Arable agriculturists commonly remained three or four years in an area before a move was necessary while pastoral agriculturists may have had to move daily or weekly to find new pasture. This **shifting** type of production gradually gave way to **sedentary agriculture** as skill and knowledge of farming were acquired. Beyond seed collection and selection, farmers learned that yields could be maintained if crops were rotated or planted in a sequence. They found that cereals planted after legumes produced higher yields than when planted after the same or another cereal grain. Wood ashes, animal manures, and crop residues were found to have fertilizing effects on crops. With these discoveries people found that they no longer needed to move to new land areas but could settle permanently in one area. Also, the increased levels of production meant that not everyone was needed for food production. These factors enabled people to begin settling into villages and cities and agricultural trade began.

It is likely that about this time agriculture changed from a **subsistence** to a **commercial** level. Subsistence farmers produced only enough food for the needs of their own family, or the extended family (tribe). Although engaged in arable agriculture, the environment or the circumstances dictated that the family lived well in good years and poorly in bad years. In subsistence agriculture, production was usually planned to meet the family's needs, and if a surplus was produced, it was viewed as a bonus, not an expectation. Probably, if one farmer produced a surplus in one year, most of the farmers in the area also produced a surplus in the same year. Without adequate roads and markets, the surplus production was of little value to the community who consequently saw little reason to progress beyond subsistence farming.

On the other hand, the objective of commercial agriculture is the deliberate creation of surplus to be sold for a profit. The development of commercial agriculture depends on a system of markets, a network of transportation, and some form of storage and processing. The produce of commercial farming is moved regularly and systematically to food consuming or food processing centers at a profit to the producer and the processor. Without profits or rewards the system fails.

Commercial agriculture may be either **extensive** or **intensive**. The objective of extensive agriculture is to obtain maximum returns from each unit of labor invested. Yield per unit of land area may be low but so are production costs. Wheat farming in the Great Plains of the United States, or cattle and sheep grazing on low producing rangelands are examples of extensive agriculture. Conversely, intensive agriculture expects to maximize profits through investments of large amounts of purchased inputs and labor. Yields per unit land area are high and management techniques are very sophisticated. Fruit and vegetable production are the most commonly given examples of intensive agriculture, but many other types of crop production are also very intensive. Production of tobacco, cotton, and irrigated rice and corn are intensive as is a corn soybean production system of the Central Corn Belt. Livestock production using a system of rotational irrigated pastures would also be classified as intensive. As with any system of classification, the assigned category is one of degree and is often dependent upon the whim of the classifier as we will discuss in later chapters. However, good system management is essential regardless of the level of production.

These steps involved in the development of agriculture have all occurred in many developed nations of the world, but each stage still exists somewhere in the world today. There are tribes in the bush of Africa who are hunters and gatherers. Natives in the Amazon region of South America practice "slash and burn" shifting agriculture in which they clear an area in the tropical forest for cultivation and then clear another area when that area can no longer produce a crop. The original area then grows back to native vegetation. Subsistence agriculture is common in many areas of Asia, Africa, and South America, while commercial agriculture is typical of Europe and many areas of North and South America and the developed areas of the Orient.

The development of agriculture in any particular region of the world depends on a combination of climatic and **edaphic** (soil) factors. Figure 1-1 shows the native vegetation for North America. The most productive grain crop production occurs in those regions that originally developed into grassland or prairie. The combination of climate, soil, and topography which results in grassland development are

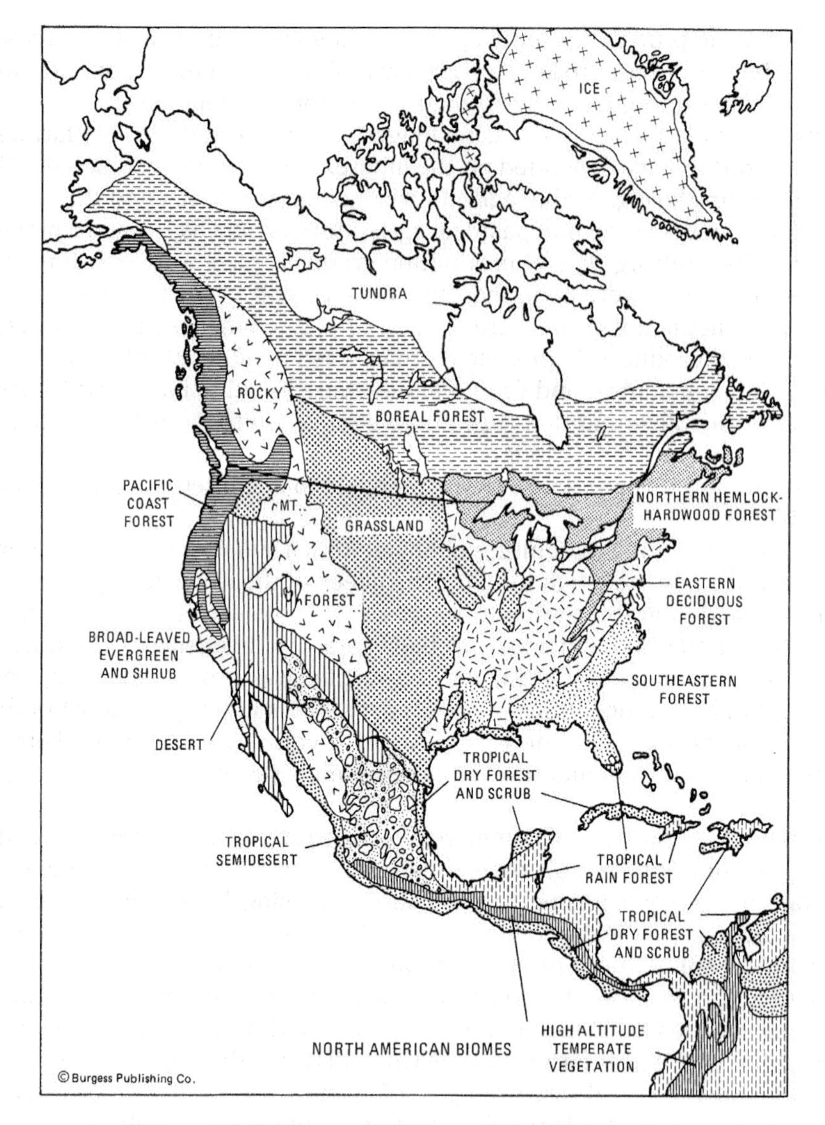

Figure 1-1 Native vegetation of North America.

also the most desirable for grain crops as they are also grasses. Other regions can be developed for crop production if climate and topography allow. However, they may require special practices such as irrigation, drainage, or removal of trees and shrubs.

WORLD AGRICULTURE TODAY

World-wide, agriculture remains an important economic activity, providing employment for a significant portion of the world's population as shown in Figure 1-2. Note that a significantly higher percentage of the population is directly involved in agriculture in lesser developed areas while only a small percentage is involved in regions with more developed agriculture. The percentages of agricultural laborers are declining world-wide.

Figure 1-3 shows that only about 11 percent of the total earth land surface have the combination of precipitation, temperature, topography, and soil factors that make it economically suitable for field crop production. As human population continues to increase, food is being produced by fewer persons on less land. Without agricultural technology our planet would soon experience world-wide food shortages.

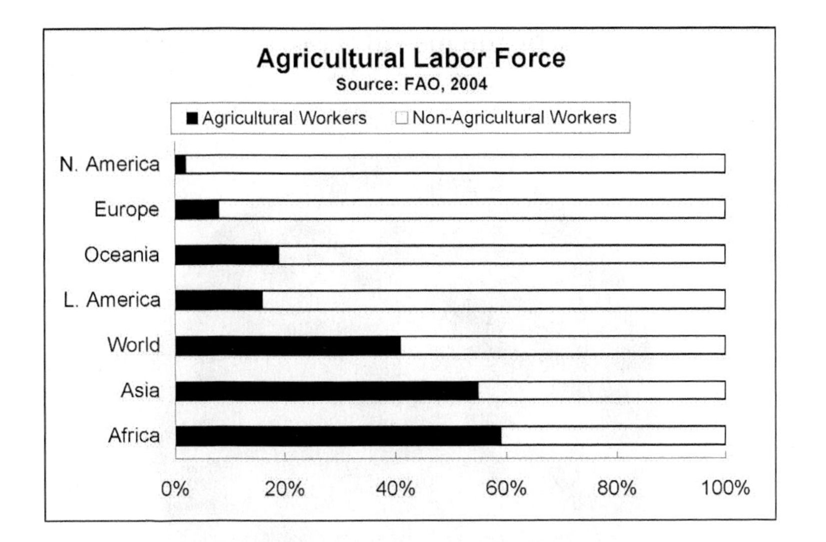

Figure 1-2 Percent of labor force active in agriculture.

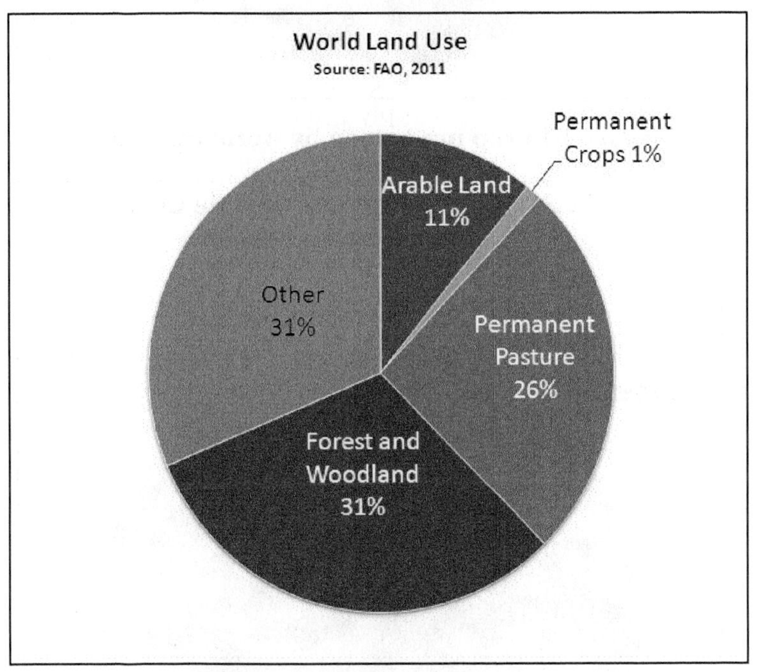

Figure 1-3 World land use.

Cropland is unequally distributed among the regions of the world with nearly one-half occurring in the countries of China, India, Russia and the United States. This cropland is not equally productive due to differences in climate, soils, and stage of agricultural development. Figure 1-4 shows the contribution of each region to world crop production.

In the United States, agriculture is also a very important economic enterprise. Although less than 3% of the population is directly involved in farming or ranching, about 25% of the population is indirectly involved in the processing, transportation, and marketing of agricultural commodities, and providing services and supplies to agricultural producers.

World Crop Production

Rice, wheat, and corn are the major world crops both in land area harvested and tons of production as shown in Figure 1-5. Note that about 50% of total crop production is wheat, rice, corn, sorghum, millet, and soybean.

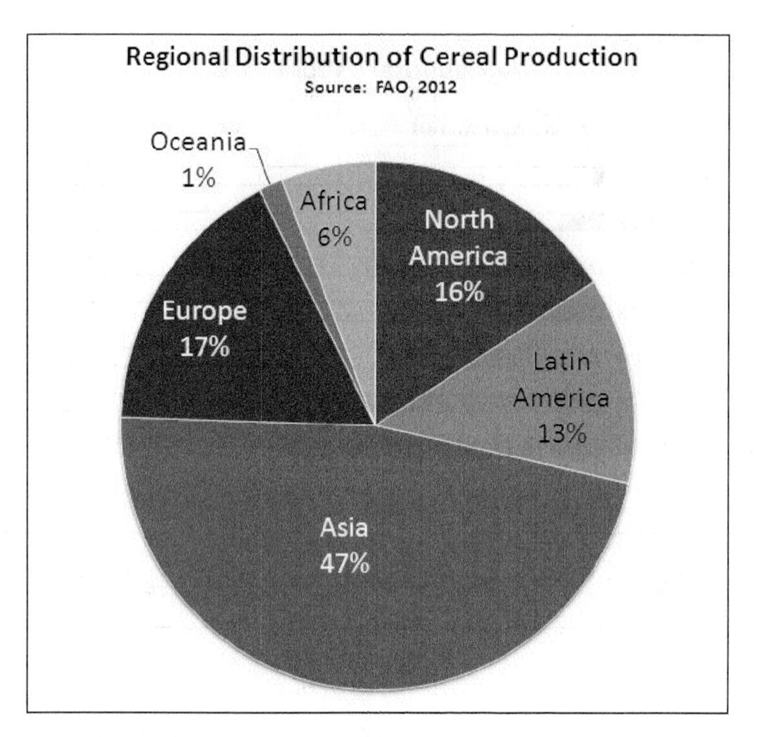

Figure 1-4 Crop production by world regions.

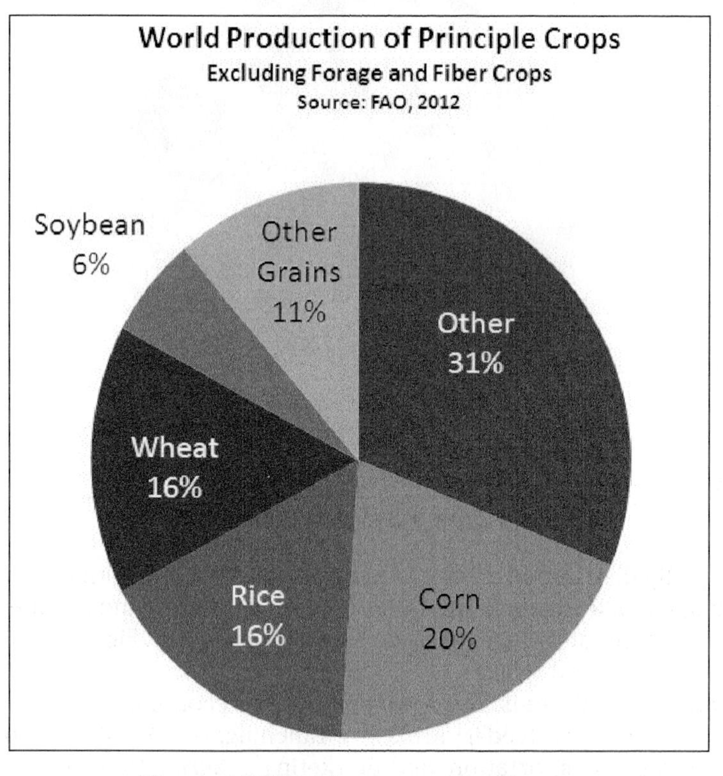

Figure 1-5 World crop production.

Rice is an important food grain and is the staple food of the greatest percentage of the world's population, primarily in Southern, Eastern, and Southeastern Asia. Rice is consumed by more humans than wheat and corn combined. China produces one third of the world total, with the combined production of India and China accounting for over one half of the world production of rice. About 75 percent of the rice grown in Asia is irrigated.

Wheat is grown extensively in the temperate zones of the world and ranks first in the total land area seeded to cereal grains. It has been estimated that wheat is being harvested somewhere in the world every month of the year. Although the region that was the Soviet Union seeds about one fourth of the land area occupied by wheat, the yields are commonly low and only adequate for domestic needs. The United States is the fourth largest wheat producing country, behind China, India, and Russia.

Corn is grown throughout North America, South America, Southern Europe, and some areas of Africa and Asia. Although the crop is used primarily in the United States as a feed grain for livestock, corn is an important food grain in many parts of the world. The U.S. produces nearly one-half of the world's total on one-fourth of the world's land planted to corn. Corn production in the U.S. is a highly specialized enterprise and yields are high. China and Brazil rank second and third in world production. Corn production is also an important crop to Argentina, South Africa, and many Balkan countries of Europe.

Although rice, wheat, and corn are the major crops from a world trade standpoint; sorghum, millet, rye, oats, barley, and potatoes are important food and feed crops in many countries. **Sorghum** and **millet** are important grain crops in Asia and Africa. Sorghum is a major food crop for India, Nigeria, Niger, Mali, and Sudan and is also grown extensively in the U.S., and Mexico. Millet is a short season, water efficient crop that is grown in the drier areas of India, China, and many countries in Africa. **Barley** is also grown in the drier areas of subhumid and semiarid regions of the world. Barley is planted throughout North Africa, the Soviet Union, and China but on a very limited area in the United States.

Oats are best adapted to the cool, moist areas of the world. Russia plants about 40% of the total land area in oat production, followed by Canada and Poland. In the United States, over 60% of the oats is grown in the North Central States of South Dakota, Minnesota, North Dakota, and Iowa. Oats is used primarily as feed for horses, swine, poultry, and cattle. Only a small percentage is used for human food, primarily breakfast foods. **Potatoes** are generally adapted to the same climatic areas as oats and are an important food source in many northern European countries. Besides white potatoes, sweet potatoes, yams, and cassava are grown for food in the subtropical and tropical areas of Africa, Asia, and South America.

Oilseed, oil nut, sugar crops, cotton, and tobacco are important in many local agricultural economies and form an important sector of world trade. Fats and oils from oilseed and oil nut crops now account for over one-half of the world's supply. Although the world demand for fats and oils has been steadily increasing, oils from plant sources are slowly replacing some animal fats and oils. **Soybean, peanut, cottonseed**, and **palm oils** are the major world sources. However, **sunflower, sesame, canola** (rape seed), **linseed** (flax), **castor**, and **coconut oils** are locally important in some areas of the world. A significant by-product of the oilseed industry, used for livestock and human food, is a high protein cake or meal that is left as a residue after the oil has been extracted.

Sugar crops include sugarcane, sugarbeet, and sweet sorghum. **Sugarcane** is produced in the tropical and subtropical climates of the world and supplies most of the world sugar supply. In the United States, about 80% of the sugarcane is produced in Louisiana and Florida. **Sugarbeet** is most productive in the northern half of the temperate zone. Since it has a high water requirement, the crop is commonly grown under irrigation. Minnesota leads in the production of beet sugar.

Cotton is the most important fiber crop based on world production. It requires a long growing season so production is limited to subtropical and tropical climates. India and China are the major world cotton producers, with the U.S. ranking third. India has the largest land area devoted to cotton.

Crop Production in the United States

The three major crops grown in the United States are corn, wheat, and soybean. As shown in Figure 1-6, corn is the number one crop in production followed by soybean and wheat.

Corn is extensively grown in the Midwest, from Nebraska to Ohio. Some corn is grown in practically all major agricultural states. Soybean is grown extensively in the Midwest and throughout the Southern and Southeastern states. Wheat is grown extensively in the Plains States from Northern Texas through North Dakota, and in parts of Washington and Oregon.

Sorghum is primarily grown in the drier areas of the Southern Plains from Texas through Kansas. Oats is grown in the Northern Plains from Nebraska through North Dakota. Barley and rye are grown mostly in the Plains States.

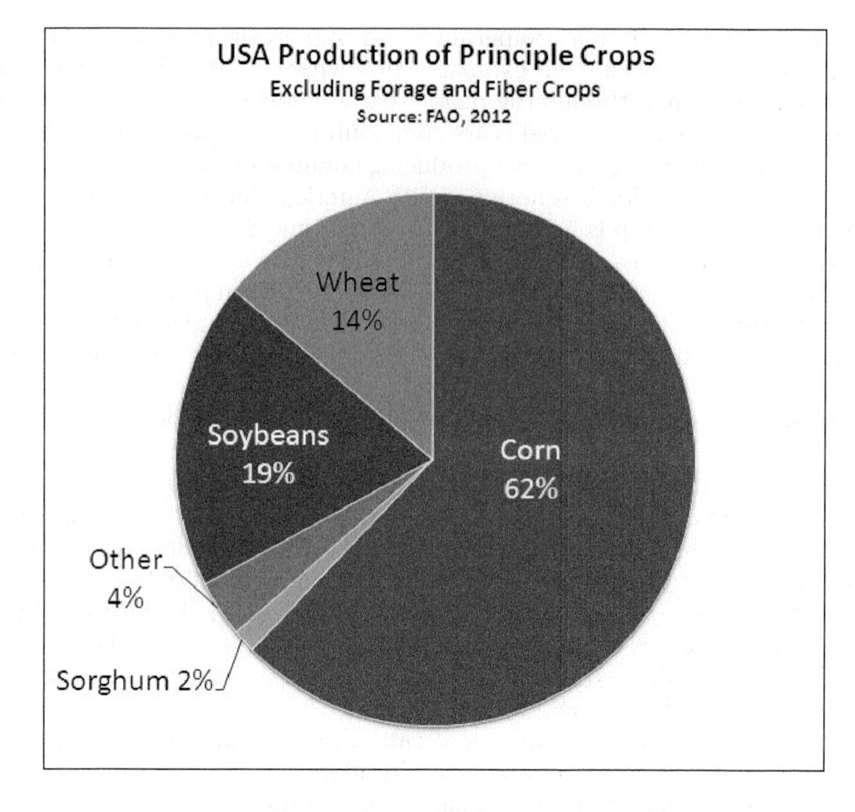

Figure 1-6 Crop production in the U.S.

MEETING THE NEEDS OF THE FUTURE

As our world's human population continues to increase agriculture will need to greatly increase food production. The most common estimates predict that grain and seed production will need to increase about 70 percent by 2050 to keep up with a combination of increasing population and better diets through better standards of living. If we are to meet this increased demand for food and prevent widespread food shortages, agronomists will need to look at new ways to increase productivity. The most promising possibilities are improved genetics of existing major crop plants using biotechnology, and looking to alternative crops to make up the difference.

Improved plant genetics using biotechnology such as genetic engineering offers excellent opportunities to increase food production. With biotechnology we can design plants to fit specific conditions within a region, increase natural defenses against pests, and improve food value by increasing specific dietary requirements. Genetic engineering is discussed in more detail in Chapter 11.

Even with the potential offered by biotechnology, we will not likely meet the needs of people in the poorer, developing countries. The answer may lie in increased research on native crops within a region. For example, there are many grain crops native to Africa that are being grown by people there that have not been the subject of much research. Examples are pearl millet, kenaf, tef, and African rice. Since most research is directed toward the "mainline" staple crops of rice, wheat, corn, and a few other grains these alternative crops still remain to be exploited. One crop native to Africa that shows great potential is pearl millet. Its agronomic and genetic development today is similar to that of corn 50 years ago.

With continued research it will be possible for crop production to continue to keep pace with population growth well into the 21st Century. But, it will be necessary to look to different crops and new technology to meet the demand.

REVIEW QUESTIONS

1. In what ways do we directly and indirectly depend on plants for our life? Are there any substitutes?
2. How did early agriculture differ from agriculture today? What practices are still the same? Can we change these?
3. Which is better, intensive or extensive cropping? Why?
4. What factors determine where a particular crop is grown? Which of these can we control or influence?
5. What crops are commonly grown in our area? What other crops could be grown? Why aren't they?
6. How can crop production keep up with increasing demand from higher population and better living standards?

CHAPTER TWO

Crop Terminology

INSTRUCTIONAL OBJECTIVES

Upon completion of this chapter you should be able to:

1. Define eight types of classification systems for crops. Give an example of each.
2. Define grain crop, cereal crop, small grain, feed grain, and food grain. Give an example of each.
3. List at least three different classifications each for corn, soybeans, and wheat.
4. Define forage crop, pasture crop, range crop, hay crop, silage crop, and soiling crop. Give an example of each.
5. Define oil crop, pulse crop, root crop, tuber crop, sugar crop, and fiber crop. Give an example of each.
6. Define horticultural crop, vegetable crop, orchard crop, small fruit crop, and ornamental crop. Give an example of each.
7. Define cover crop, catch crop, companion crop, green manure crop, and trap crop. Give an example of each.
8. Outline the life cycle of a summer annual, winter annual, biennial, and perennial plant. Give an example of a crop for each.
9. Define tropical, subtropical, temperate, and boreal plants. Name the climatic adaptation of most crops.
10. Define warm-season and cool-season plants. Give an example of each.
11. Define row crop, close-seeded crop, and pasture or hay crop. Give an example of each. Discuss the conservation implications of each.
12. Define herbaceous and woody stems. Discuss differences in their growth habit and digestibility.
13. Define hydrophyte, mesophyte, xerophyte, and halophyte. Give an example of each. Name the water adaptation of most crops.
14. Define genus, species, and cultivar. Explain their significance in plant classification.
15. Name the two most important botanical families of crops. Give two examples of each.
16. Define monocot and dicot. Give an example of each.

INTRODUCTION

The terminology associated with crop plants is based on different types of classification. Plants can be classified using many different systems, depending on the need and purpose of the person or group doing the classifying, or using the classification. Probably the most common and well-known classification system is the **botanical classification system** that is based on morphological differences in stems, leaves, and/or flowers, or flower arrangement. However, for the sake of convenience, crop plants are often categorized by agronomists and farmers according to crop use, cultivation practices, environmental adaptation, or life cycle.

In this chapter we want to study the different types of **agronomic classifications** as well as the botanical classification system. The agronomic systems are not as exact as the botanical system but they are very convenient and serviceable to the user. Also you will notice that there are many exceptions to the rules in the agronomic classification systems. These exceptions are not meant to confuse you but do illustrate the versatility of crops and cropping systems.

CROP USE

Grain Crops

Grain crops are those crops that are grown for their dry edible seeds or fruits and are sometimes also called cereal crops or cereal grains. Technically, all grain crops are grasses, but sometimes crops, such as flax and soybeans, are also called grain crops.

A **cereal crop** is any grass grown for its edible fruit or grain. The term can refer to either the entire plant or to the grain itself. Although buckwheat is not a grass, it is usually classified as a cereal since its grain is used like the true cereals; an example of the exceptions mentioned before.

A **small grain** is a collective term used to include wheat, rye, oats, and barley. These crops are relatively short plants with small seeds as compared to other grain crops such as corn and sorghum. Small grains can be further classified as **fall-seeded small grains** such as winter wheat, winter barley, winter oats, or rye or as **spring-seeded small grains** such as spring wheat, spring barley, or spring oats, depending on their life cycle.

Feed grains are cereal crops grown for their seeds or fruits to be fed to livestock. These crops are sometimes called **coarse grains** in the commodity market. The most common feed grain in the U.S. is corn followed by grain sorghum. Other common feed grains are oats, barley, rye, buckwheat, and millet. Although these are the most common feed grains in the U.S., keep in mind that any grain crop fed to livestock can be classified as a feed grain. Sometimes market and economic conditions will result in different crops being fed to livestock when they normally would be used for other purposes. Wheat is a good example of this.

Food grains are cereals grown for the human consumption of their edible seeds or fruits. The most common food grain in the U.S. is wheat. Other common food grains are rice, rye, corn, and oats. The use of food grains varies throughout the world. Rice is used in the greatest quantities as a food grain over the world, but wheat is grown on the most acres, and is the most common in many parts of the world particularly the Orient and Asia. Corn is the most common food grain in Latin America, and sorghum and millet are common in Africa.

Forage Crops

Forage crops are grown for their vegetative matter (stems and leaves) as a feed for livestock, primarily ruminants. They should not be confused with feed grains in which only the grain or seed of the plant is used for animal feed.

A **pasture crop** is forage that is harvested directly by livestock through grazing. This classification usually includes only those crops that have been seeded into an area and may include annuals, such as sudangrass or perennials, such as smooth bromegrass. Pasture crops can be grasses, or legumes, or mixtures of both, such as an alfalfa-bromegrass pasture. Pasture crops may be either permanent or temporary vegetation on the land they occupy. A **range crop** is similar to a pasture crop except that it is composed of only native perennials that are permanent in nature. Management practices such as fertilization or controlled grazing may or may not be used on pasture or range crops.

A **hay crop** is forage that is cut while still green, allowed to dry in the field, processed, stored, and fed to livestock. Processing can include packing the hay into round or square bales, or gathering it loose and storing it directly into stacks or piles. Square bales usually are stacked to prevent excessive spoilage. It is necessary to dry hay to about 10% moisture or less to retain its quality and prevent losses from molding. When gathering and processing hay, it is important to retain as many of its leaves as possible to maintain nutritional quality.

A **silage crop** is forage that is harvested in a green, succulent condition and then stored under anaerobic conditions where controlled fermentation occurs. In order for anaerobic conditions to occur, oxygen must be excluded from the crop during storage. This can be accomplished by either storing it in an airtight silo or by packing it to remove the air. A silage crop is usually finely chopped during harvest to aid packing and develop anaerobic conditions during storage. When properly processed and stored, silage is an excellent animal feed.

There are two different methods of silage harvest. It can either be harvested and directly stored as **high moisture silage**, or cut and allowed to partially dry before being chopped and stored as **low moisture silage**. High moisture silage is usually about 60-70% water and low moisture silage is usually about 40% water. Any forage crop can be used for silage but the most common is corn, followed by alfalfa.

A **soiling crop** is forage that is harvested in a green, succulent condition and fed directly to livestock. It is very similar to silage except that it is not fermented and stored. Any crop that can be used as a silage crop can also be used as a soiling crop. Usually the same harvesting equipment used for silage is also used to chop a soiling crop. Another common name for a soiling crop is **green chop**.

Oil Crops

An **oil crop** is a crop grown primarily for its oil content. Plant oils are primarily used for food processing, and in vegetable oils and shortening. Other common uses are lubricants and industrial processing. Some plant oils have a higher smoke and flash point than conventional petroleum-based lubricants which makes them better suited for some industrial applications.

The most common oil crop in the U.S. is soybean. It accounts for more oil than all the other oil crops combined. Soybean oil is the main ingredient in most vegetable oils sold in grocery stores. Another common oil crop is sunflower, which is the main oil crop in some parts of the world, particularly Russia. Peanut oil is used mostly as industrial cooking oil because it has a higher smoke point than soybean or sunflower oil. Canola oil use is growing rapidly due to its health benefits. Other common oil crops in the U.S. include cottonseed used in margarine, and corn and safflower.

Linseed oil is obtained from flax seed and is used in paints and varnishes. Castor oil has a high heat resistance and is used in racing cars and in industry. Sesame oil is reddish in color and is an important cooking oil in Latin America and India. Other oil crops are oil palm, mustard, and rapeseed and canola.

Pulse Crops

Pulse crops are **large-seeded legumes** that are used as a source of protein for humans and livestock. They are also called **seed legumes**. Large-seeded legumes should not be confused with small-seeded legumes such as alfalfa and the clovers that are grown primarily as forage crops.

The most common pulse crop in the U.S. is soybean, followed by peanut. Other pulse crops are the field beans, which include pinto beans, lima beans, great northern beans, kidney beans, mungbeans, and broadbeans; and the field peas, which include chickpeas, pigeon peas, and lentils.

Root Crops

Root crops are crops from which the true root is harvested for human food or livestock feed. They include such edible root vegetables as radish, carrot, turnip, beet, sweet potato, and rutabaga. Sugarbeet is an agronomically important root crop grown for its roots high in sugar. Other root crops are mangel and cassava.

Tuber Crops

Tuber crops are grown for their underground tubers. A **tuber** is not a true root, but is a thickened, modified underground stem as discussed in Chapter 7. Tubers are used mostly for human consumption, but can be fed to livestock. The most common tuber crop is the Irish potato. Another tuber crop is Jerusalem artichoke.

Sugar Crops

Sugar crops are grown for their sweet juice from which **sucrose**, or common table sugar, is extracted and refined. The most common sugar crop in the U.S. is sugarbeet. Sugar is extracted from its large fleshy taproot. Another important sugar crop in the world is sugarcane. Sweet sorghum is a sugar crop that has gained in interest in recent years for use in alcohol fuel production. In these latter two crops, the sugar is extracted from the stalk or stem. **Fructose** (corn sugar) is extracted from the grain of corn and sorghum.

Fiber Crops

Fiber crops are grown for the fiber produced in the fruits or stems, which is used for making textiles, rope, twine, bags, and similar products. The most common fiber crop in the U.S. is cotton, which has its fibers attached to the seed. Another important fiber crop is flax that is used for making linen from the fibers contained in its stem. Other fiber crops are hemp, hennequen, sisal, jute, ramie, kenaf, and broomcorn. These are mostly grown in other parts of the world.

Horticultural Crops

Horticultural crops usually require very intensive cultural practices as compared to **field crops** such as wheat or corn. They include crops grown for their edible roots, leaves, stems, or fleshy fruits as well as plants grown as ornamentals or as fresh flowers. Horticultural crops are not considered agronomic crops in most areas, but can be in areas where they are important such as the Imperial Valley of California. However, it is important to include them in any discussion of crop classification to present a complete picture of the uses of crops.

Vegetable crops are those that are grown for their edible vegetative parts, or fleshy fruits. Most vegetable crops are herbaceous annuals or biennials but a few, such as asparagus, are herbaceous perennials. The fleshy fruits of woody perennials are not usually included in a classification of vegetable crops and will be discussed later as tree fruits. Examples of vegetable crops include carrots, beets, and rutabagas grown for their edible roots; lettuce, cabbage, and spinach, grown for their edible leaves; and asparagus grown for its edible stems. Other vegetable crops are broccoli, cauliflower, and Brussels sprouts grown for their edible flowers; and peas and sweet corn grown for their edible seeds; and tomatoes, cucumbers, and peppers grown for their edible fruits. Some vegetable crops are harvested for their specialized vegetative structures, such as onion and garlic, which are grown for their fleshy bulbs.

Orchard crops include woody perennial trees grown for their edible fruits. These are also called **tree fruits**. They include apple, peach, pear, and citrus grown for their fleshy fruits; and walnut, pecan, and almond grown for their dry fruits.

Small fruits are perennial crops grown for their edible fruits. They include **cane plants** such as blackberry, raspberry, and gooseberry; and other small woody plants such as grape; and herbaceous perennials such as strawberry.

Ornamental crops are those plants used for landscaping which are grown for sale. They include perennial shrubs and small trees, also flowers. **Floricultural** crops are those plants that are grown for their flowers that are cut and sold individually and in floral arrangements.

Other Crops

A **beverage crop** is grown for its seeds or leaves that are processed and used in making beverages. The most common are coffee, tea, cocoa, and cola nuts that are grown in tropical regions of the world. The only beverage crop grown in the U.S. is coffee grown in Hawaii.

A **medicinal crop** is grown for a specific drug that it contains or produces. The only medicinal crop grown legally in the U.S. is tobacco that is grown for its nicotine, and hemp which is allowed only in certain states. Other medicinal crops grown in the world are belladonna, digitalis, poppy, hemp, and sassafras.

Rubber crops are grown for the latex they produce which is used in making natural rubber products. None are grown extensively in the United States. Examples are the rubber tree, guayule, and koksagyx (Russian dandelion).

SPECIAL USE CROPS

Cover Crop

A **cover crop** is grown to protect the soil from wind and/or water erosion, grown usually when the soil would not normally have the protection of a growing crop. Cover crops may or may not be harvested, and are not usually allowed to mature. Examples of cover crops are small grains such as oats or rye. These are seeded in the fall after the harvest of spring-sown row crops such as corn, sorghum, or soybeans, to provide soil erosion protection during the winter. The cover crop will then either die during the winter or be destroyed early the next spring during soil preparation for the subsequent spring-seeded row crop. Cover crops are usually used on highly erodible lands with sandy soils (wind erosion), or steep slopes (water erosion).

Catch Crop

A **catch crop,** or **emergency crop**, is seeded as a "fill-in" when the regular crop has failed or when seeding has been delayed too long for the regular crop to mature. Examples include seeding millet or sorghum after wheat has been destroyed by hail; or seeding soybeans when it's too late to seed corn due to excessive spring rains. A catch crop will not have the yield potential of the regular crop due to its shortened growing season, but it does allow the producer to recoup some losses that occurred.

Companion Crop

A **companion crop,** or **nurse crop,** is an annual crop that is seeded with a perennial crop to aid in the establishment of the perennial crop. Companion crops are usually spring-seeded small grains. An example is spring oats seeded with alfalfa. The oats are harvested in the summer for grain or hay leaving the established alfalfa.

Companion crops aid in establishing the main crop by providing protection for the seedlings of the main crop that are usually much slower to germinate and become established. The companion crop can aid in weed control by providing competition to weed seedlings that are also germinating after seeding, and by shading the soil to reduce soil evaporation and soil temperature. They can also provide erosion protection during this critical period. In addition, they provide some return from the land the first year when the main crop usually produces very little.

Companion crops do, however, provide competition for the perennial seedlings so their management is very important. It is sometimes better to provide weed control with herbicides and soil protection with crop residue management than to use companion crops.

Green Manure Crop

A **green manure crop** is grown and then tilled into the soil while still in the green, succulent condition to improve the soil. The crop can improve the soil by providing additional organic matter that can improve its structure. It can also increase plant nutrient availability by immobilizing soil nutrients that are then released when the crop rapidly decomposes, and it can increase soil nitrogen if the crop is a legume. Chapter 4 discusses soil nutrient cycles in more detail.

Green manure crops are usually short-lived legumes such as sweetclover, red clover, cowpea, or vetch; but they can also be nonlegume crops such as small grains or sudangrass. Sometimes the green manure crop is seeded in the fall after the principal crop is harvested, and then tilled under the following spring in which case it doubles as a cover crop.

Trap Crop

A **trap crop** is seeded to attract certain insects or parasitic weeds and then is destroyed when it has served its purpose by aiding in the control of the pest. Trap crops do not always work well and better general control can usually be obtained with pesticides; besides the fact that the land devoted to the trap crop is usually lost to production. Sorghum seeded in a strip next to a wheat field to attract chinch bugs is an example of a trap crop.

LIFE SPAN

Crop plants are grouped based on the length of their life span, also the seasons in which they grow. Knowledge of crop life span is essential because the cultural practices used are influenced by the crop growing seasons and the expected life span. Also, weeds that have the same life cycle as the crop are most likely to cause problems in a particular crop.

Annual Crops

Annual crops are those that complete their life cycle in one year or less. Annuals always reproduce by seed. They can be further classified as winter annuals and summer annuals.

Winter annuals germinate in the fall, live through the winter, usually in the dormant state, and then mature and produce seed in the late spring or early summer. Most winter annual crops are small grains such as winter wheat, winter rye, winter barley, and winter oats. Winter annual weeds include downy brome, hairy chess, pennycress, shepherd's purse, and the wild mustards.

Summer annuals germinate in the spring, grow and develop during the summer, and mature and produce seed in the fall before frost. Most crops are summer annuals and include corn, sorghum, soybean, cotton, peanut, and spring-seeded small grains such as spring wheat, spring barley, and spring oats. Summer annual weeds include crabgrass, foxtail, pigweed, and velvetleaf.

Biennial Crops

Biennial crops require two years to complete their life cycle. Biennials grow vegetatively the first year, usually with leaves in a rosette pattern. A **rosette** is a group of leaves growing from a compact stem with little internode elongation. During the first growing season, food reserves are stored in the roots that are usually large and fleshy. The second year a flower stalk is produced from the center of the rosette and the plant matures and produces seed before frost in the fall.

Biennial crops and weeds are less important compared to annuals and perennials. Examples of biennial crops are sugarbeet, carrot, and sweetclover. Biennial weeds include common mullen and many thistles.

Perennial Crops

Perennial crops have an indefinite life span which can range from a few to many years. They also can vary in their growth habit that can affect stem structure and growth. Some have **herbaceous stems** that die back to the soil surface every winter. These plants resume growth in the spring from a crown or taproot. Examples of herbaceous perennials are pasture grasses and alfalfa. Other perennials, such as trees and shrubs, add new growth every year from **woody stems**. Herbaceous and woody stems are discussed in more detail later in this chapter.

Some crops are perennials in one climate and annuals in another. Examples are cotton, castor, and sorghum. In tropical regions these plants may grow for several years producing new growth each year. In temperate regions, they are killed by frost each fall and are grown as annuals.

CLIMATIC ADAPTATION

Tropical Crops

Tropical crops are grown in warm climates where freezing never occurs. Tropical crops are mostly long-lived perennials, but may be annuals or biennials. In the U.S., tropical crops are grown only in Hawaii. Examples include banana, pineapple, and coffee bean.

Subtropical Crops

Subtropical crops are grown in warm climates with long growing seasons and mild winters where freezing rarely, if ever, occurs. As with tropical crops, they are mostly perennials. Examples include most citrus crops such as orange, grapefruit, and lemon. Subtropical crops grow in the southernmost U.S. particularly the southern parts of Florida, Texas, and California.

Temperate Crops

Temperate crops are grown in climates where there is a marked winter season with prolonged periods of freezing temperatures. Most crops are temperate and include corn, soybean, and small grains. Temperate crops are grown over all the United States. They are also commonly grown in tropical and subtropical regions.

Perennial temperate crops can be further classified as cool-season or warm-season. **Cool-season crops** grow primarily during the spring and fall, and are usually dormant during the summer and winter. Examples include bluegrass, wheatgrass, and bromegrass. **Warm-season crops** grow primarily during the summer months and remain dormant during the rest of the year. Examples include bluestem, buffalograss, and switchgrass. Although these terms are used mostly with perennial forage grasses, they can also be used with any plants that exhibit these types of growth.

Boreal Plants

Boreal plants grow in climates where temperatures are below freezing most of the year except a very short growing season. They are found near the Polar Regions and in the high altitudes of mountains in lower latitudes. Few crops can grow in these regions but some cool-season temperate vegetable crops such as cabbage can produce well at high latitudes because of the extremely long daylength that occurs during the short growing season.

CULTURAL PRACTICES

The classification of crops by cultural practices is based primarily on the row spacing used when the crop is seeded. Row spacing greatly affects the soil conservation practices that are needed to prevent soil erosion. Soil erosion is much more likely to occur in those fields that are seeded to crops with wide rows than with narrow rows. Therefore, it is necessary to maintain more plant residues on the soil surface, or to use special practices such as terracing or contour farming.

Row Crops

Row crops are seeded in rows that are spaced wide enough to allow inter-row tillage for weed control and soil management. Row spacing will vary with the crop grown and the machinery used, but will usually range from 50-100 cm (20-40 inches), with 75-100 cm (30-40 inches) the most common. Examples of row crops are corn, sorghum, soybeans, cotton, sugarbeet, and many vegetables.

Close-Seeded Crops

Close-seeded crops, or **drilled crops**, are seeded in rows that are too narrow for inter-row cultivation. Sometimes these crops are **broadcast,** that is, not seeded in rows but merely spread over the soil surface and incorporated with shallow tillage. Row spacing of drilled crops ranges from 18-50 cm (7-20 inches). Small grains, millet, and buckwheat are usually drilled. Soybeans and sorghum can also be drilled but are usually seeded as row crops.

Pasture or Hay Crops

Pasture crops or **hay crops** are usually perennial, close seeded crops in which no cultivation occurs after initial seedbed preparation. Since they protect the soil throughout the year, they are the best for controlling erosion especially in moderate to highly erodible soils. The uses of pasture and hay crops were discussed earlier in the chapter, but are mentioned here as a unique cultural system.

GROWTH HABIT AND LEAF RETENTION

Herbaceous Plants

Herbaceous plants have soft and succulent stems with little secondary tissue produced by the cambium. The **cambium** is a ring of cells near the outside of a stem that produces lateral growth. The cambium may be present in a herbaceous plant but it has little function. Plant parts are likely to be edible since they are low in cellulose and fiber. Herbaceous plants may be annual, biennial, or perennial, and include practically all agronomic crops. All annual and biennial crops are herbaceous. Alfalfa is an example of a herbaceous perennial plant.

Woody Plants

Woody plants have an active cambium that develops large amounts of secondary tissue and abundant xylem high in cellulose, lignin, fiber, and other poorly digested compounds. Woody tissue is necessary to support large plants such as trees. Most woody plants are perennial and can be further classified as evergreen or deciduous.

Evergreen plants, such as pine and spruce, maintain their leaves throughout the year. No field crops common in the U.S. are in this category. **Deciduous plants** lose their leaves every year, usually in the fall. Most tree fruits, such as apple, pear, and citrus are deciduous woody plants.

WATER REQUIREMENT OR ADAPTATION

Hydrophytes

Hydrophytes are plants that are adapted to live in water or in soil saturated with water. These plants usually use large quantities of water in their growth and development. The only major crop classified as a hydrophyte is paddy or lowland rice, but all plants that grow in marshes, ponds, and other poorly drained areas are hydrophytes.

Mesophytes

Mesophytes are intermediate in water use and needs. They can tolerate a water saturated soil for only a brief time, and are also very sensitive to extremely dry soil conditions. Mesophytes prefer moist, well-drained soils. Most crops are mesophytes.

Xerophytes

Xerophytes can survive in areas where soils are very dry for long periods of time. These plants usually have special features which enable them to store or conserve water, such as fleshy leaves, or stomata that open only at night. Others have an extremely short life span that allows them to produce seed before the soil is dry. No common crops are xerophytes. Examples are cactus and mesquite.

Halophytes

Halophytes can grow in soils that have a high concentration of salts. These salts make it very difficult for the plant to take up soil water so most plants cannot survive under these conditions. Some varieties of barley are halophytes, but few other crops.

BOTANICAL CLASSIFICATION

The **botanical classification system** is a method of classifying plants that gives every type of plant a unique position in the system. As mentioned at the beginning of this chapter there is no repetition as shown in the other systems studied in this chapter which makes it very valuable for scientific purposes. To illustrate this system we will use Arapahoe wheat as an example.

KINGDOM-Plantae (plants)
DIVISION-Spermatophyta (seed plants)
CLASS-Angiospermae (seeds in fruits)
SUBCLASS-Monocotyledonae (one seed leaf)
ORDER-Graminales (grasses and sedges)
FAMILY-Poacae (grasses)
GENUS-Triticum (wheats)
SPECIES-aestivum (common wheats)
CULTIVAR-Alliance

The **genus** and **species** form the **scientific name** for any plant. This name is unique for that particular type of plant and aids in identification of plants when different languages are used. For instance, *Zea mays* is called corn in the U.S., but is called maize in many other parts of the world. The example given earlier, *Triticum aestivum*, is called wheat in the U.S. but is called corn in other places. **Cultivar,** or **variety**, is used to designate a group of plants within a species that have the same genetic background.

The two most important **families** of crop plants are the *Poaceae* or grasses, and the *Fabaceae* or legumes. Grasses include small grains, corn, sorghum, millet, sugarcane, and rice. Legumes include soybeans, alfalfa, clover, vetch, peas, and beans. Other families that contain crop plants include *Polygonaceae* (buckwheat), *Chenopodiaceae* (sugarbeet), *Brassicaseae* (mustard, rape, and kale), *Solanaceae* (potato and tomato), *Malvaceae* (cotton), and *Asteraceae* (sunflower, safflower, and Jerusalem artichoke).

Another important division of the botanical system is the **subclass**. Most crops are classified as **monocots** (*Monocotyledonae* subclass), or as **dicots** (*Dicotyledonae* subclass). Monocots have one seed leaf in the embryo and include grasses and sedges. Dicots have two seed leaves in the embryo and include legumes and most other plants. There are many differences between monocots and dicots in their seeds and seedling emergence, as well as other factors, which are discussed in later chapters.

REVIEW QUESTIONS

1. What are the different ways that crops can be categorized by agronomists and farmers?
2. What are the different ways that grain crops can be classified?
3. What are the differences in the harvesting and storage of the different types of forage crops?
4. What are the major uses of oil crops? What is the most common oil crop grown in the U.S.?
5. What type of crop is the major source of protein from plants?
6. How does a tuber differ from a root?
7. What is the major fiber crop, sugar crop and tuber crop grown in the U.S.?
8. How does a horticultural crop differ from a field crop?
9. Can a crop double as a cover crop and a green manure crop?
10. What are the advantages and disadvantages of a companion crop?
11. What is the life cycle of a summer annual, winter annual, biennial, and perennial? When would each be seeded and harvested?
12. What is the climatic adaptation of most crops?
13. How does row spacing affect conservation practices for erosion control?
14. How do herbaceous and woody stems differ in growth and digestibility? Why are annual plants never woody?
15. What is the water adaptation or requirement of most crops?
16. How is the botanical classification system different from the agronomic systems?
17. What is the scientific name of a plant?
18. What are the important subclasses of crops? What are the two most important families of crops?

CHAPTER THREE

Agroecology

INSTRUCTIONAL OBJECTIVES

Upon completion of this chapter, you should be able to:

1. Define ecology, agroecology, ecosystem, and agroecosystem.
2. Define symbiosis, mutualism, competition, and allelopathy. Give an example of each.
3. Define open ecosystem, and closed ecosystem. Give an example of each.
4. Define managed ecosystem and list four ecological factors that the crop producer can control in an agroecosystem. Discuss how these factors are affected differently in a natural ecosystem.
5. Define monoculture and ecological niche. Explain how a monoculture crop field cannot be ecologically stable.
6. List five factors that affect plant competition. Give an example of each.
7. Define biotic and abiotic factors.
8. Define limits of tolerance, optimum range, range of tolerance, and stress zones. State what ultimately determines yield in an agroecosystem.
9. Discuss differences in nutrient cycling in natural ecosystems and agroecosystems.

INTRODUCTION

Ecology is the study of the relationships of living things to their environment and to each other. **Agroecology** is the study of the specific relationships of crops and farm animals to their environment and to each other. Agroecology is also spelled **agro-ecology**. An **ecosystem** is the total interaction between all organisms that live in an area with each other and their environment. A pond, forest, or rangeland will each have its own ecosystem. An **agroecosystem** is the total interaction between crops and farm animals and their environment within a farming or ranching operation.

The concept behind agroecology is that species controlled by the farmer, such as crop plants and livestock, are directly and indirectly influenced by other living organisms that are present within the boundaries of the farm. These other organisms can include other plants (weeds), soil microorganisms, and other animals, as well as weather and climate that occur in the immediate area. Some studies of agroecosystems also include social and economic factors that affect the choice of crops and livestock the farmer will utilize; but in this chapter those factors will not be discussed. Also, this chapter will concentrate on the ecological factors that affect crop plants, and will not discuss to any extent the effects that livestock found on the farm would have on the overall ecology of the agriculture system.

PLANT INTERACTIONS

Plants never exist without influencing, and being influenced by other plants and organisms that are also growing in the area. These interactions can be beneficial, detrimental, or have no effect on the organisms involved. In crop production, we can greatly influence these interactions by determining plant densities, types of crops or weeds present, or managing soil factors that influence plant growth and development. Understanding plant interactions and creating the optimum combination of plants can lead to increased yields. There are three types of interactions between organisms: symbiosis, competition, and allelopathy.

Symbiosis means "living together". It is a general term signifying the ways that organisms relate to each other. **Mutualism** is an interaction between two different organisms which is mutually beneficial to each. The most common example of mutualism is the relationship between legume plants and *Rhizobium sp.* bacteria. The bacteria fix nitrogen from the air and make it available to the legume plant; and the legume provides the bacteria with food and shelter. Symbiotic nitrogen fixation is discussed in more detail in Chapter 5.

Competition is a mutually negative interaction between organisms which utilizes a common resource that may be in short supply. The most common resources for which plants compete with each other are light, water, nutrients, and soil volume for roots. Usually when we think of competition between plants, we think of weeds that compete with crop plants. However, crop plants that are growing together also compete with each other; and crops also compete with weeds. Proper management of a cropping system can give the crop plants a competitive advantage over weeds, and minimize competition between the crop plants themselves. For example, seeding a crop at the optimum time enables it to rapidly germinate and emerge, giving it a competitive advantage over weeds that may germinate later. Proper seeding rate minimizes competition between crop plants while maximizing competition with weeds.

Allelopathy, also called **amensalism**, is an interaction which has a negative effect on one organism while affecting the other organism very little. An example is the release of a metabolic by-product by one plant that inhibits the growth and development of another plant. There have been many studies which show that some crops have allelopathic effects on other crops which are growing at the same time, or which follow in a subsequent cropping season. For example, sorghum can have an allelopathic effect on crops which follow it in a crop rotation system, especially small grains. Knowledge of potential allelopathic effects is important when planning a crop rotation system.

ENVIRONMENTAL FACTORS AFFECTING ECOSYSTEMS

Environmental factors that affect any ecosystem are either biotic or abiotic. **Biotic** factors are all living organisms within the ecosystem plus the products of those organisms. Biotic factors include all plants, animals, and microorganisms. The products produced by these organisms, including waste products, are also part of the biotic environment. **Abiotic** factors are all non-living factors that affect an ecosystem, including light, temperature, water, soil pH, and nutrient supply.

The range of biotic and abiotic conditions that species in an ecosystem can tolerate is called the **range of tolerance**. Each species has a set of conditions within which it does best called the **optimum range**. The high and low extremes are the **limits of tolerance**. Environmental conditions between the optimum range and the limits of tolerance are the **stress zones**.

A crop producer has some control over biotic and abiotic factors in the field, but cannot control them completely. Management practices try to maintain environmental factors as close to the optimum range as possible, however, variation in crop yield from year to year attests to the fact that not all environmental factors are being controlled. The yield of a crop field is ultimately determined by the environmental factor that is most limiting, whether it is nutrient supply, temperature, moisture supply, pests, etc.

AGROECOSYSTEMS VS. NATURAL ECOSYSTEMS

The basic difference between an agroecosystem and a naturally occurring ecosystem is that the crop producer affects, and ultimately changes, the ecosystem within the crop field, while a natural ecosystem is only affected by the environment in which it thrives. A natural ecosystem is basically a **closed ecosystem** in which the factors needed for plant growth, particularly nutrients, are continually recycled within the ecosystem. Although animals can remove some vegetation from the area, they will likely deposit waste material which will cycle into the system. Sunlight and precipitation are inputs into the system, but their effects are part of the overall ecosystem.

An agroecosystem, however, is basically an **open ecosystem**. The crop producer removes crop produce, such as grain or vegetation, from the system; and inputs resources such as plant nutrients in the form of soil fertilizers. An agroecosystem is a **managed ecosystem** because the producer is altering the system through management practices. Table 3-1 lists some differences between natural ecosystems and agroecosystems. Notice in this table that although the crop producer has considerable influence over the ecosystem in a crop field by controlling the types of plants that are present and some environmental factors such as nutrient supply; an agroecosystem is not stable and can exist only through continued intervention by the crop producer. If cropland is abandoned, or left alone, it will eventually progress to a stable ecosystem, although it will take many years to do so. The climax vegetation will likely be different, however, unless the climax plant species are introduced by humans or animals; and even then it can only approximate the diversity associated with the original or native ecosystem.

Table 3-1 Differences between natural ecosystems and agroecosystems.

Characteristics	Agroecosystem	Natural ecosystem
Net productivity	High	Low
Species diversity	Low	High
Genetic diversity within species	Low	High
Nutrient cycles	Open	Closed
Stability	Low	High
Human control	High	Not needed
Permanence	Short	Long
Flowering, maturity, etc.	Synchronized	Seasonal
Ecological maturity	Early, very immature	Mature, climax

Plant Diversity

An agroecosystem is much less diverse than a natural ecosystem. It may have only one species of plant, called a **monoculture**, while a natural ecosystem will likely have hundreds of species living together. An agroecosystem is also much less stable because of its lack of diversity. In a stable natural ecosystem, each organism, whether plant or animal, occupies an **ecological niche** in which it thrives with minimal competition with the other organisms in the system. In its niche, a plant receives adequate light, water, nutrients, and space to survive. Competition exists, but not at a level which results in the elimination of the species. An agroecosystem, by its very nature, is not stable. In most cases, there is a monoculture of only one species of plant (the crop) which occupies only a fraction of the ecological niches present in the field. If not controlled, weeds will germinate and occupy the empty niches in the field. Also, the crop producer is modifying the environment in the field through fertilizers, pest control, weed control, irrigation, etc. which further destabilizes the ecosystem. Without the diversity and stability of a natural ecosystem, the producer must constantly modify the system to protect the crop.

Another difference in the diversity of plants in agroecosystems and natural ecosystems is the life cycle of the plants in the system. A natural ecosystem has a mix of annual, biennial, and perennial plants. An agroecosystem, however, usually contains only annual plants, such as grain crops. When grown in a monoculture, the plants progress through their life cycles together and significant growth stages, such as flowering and maturity, occur simultaneously. This synchronization of growth stages is desirable for the producer as it enables farm operations, such as harvest, to be orderly and efficient. However, this further destabilizes the ecosystem, making it more prone to damage from weather or pests.

A good crop rotation system, as discussed in the next chapter, will contain a mix of annual and perennial plants which will aid in improving ecological stability. A rotation system, however, will still not be nearly as stable as a natural ecosystem; especially if each crop in the rotation is grown in a monoculture. Intercropping systems, also discussed in the next chapter, can further increase stability in the system. A combination of crop rotation and intercropping can provide the most stable agroecosystem. However, this combination is rarely practiced in industrialized agriculture systems, such as found in the United States, because of problems associated with mechanization of tillage, seeding, and harvesting.

Plant Competition

As previously discussed, plants compete with other plants in their immediate vicinity. The degree of competition depends on several factors including plant density, leaf characteristics, plant height, type of root system, and water and nutrient supply. A producer takes these factors into consideration when determining the **yield goal** for a crop, and manipulates these factors in the management of a crop. Many of these factors are discussed more fully in later chapters, but the general concepts will be discussed here.

An important rule to remember when studying plant competition is that any given area can only produce a finite amount of plant material or dry matter. The maximum amount is primarily determined by the supply of light, water, and soil nutrients. Other factors, such as type(s) of plant(s), soil pH, soil structure, pests, and diseases can also affect maximum production.

Plant density, or the number of plants per area, affects the spacing of plants, and ultimately the degree of competition between individual plants in any part of the field. If plant density, as determined by seeding rate, is increased, each individual plant will produce less dry matter and less yield. With grain crops, this means that each plant will produce fewer seeds. For example, corn will not completely fill its ears, soybeans will produce fewer pods and/or fewer seeds per pod, and wheat will reduce tillering and not fill all florets in its head. Seeding rate is discussed more fully in Chapter 6.

Leaf characteristics and plant height affect the efficiency with which plants intercept sunlight. This factor is not as important in a monoculture where all leaves are of the same type as it is with intercropping where there may be different types of leaves and different plant heights. For example, a tall plant with large, wide leaves will intercept more light than a shorter, smaller leaved plant. Light interception is discussed more fully in Chapter 8.

The type of root system affects the density of roots in the soil, as well as their lateral and vertical spread. A plant with a more vigorous root system will have a competitive advantage over another plant with a smaller root system. Again, with a monoculture, this is not a problem if plants are spaced properly; but it can be with intercropping. A carefully designed intercropping system will use plants with different types of root systems to utilize different depths of the soil. Roots are discussed in Chapter 7.

Water and nutrient supplies greatly affect the degree of competition between plants in both monoculture and intercropping systems. When water and nutrients are in abundant supply, the root systems will grow to their maximum extent. However, when either water or nutrients become limiting to crop growth, the competition for them increases and root growth decreases, further limiting the ability of the plant to absorb them. As discussed before, different root systems in an intercropping system can lessen root competition between different types of plants.

Cycling of Nutrients and Organic Matter

Cycling of nutrients through the decomposition of soil organic matter is an important function of any ecosystem. This cycling process, called **mineralization**, is discussed in detail in Chapter 5. In this chapter, we will discuss differences in nutrient cycling between agroecosystems and natural ecosystems. In a natural ecosystem, most of the nutrients are cycled within the system. Some nutrients will be removed by animals, but the total is small, and some of the nutrients are replaced in the animals feces and urine. As mentioned earlier in this chapter, a natural ecosystem is basically a closed system.

In an agroecosystem, which is an open system, nutrients are removed from the system by harvesting of grain or vegetation. These nutrients must be replaced by the crop producer to maintain productivity. The most common method for replacement of nutrients is the application of soil fertilizers which come from outside the system. Figure 3-1 shows nutrient cycling in an agroecosystem and Figure 3-2 shows nutrient recycling in a natural ecosystem. Note that the thickness of the arrows indicates the relative amounts that are being removed (outputs) and supplied by the producer (inputs). Figures 3-3 and 3-4 show the overall cycling of nitrogen and carbon in natural ecosystems. These figures show mineralization within individual ecosystems and the cycling of nitrogen and carbon on a more global scale. These cycles, together with the cycling of other important elements such as oxygen maintain life in our world. Some processes shown in Figures 3-3 and 3-4, such as photosynthesis and nitrogen fixation are discussed in later chapters.

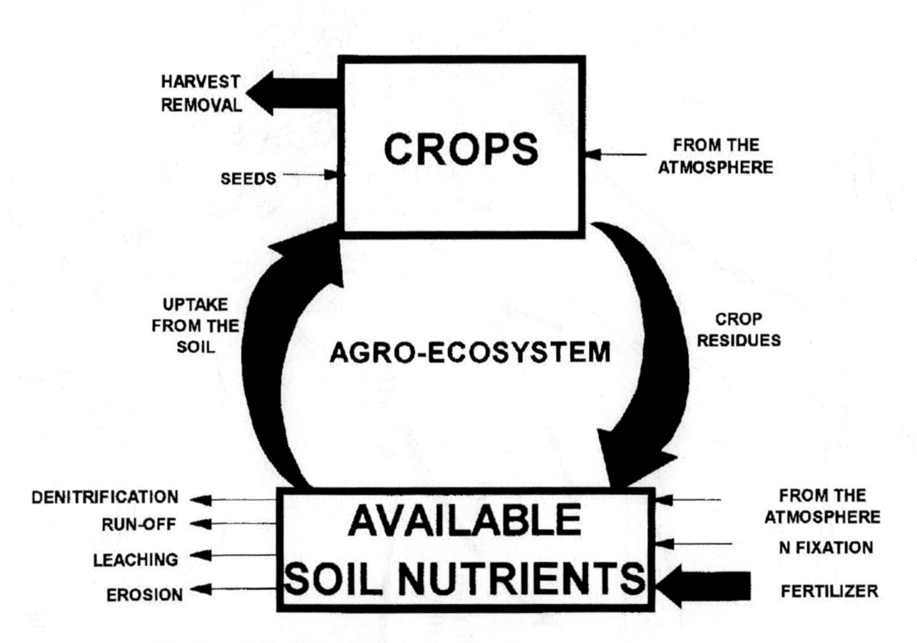

Figure 3-1 Nutrient recycling in an agroecosystem.

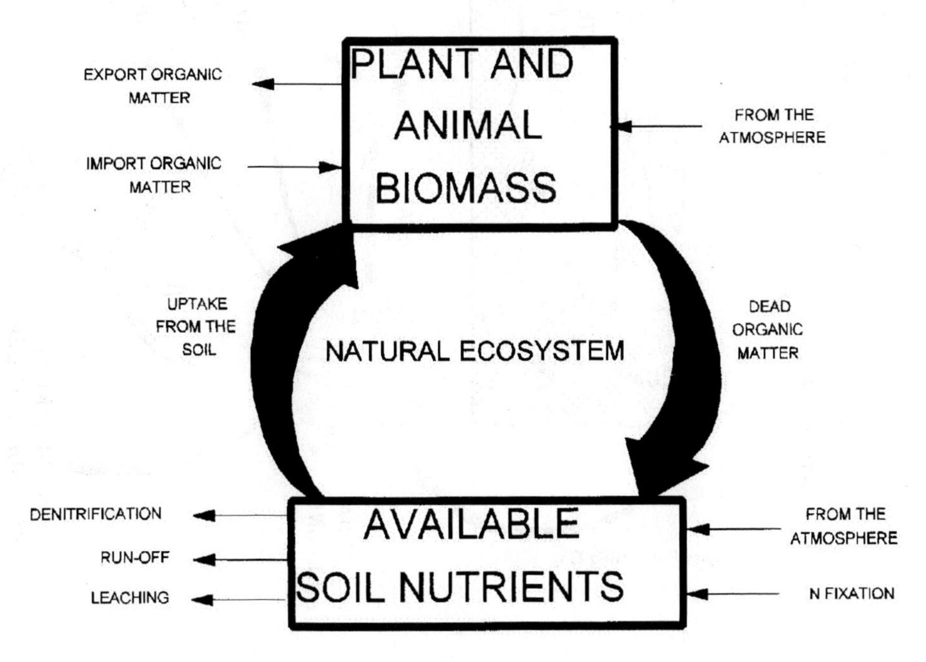

Figure 3-2 Nutrient recycling in a natural ecosystem.

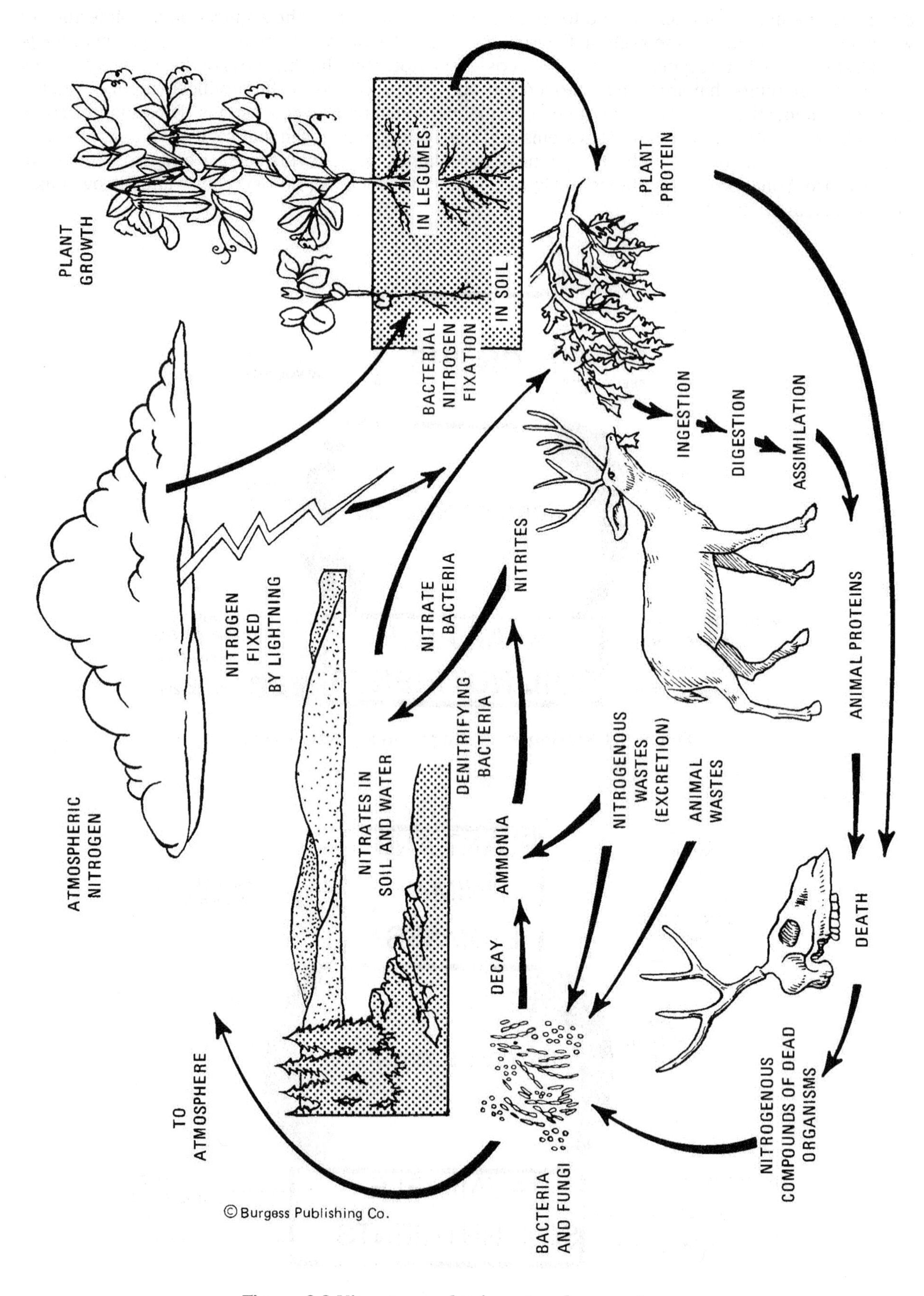

Figure 3-3 Nitrogen cycles in natural ecosystems.

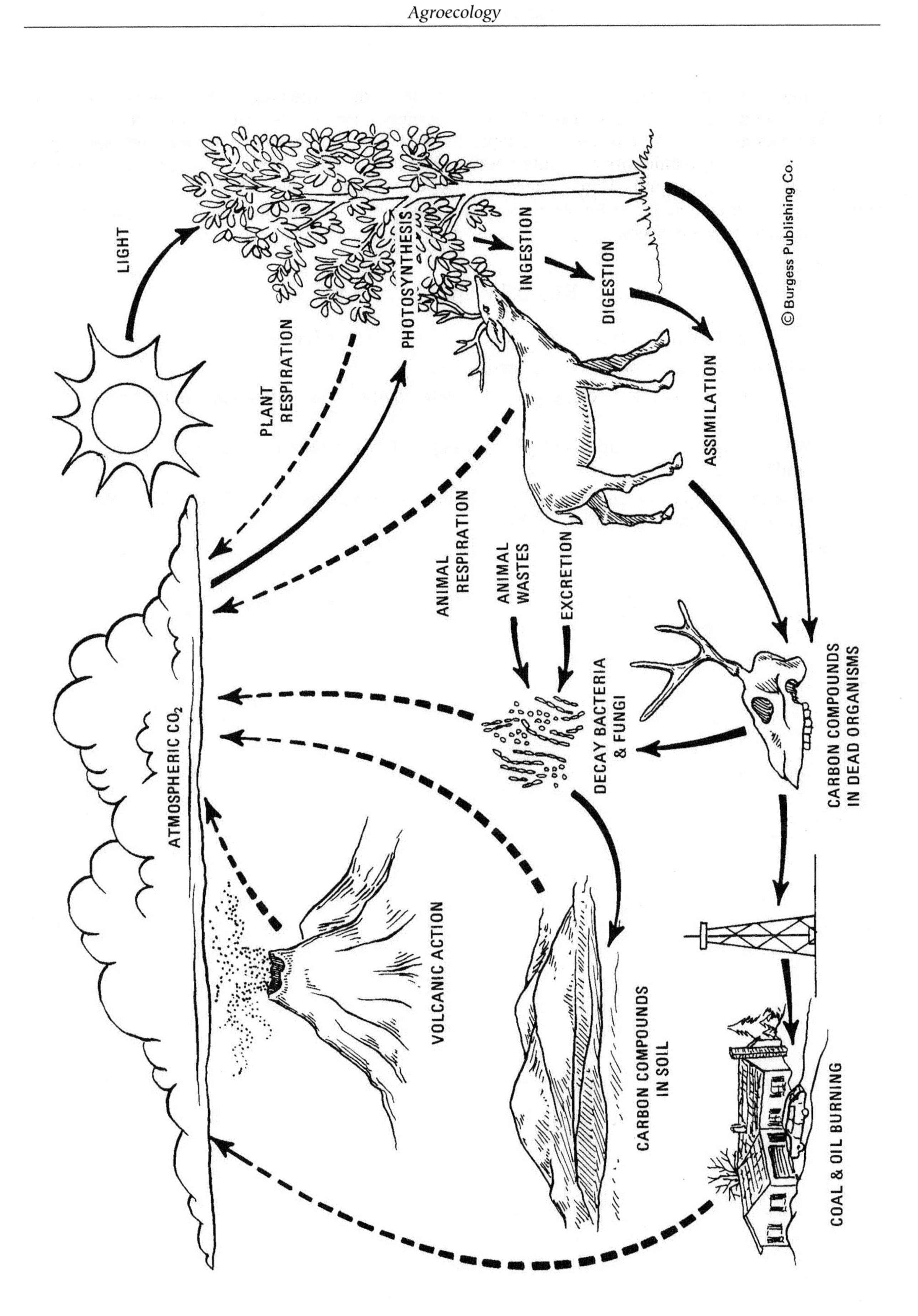

Figure 3-4 Carbon cycles in natural ecosystems.

The amount of inputs needed in an agroecosystem depends on the type of cropping system that is used. A continuous cropping system requires considerably more inputs that a crop rotation system. These systems are discussed in the next chapter. The types of crops grown, either continuously or rotated, also affect the amounts of inputs needed. For example, legumes can help supply nitrogen through symbiotic nitrogen fixation, while grains use considerable amounts of nitrogen. The goal of any crop producer is to minimize inputs while maximizing yield. Various methods for doing this are discussed in subsequent chapters.

REVIEW QUESTIONS

1. Why would a monoculture never develop under natural conditions?
2. What role do weeds play in an agroecosystem?
3. Why is it not possible to completely eliminate competition between plants in a crop field?
4. What are some advantages and disadvantages of synchronized life cycles in a crop field?
5. How can a crop producer increase the stability of the agroecosystem in a field? Would that be practical or economically feasible?

CHAPTER FOUR

Crop Production Systems

INSTRUCTIONAL OBJECTIVES

Upon completion of this chapter, you should be able to:

1. Define cropping system, crop rotation, continuous cropping, fallow, and organic farming. Give an example of each.
2. Define multiple cropping, double cropping, intercropping, strip intercropping, and relay intercropping. Give an example of each.
3. List the criteria that should be met for a crop rotation system to be successful.
4. Compare crop rotation with continuous cropping. List the advantages and disadvantages of each.
5. List the advantages and disadvantages of multiple cropping.
6. List the advantages and disadvantages of double cropping.
7. List the criteria for a successful fallow system. Define fallow efficiency.
8. Define organic farming and compare it to conventional farming.
9. Define sustainable agriculture and compare it to conventional farming.

INTRODUCTION

Crop production involves a series of distinct yet related operations and decisions by the producer. These operations and decisions include crop selection, hybrid and/or variety selection, the number of tillage operations for weed control and seedbed preparation, method and time of seeding, time and method of fertilizer application, method and amount of irrigation if used, harvest and storage procedures, etc. In this book, we will study these various operations individually and in detail. However, it is important to understand that these individual operations go together into a system of producing a crop or crops on a particular farm.

In this chapter we will discuss the different types of crop production systems commonly found in agriculture. With this information you will have a better understanding of how the various operations studied later fit into the total scheme of crop production. These systems are found in all types of agriculture, from the primitive to the highly developed, and with all types of climates and crops. The only differences are the specific operations used in the system.

CROP ROTATIONS

Crop rotation is a regular recurrent succession of different crops on the same land. Crop rotations were previously mentioned in Chapter 1 as one of the early techniques used to sustain economic levels of crop yields.

A farm using a crop rotation system will be divided into several fields that are planted to two or more different crops in any given year. These crops are then rotated around the different fields in subsequent years. Table 4-1 gives a very simple example of a crop rotation. When studying Tables 4-1 and 4-2, note that the sequence of crops over years is found in the table columns, and the number of crops in each year is found in the table rows.

Table 4-1 A simple crop rotation system.

	FIELD			
YEARS	A	B	C	D
1	Corn	Corn	Wheat	Oats
2	Oats	Corn	Corn	Wheat
3	Wheat	Oats	Corn	Corn
4	Corn	Wheat	Oats	Corn
5	Corn	Corn	Wheat	Oats

Notice that in this example three crops are repeated or rotated on four fields over a four-year period, and the same sequence is used on all the fields. Since this is a four-year rotation, the sequence repeats itself beginning the fifth year as shown in the table. The sequence on all the fields is two years of corn, followed by one year each of oats and wheat.

Table 4-2 is an example of a more complicated rotation that involves not only annual crops but also a perennial crop of alfalfa.

Table 4-2 A crop rotation system involving both annual and perennial crops.

	FIELDS				
YEARS	A	B	C	D	E
1	Sorghum	Corn	Corn	Soybean	Alfalfa
2	Corn	Sorghum	Soybean	Corn	Alfalfa
3	Soybean	Corn	Corn	Sorghum	Alfalfa
4	Corn	Soybean	Sorghum	Alfalfa	Corn
5	Sorghum	Corn	Corn	Alfalfa	Soybean

Notice that in this rotation system some fields are being rotated with annual row crops while the perennial alfalfa is allowed to grow for three years on a single field. The four-year sequence of the annual crops is sorghum, corn, soybeans, corn. The fifth year is the same as the first except that the soybeans have been shifted to Field E and the alfalfa to Field D. In subsequent years the alfalfa could be rotated to Fields A, B, and C, or it could be planted back onto Field E depending on how well adapted the various fields are to alfalfa, or their locations. While it is desirable to maintain about the same acreage of the various crops grown from year to year in a rotation, it is not always possible to grow all crops on all the fields. Differences in soils, topography, and location can make some fields better adapted to some crops than to others. For example, Fields D and E might be the most erodible and should be left in perennial crops as much as possible.

In order for a crop rotation system to provide the maximum benefits and minimum problems, certain criteria should be met. First, the area of each crop should be nearly equal every year unless there is a good reason for changing it. This requirement does not mean that each crop grown should have equal area, but that each crop should have about the same acreage every year. Second, there should be as large an area of the most profitable crop or crops as can be easily fit in the rotation. Third, the rotation should provide roughage and pasture for any livestock kept on the farm. Fourth, the rotation should include a perennial crop, preferably a legume, if possible. There should also be at least one tilled crop in the rotation for the elimination of weeds. Finally, the rotation should provide for maintaining the organic matter and other properties of the soil.

ADVANTAGES AND DISADVANTAGES OF ROTATIONS

Crop rotations can be very effective in controlling crop pests such as insects, diseases, and weeds, because the different cultural practices associated with each crop can alter the life cycle of the pest. Crop rotations are an excellent way of controlling weeds. Of some twelve hundred species of weeds, less than 30 can survive indefinitely when crop rotation is used. Each crop has a particular growth habit and life cycle and weeds that most closely match the crop will cause the most problems in that crop. By growing different crops, cultural practices will be varied from year to year thus upsetting the life cycles of the weeds. For example, winter annual weeds, such as wild mustard and goatgrass, are a problem in winter annual crops such as winter wheat and winter barley. By rotating to a summer annual crop such as corn or soybeans, tillage and seedbed preparation will occur during the early spring when these weeds are trying to flower and produce seed.

Crop rotations are also effective in managing many insects and diseases. Insects such as white grubworms and cutworms feed on roots of crops of the grass family. Legumes are unfavorable for their development. Corn rootworm does little damage except in corn fields. If corn is grown continuously, its numbers will increase rapidly. Rotation to other crops will effectively control it. Nearly all diseases are controlled to some extent and many of them are completely controlled by proper crop rotation. Scab of cereals, flag smut of wheat, anthracnose blight of beans, and Texas root rot of cotton can all be controlled by simply rotating to a nonsusceptible crop. Often rotations are the only economical way of controlling a pest. Nematodes that usually cannot be economically controlled with chemicals is an example.

Rotations are very effective in maintaining crop yields because of their effects on soil properties. A good rotation can help maintain soil organic matter levels and soil nitrogen levels, especially if the rotation includes a legume. Legumes can increase soil nitrogen through a symbiotic relationship they have with Rhizobia bacteria (see Chapter 5). Two to three years of vigorous alfalfa can provide essentially all of the nitrogen needs of a succeeding corn crop. Rotations can also provide good protection against soil erosion from wind and water, especially if the rotation includes a perennial crop. Since different crops vary in the amounts of nutrients they remove from the soil, a rotation permits a better balance of nutrient removal which results in less need for added fertilizer. For instance, corn is a heavy user of nitrogen, but alfalfa can increase the soil nitrogen level. Alternating these crops can sometimes eliminate the need for any additional nitrogen fertilizer. Crops also differ in their rooting habits which changes the extent to which water and nutrients are extracted from the soil from year to year.

Rotations can result in a more uniform labor distribution in the farming operation since different crops are planted, cultivated, fertilized, irrigated, and harvested at different times of the year. Generally, a crop rotation will result in fewer inputs for energy, fertilizer, and labor when compared to a continuous cropping system.

However, crop rotations may not always be the best system to use. Frequently government programs, as well as economic, climatic, and/or soil factors make it more profitable to grow only one crop in a particular area. Machinery requirements are greater when several crops are grown thus increasing total capital investment. Markets may not be available for a variety of crops. Finally, crop failure can upset the rotation especially when a catch crop is used.

Before the availability of economical fertilizers, crop rotations were the primary methods of maintaining soil productivity, because legumes were needed to help maintain soil nitrogen. As the costs of fertilizer, energy, and other inputs rise, there will be an increase in the number of acres in crop rotations. Although production levels may drop when a rotation is used as compared to an intensive continuous cropping system, total profits may be higher due to reduced inputs.

CONTINUOUS CROPPING SYSTEM

A **continuous cropping system** is the practice of growing only one crop year after year on the same field. It is also called a **monoculture system**. Continuous cropping is the opposite of crop rotation and many comparisons have been made between them, some of which were discussed in the previous section. Continuous cropping is a very predominant cropping system in industrialized agriculture where energy inputs are cheaper.

It is possible to maintain high levels of production with continuous cropping provided soil productivity is maintained with the adequate use of fertilizers and erosion control practices. Sometimes, the total level of productivity may be higher with a continuous cropping system compared to a crop rotation system.

ADVANTAGES AND DISADVANTAGES OF A CONTINUOUS CROPPING SYSTEM

Often, where external inputs (fertilizers, pesticides, etc.) are cheaper, a continuous cropping system is more profitable when the crop grown is high yielding and well suited to the soil and climate of the area. Only one kind of machinery is required which lowers the capital investment of the producer. When only one crop is grown the producer becomes a specialist in that crop resulting in maximum yields.

A big disadvantage of a continuous cropping system is the increase of weeds, pests, and diseases that affect the crop grown. As mentioned before, the presence of some pests can prevent the producer from using a continuous cropping system although it would give the highest returns. On sloping lands, there is an increased risk of water erosion with continuous row crops, especially if crop residues are small such as with soybeans. As mentioned before, applications of fertilizer are usually greater because the crop is using great quantities of certain nutrients. The net result of these problems may result in decining crop yields and profits unless increasing levels of inputs are maintained under good management.

MULTIPLE CROPPING SYSTEMS

Multiple cropping systems involve the production of more than one type of crop in an individual field during one growing season. Multiple cropping is the opposite of **monoculture**, which occurs when a single species of crop plant is seeded in a field. Here, the term *monoculture* is different from *monoculture system* discussed earlier. There are many types of multiple cropping systems. A few of the most common systems will be discussed here.

INTERCROPPING

Intercropping is the practice of growing two or more different crops in the same field simultaneously. Usually these crops are grown in alternating rows or groups of rows because of the machinery being used. An example of intercropping is the practice of alternating rows of soybeans between rows of corn. Intercropping can provide maximum productivity from the land because the different crops complement each other if properly selected and managed. For instance, the soybeans can provide nitrogen to the corn, and the corn can provide protection against heat and wind for the soybeans.

As discussed in the previous chapter, the planting of several different plant types provides for a more stable ecosystem than is possible with a monoculture of one species. If properly designed, each type of plant will fill a particular ecological niche and competition between the species will be minimized. Although each individual crop will yield less than it would have under a monoculture, the total yields of the entire field can be higher.

Very little intercropping is practiced in the U.S. because of the problems associated with the mechanical harvest of the crops growing together. However, intercropping is used in many parts of the world where abundant manual labor is available for harvesting and separating the crops.

Strip Intercropping

Strip intercropping, shown in Figure 4-1, is the practice of alternating strips of four to eight rows of two or more crops across a field. Strip intercropping allows the use of mechanical seeders and harvesters while retaining some advantages of intercropping. A common strip intercropping system alternates strips of soybeans and corn. By rotating the strips every year, many advantages of crop rotations can be attained. In addition, the corn benefits from its border association with the beans on the edges of each strip, and there is less spread of pests and diseases between the strips.

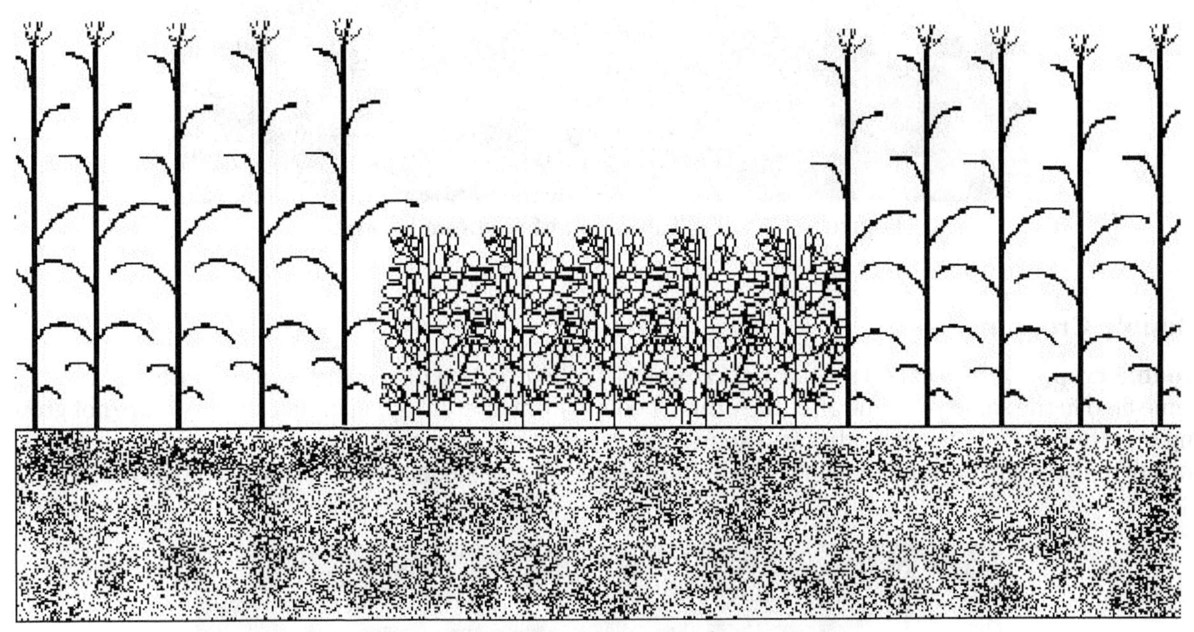

Figure 4-1 A strip intercropping system.

Relay Intercropping

Relay intercropping, shown in Figure 4-2, is a variation of intercropping in which the second crop is seeded in the first before the first crop is harvested. For example, soybeans are seeded into winter wheat in late April or May before the wheat has matured. The wheat will be harvested in late June or early July, leaving the soybeans to finish growth and mature. In China, corn is transplanted into wheat and then the wheat is harvested manually leaving the corn to mature later. Relay intercropping combines elements of both intercropping and double cropping.

The first crop will have a definite competitive advantage over the second crop. Its root system will be fully developed while the new seedlings will have a very limited root system and will find it very difficult to compete for water and nutrients. Careful selection of varieties of both crops is necessary. The first crop should not develop too dense a leaf canopy, and the second crop must be able to survive with minimum light. Application of fertilizer near the seedlings and irrigation can greatly aid the second crop, and may be necessary.

Figure 4-2 A relay intercrop of soybean in wheat.

Double Cropping

Double cropping, also called **sequential cropping**, is the practice of growing two or more crops on the same field in the same year, one after the other. It differs from intercropping in that the crops are not growing together but follow each other in a sequence. An example, shown in Figure 4-3, is soybeans planted in June or July after winter wheat has been harvested. The soybeans are then harvested in the fall.

Figure 4-3 Double crop of soybean after wheat.

Double cropping is practiced in some parts of the U.S. where moisture is adequate and the growing season is long enough to allow the second crop to mature. The big advantage of double cropping is that it enables the production of two crops instead of one. However, frequently, the yield of the second crop is lower because of restrictions in water supply and growing season, and the producer must use good management to produce a profit. Sometimes, weed control is also more difficult with double cropping. When the second crop is planted later in the growing season, weed seedlings can vigorously compete with the crop seedlings because the soil is warmer. Herbicide carryover is also more critical when one crop immediately follows another and may restrict the use of certain herbicides on the preceding crop. Usually soil water is depleted by the previous crop and rainfall or irrigation is needed to assure good germination and subsequent growth of the succeeding crop. In spite of these problems, however, double cropping is becoming more widespread as crop producers endeavor to maximize productivity.

A diagram of multiple cropping systems is shown in Figure 4-4.

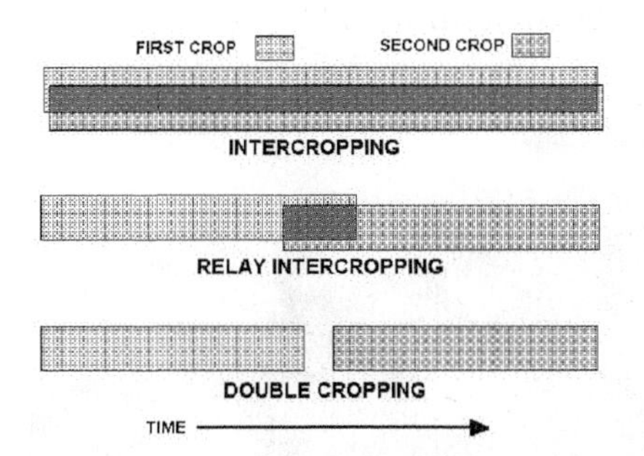

Figure 4-4 A schematic of multiple cropping systems.

FALLOW SYSTEMS

A **fallow system** is the practice of keeping the land free of all growing vegetation for at least one growing season. The purpose of fallow is to store the precipitation that occurs during the fallow period for use by the subsequent crop. Fallow is commonly practiced in semiarid regions that receive less than 20 inches of annual precipitation. Following the fallow period, crops are planted for one to three years, and then the land is fallowed again. A common practice in the High Plains of the U.S. is to alternate fallow with winter wheat. Figure 4-5 shows a wheat-fallow system. In this system, wheat is grown every two years on a field, but the wheat yields are more than doubled so total production is greater than would occur with continuous wheat. Another common practice in this region is to follow the wheat with a row crop such as sorghum or corn and then fallow. Often, it is best to precede the wheat with fallow as it is very efficient at using the water stored during the fallow period, the land will have the protection of a growing crop during the winter and early spring when the erosion hazard is greatest, and the fallow period will include the summer months when precipitation is usually greatest.

Figure 4-5 A wheat-fallow system.

In order for fallow to be successful, the soil surface must maintain high infiltration rates for the precipitation that occurs, the soil must be protected from erosion, and all weeds must be controlled. These three goals are primarily accomplished with good tillage practices that maintain adequate crop residues on the soil surface and control weeds before they achieve much growth. The timing and method of tillage are very important since tillage itself can cause loss of soil water and crop residues.

Wind and water erosion are major problems with fallow systems because the soil is without the protection of growing crops for a considerable time. To minimize the erosion hazard, the soil surface structure should be maintained to provide stable aggregates large enough to resist the force of wind and water. Also, microbarriers to the wind should be maintained at the soil surface such as crop residue, surface roughness, and/or large clods; and minimum field width perpendicular to the prevailing wind direction or slope helps to reduce erosion. Figure 4-6 shows a stubble-mulch fallow system contrasted with fallow in which no residue has been maintained on the soil surface.

Figure 4-6 An example of stubble-mulch fallow on the right, fallow with no crop residues on the left.

The minimum amount of residue to maintain on the soil surface is partly decided by soil texture and the type of residue present. Sandy soils need more residue cover than silt or clay soils, and finer residues such as wheat straw provide better protection than coarser residues such as corn and sorghum. Besides soil surface microbarriers, non-woody windbreaks such as grass strips are effective measures to keep soil erosion in check, especially in situations where surface barriers are difficult to manage.

In recent years there has been an increased use of herbicides for weed control in fallow thereby reducing the number of tillage operations needed. This is commonly called **ecofallow**. For best results, the herbicide should be applied soon after wheat harvest so that the residues can be left undisturbed over the following winter. This practice usually results in better **fallow efficiency**, that is, the proportion of the precipitation that is stored in the soil for use by the subsequent crop. However, even under the best management, only about 30-35% of the precipitation that falls during the fallow period is stored for the subsequent crop and the fallow efficiency can be as low as 10-15% when poorly managed.

SUSTAINABLE FARMING SYSTEMS

In the last two decades the rising costs of petroleum energy, and of the agrichemicals derived directly and indirectly from petroleum, have greatly increased interest in discovering and using alternatives to these crop production inputs. The possible impact to the environment, the potential health hazard to people and animals, and the effect on food safety from continual and/or heavy uses of pesticides have prompted some crop producers to explore and adopt other methods of disease, pest, and weed control. Of equal concern is the constant decline in soil productivity due to soil erosion, decreasing soil organic matter, and the pollution of surface water from soil sediments and agrichemicals. These concerns and others of a more ethical and philosophical nature have lead to the development of alternative agricultural systems.

Organic farming is a production system that prohibits or severely restricts the use of manufactured (not naturally produced) agrichemicals, including all pesticides, fertilizers, growth regulators, and livestock feed additives. **Sustainable agriculture** is a production system that strives to greatly reduce the level of purchased inputs and minimize the impact of cropping practices on the environment. Sustainable agriculture does not prohibit the use of chemical fertilizers and pesticides, but it does reduce their use.

A key premise behind a sustainable agricultural system is the maintenance of long-term productivity within the physical and biological limits of the environment, while maintaining the economic viability of the producer within the socio-economic constraints of society. With proper management practices, crop yields under sustainable agriculture systems can be comparable to yields under more conventional cropping systems that use higher levels of inputs.

In both organic and sustainable systems, soil productivity is maintained through crop rotations, legumes, animal wastes, green manures, other organic wastes produced both on and off the farm, management of crop residues, and the extensive application of soil and water conservation practices. Weeds, pests, and diseases are controlled by crop rotation, tillage and cultivation, adjustment of planting date, and the use of resistant varieties.

SOIL FERTILITY MAINTENANCE

The use of crop rotations, especially ones that include legumes and/or perennial grasses is an effective method of maintaining soil organic matter levels and enhancing soil nitrogen availability. The use of green manure crops can temporarily increase the availability of plant nutrients but this practice will not maintain or increase the inherent soil supply. In a completely organic system, nutrients other than nitrogen must be added through animal manure or other organic sources high in the desired nutrient. Careful management of crop residues is an effective method of nutrient recycling, and soil conservation practices avoid nutrient losses due to erosion; but neither of these activities increases the supply. Without some procedures to add nutrients like phosphorus, potassium, sulfur, calcium, magnesium, and the micronutrients as needed, it is unlikely that mineralization from organic matter and release from soil minerals will be adequate to produce optimum crop yields.

WEED CONTROL

Nonchemical methods of weed control are used by both conventional and organic crop producers, but are commonly managed with more expertise by the organic farmer since these are the only weed control practices. Crop rotations are effective in pest control when they break or interfere with the life cycle of the pest as discussed earlier in this chapter. Alternating crop species and cropping seasons, such as winter annuals, summer annuals, and perennials, can reduce weed growth or increase control with tillage and cultivation. Tillage and cultivation destroy weed seedlings and frequently warm the seedbed which promote more rapid crop seedling emergence. Higher plant densities and narrow row spacing provide earlier crop canopy shading of the soil surface which reduces weed seed germination. Mulching with crop residues or other organic materials to smother weeds and reduce germination are effective practices but difficult to apply on a field basis. Biological control through parasitic insects and diseases and selections of crops that have detrimental effects on weed growth are future possibilities for nonchemical weed control.

PEST CONTROL

Organic farmers use resistant and tolerant varieties, crop rotations, crop residue management, adjustment of planting date to avoid insect damage, and weed control to destroy breeding and overwintering sites for insects. However, most of the methods are useful only with a specific insect and a specific crop, and genetic resistance on a multi-pest level is not currently available. Some biological controls, or the use of predaceous and parasitic insects, insect diseases, and sex interference techniques (see Chapter 13) have been successfully developed and used. Generally, these methods have not been universally applicable and are difficult to manage successfully.

DISEASE CONTROL

The most commonly used disease control method is the use of resistant and/or tolerant varieties and hybrids. Organic farmers frequently use the latest resistant varieties and cultural practices to control diseases.

YIELD LEVELS

Organic farming and sustainable agriculture use less off-farm purchased inputs but tend to be more labor and information intensive than conventional farming. Crop yields are generally equal to, or slightly lower than, conventional crop yields. However, gross receipts may be lower because grasses, legumes, and lower value feed grains (oats or barley as companion crops) are included in crop rotations. Total economic return, however, may be comparable to conventional farming because fewer inputs are purchased.

The future adoption rate of organic farming and sustainable agriculture is dependent on many economic, political, and social factors. Research is needed to supplement the trial and error techniques currently used in these systems.

REVIEW QUESTIONS

1. Why should a crop rotation plan provide for approximately the same area for each crop in each year?
2. What are the advantages of including a perennial legume in a crop rotation?
3. How does a crop rotation system help to maintain soil properties? Why is this important?
4. How does a crop rotation aid in the control of crop pests?
5. Why has the use of crop rotations decreased in recent years? Will it become more popular in the future? Why?
6. Can a continuous cropping system maintain crop yields as well as a crop rotation system? Why? How?
7. What is the primary factor that decides whether a crop producer uses a particular cropping system?
8. Why is multiple cropping not widely used in the U.S.?
9. What factors decide the chances of success of double cropping? Can they be controlled by a producer?
10. What are the advantages of relay intercropping and strip intercropping when compared to complete intercropping?
11. How much yield increase does a fallow system need to achieve to be economically easible?
12. What is fallow efficiency and what factors affect it? How does each factor affect it?
13. What is the difference between organic farming and conventional farming?
14. Why do some producers favor organic farming?
15. Does organic farming require a higher level of management? Why?
16. What is the difference between organic farming and sustainable agriculture?

Part II

Plants and Soils

CHAPTER FIVE

Soil and Plant Nutrition

INSTRUCTIONAL OBJECTIVES

Upon completion of this chapter, you should be able to:

1. Define soil. Briefly discuss the importance of soil.
2. List three functions of soil.
3. List four components of soil. Discuss the relative proportion of each.
4. Define soil texture, soil structure, and soil tilth.
5. Define sand, silt, and clay. List four textural classes, and list two soil textures for each.
6. List four factors that affect soil structure.
7. Define soil organic matter, litter, duff, and humus. Discuss the process of mineralization.
8. Define gravitational water, available water, capillary water, hygroscopic water, water potential, field capacity, saturation capacity, and wilting point.
9. Discuss the relationship of soil air to soil water.
10. Define cation exchange capacity and soil pH. Discuss their effects on soil nutrients.
11. Define macronutrients and micronutrients. List at least four nutrients for each.
12. List three soil nutrients used by plants in the greatest quantities.
13. List three general sources of soil nutrients.
14. Define nitrogen fixation, symbiosis, nitrification, and denitrification.
15. Define mass flow, diffusion, and root interception methods of nutrient uptake by roots.
16. Define fertilizer application methods: preplant, starter, side dress, top dress, and fertigation.
17. List the forms that N, P, K, Ca, Mg, S, Fe, and Zn are absorbed by plants.
18. List the metabolic uses of N, P, K, Ca, Mg, S, Fe, and Zn by plants.
19. Define tillage, primary tillage, and secondary tillage. List five purposes of tillage.
20. List two major categories of tillage implements, and discuss their differences in residue management.
21. Define conservation tillage. Discuss its importance to the soil.
22. Define soil compaction. List three causes of it.

INTRODUCTION

Although this book deals primarily with crops, it is impossible to study the science of crop production without including soils since soils are the medium in which plants germinate and grow. In previous chapters we already have mentioned soils often including the importance of properly managing them. In this chapter we will only scratch the surface of Soil Science, but you need to have a basic understanding of soils to fully understand how plants function.

Soil is a natural material derived from weathered rocks and organic matter that provides water, nutrients and anchorage for land plants. Soil appears to the uneducated person as an inert substance; but, in reality, it is a very dynamic medium that is chemically active, constantly changing, and supports a tremendous amount of life within itself.

Our standard of living is largely dependent on soil characteristics and the quantity and quality of plant life that the soil will support. Many civilizations have developed and advanced because the soils were excellent, such as the Egyptian Civilization along the Nile Valley. Other civilizations have fallen or declined because the soil of the area could not sustain continued crop production or because the soil was neglected or abused. An example of this was the decline of the Babylonian Civilization where the soils were ruined through irrigation with salty and silty water.

The United States, which has been blessed with an abundance of excellent soils, has been, and is continuing to abuse one of its greatest natural resources. It is estimated that over one-half of the tillable soils in the U.S. has lost at least one-fourth of the original topsoil through erosion by wind and water.

Soil is not an absolute requirement for plant growth. Hydroponics (growing plants in nutrient solutions) and sand cultures using nutrient enriched water will support luxuriant plant growth and are used quite extensively in the production of such vegetables as tomatoes and peppers in greenhouses. However, economically efficient food production is still highly dependent on soil and proper soil management.

FUNCTIONS OF SOIL

Soil serves several important functions for plants. First it is the medium in which plants are anchored. Most crop plants have an upright growth habit on vertical stems. The use of vertical stems provides the most efficient orientation and placement of leaves for maximum light interception for photosynthesis. When the soil fails to provide suitable anchorage because of root damage, the plant lodges causing yield and harvest losses. **Lodging** is discussed in more detail in Chapter 8.

Soils also function as a storehouse or reservoir of plant nutrients. These nutrients can come from natural or inherent sources such as the parent material of the soil. They can also come from the decomposition of plant and animal residues in the soil or from applied fertilizers. However, no matter the source, the soil aids in storing these nutrients and making them available to plants. Plant nutrients in the soil are discussed in more detail later in this chapter.

A soil also functions as a storehouse or reservoir of water for plants. Soils differ in their ability to store water and make it available to plants. This subject is also discussed in more detail later in this chapter.

SOIL COMPONENTS

Soil is composed mostly of mineral matter, organic matter, water, and air. Figure 5-1 shows the relative proportions of these four major components on a volume basis.

As you can see in Figure 5-1, a typical soil is composed of about 50% solids, mainly mineral matter and some organic matter. The other 50% of the soil is composed of voids or pore space. At field capacity, which occurs when the soil is holding its maximum amount of water against the force of gravity, the pore space is about evenly divided between air and water. The relative proportions of air and water are constantly changing as a soil is wetted and dried. Air and water are inversely proportional to each other. In other words, when one increases, the other decreases. If the soil is water saturated, 100% of the pore space contains water, or 50% of the total soil volume is water. If the soil is air dry, 100% of the pore space contains air and 50% of the total soil volume is air. The solid proportion of the soil remains fixed, although the organic matter fraction can vary slightly as will be shown later.

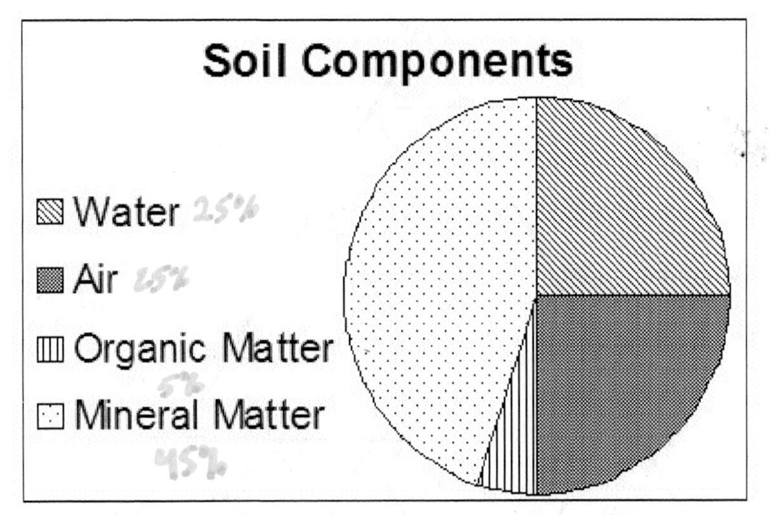

Figure 5-1 Proportion of soil components in a typical soil.

Mineral Matter

The **mineral matter** of a soil is the solid portion of a soil that was derived from inorganic or nonliving sources. Mineral matter originated from the **parent material** of the soil. Parent material can be residual, such as bedrock, or it can be transported, such as silt deposited by water or wind. Whatever the source, the parent material goes through chemical and physical changes during weathering which make it more suitable for supporting plant life.

Particle size and the proportion of these particles of mineral matter determine many physical and chemical properties of a soil. The size of particles can be grouped into three general categories. **Sand** is the largest particle size and includes all particles greater than 0.05 mm in diameter. **Silt** includes medium sized particles less than 0.05 mm but larger than 0.002 mm in diameter. **Clay** is the smallest particle size and is less than 0.002 mm in diameter. The chemical reactivity of the particles is inversely proportional to the size. Clay particles are more active in the chemical reactions occurring in the soil than sand particles.

Soil Texture

The relative proportion of sand, silt and clay determines **soil texture**. Soil texture can affect the size and number of the soil pores and the ability of water to enter the soil and be stored within it. Figure 5-2 shows how different proportions of particles sizes determine soil texture.

Soil textures can be grouped into **textural classes**. Coarse textured soils include sands, loamy sands, and fine sandy loams. Medium textured soils include very fine sandy loams, loams, silt loams, and silts. Moderately fine textured soils include clay loams, sandy clay loams, and silty clay loams. Fine textured soils include sandy clays, silty clays, and clays.

Soil Structure

Soil structure is the arrangement of the particles into stable aggregates or granules. Surface soils exhibit the strongest structural development. **Soil tilth**, which describes the ease or difficulty with which a soil is tilled is often determined by soil structure. Soil structure can be affected by natural and cultural factors that improve it by increasing aggregation, or destroy it by reducing aggregation.

Soil structure can be improved by many factors. The action of plant roots binds aggregates together through the substance that is sloughed off the root cap as it pushes through the soil. Also, when roots die and are decomposed, a channel may be left in the soil. Soil organisms, especially soil fungi and earthworms, improve soil structure by secreting substances that coat aggregates making them more stable. Additions of organic matter to the soil improve soil structure because the products of decomposi-

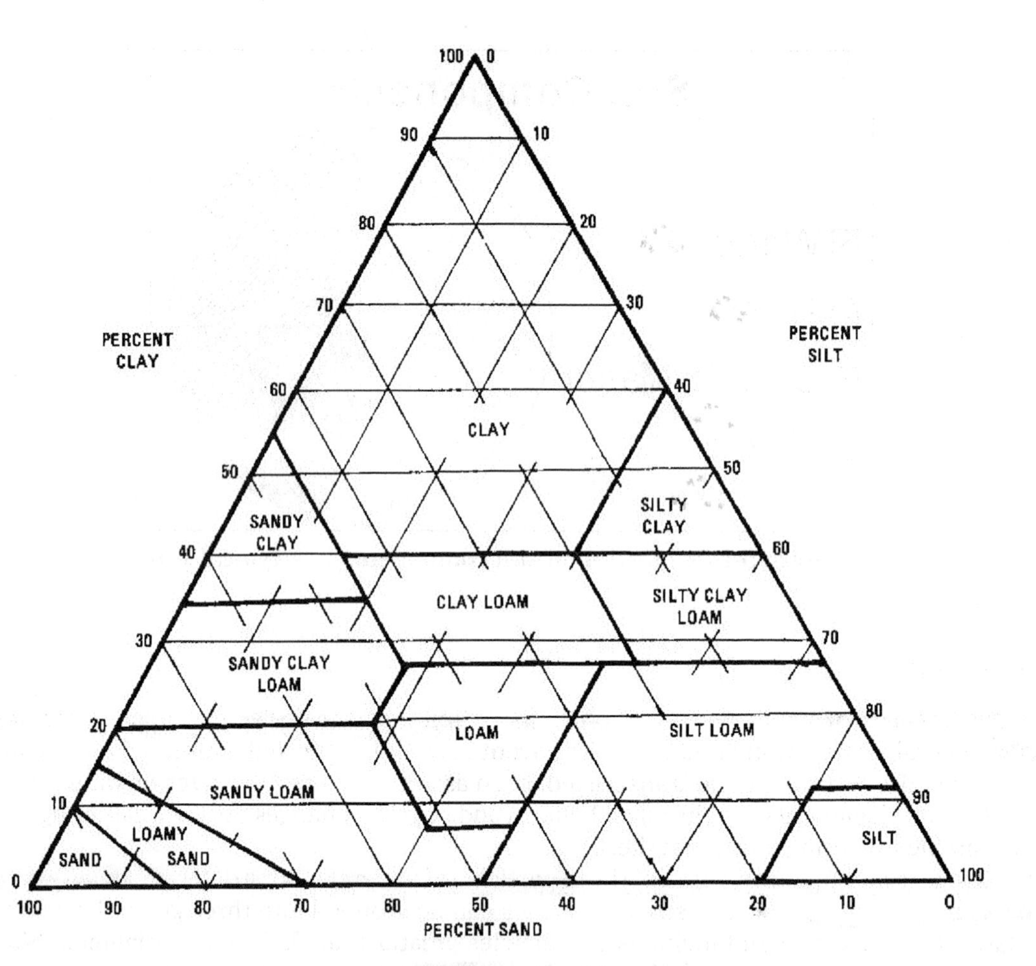

Figure 5-2 Soil textural triangle showing the proportion of sand, silt, and clay in various soil textures. (Courtesy USDA)

tion act as cementing agents. These factors that improve soil structure are almost semipermanent in nature because their effects last for a considerable time.

Climatological factors can also improve soil structure. Wetting and drying, and freezing and thawing can increase the granulation of soil aggregates particularly near the soil surface. These effects are most visible in the early spring after thawing occurs. Although valuable, these factors are temporary in nature as their effects are comparatively short-lived.

Soil structure can be destroyed by several factors, most of which are related to poor soil management. Excessive tillage destroys the natural aggregation of the soil leaving the surface very powdery. Tillage when the soil is too wet forms large aggregates or clods that become very hard when they dry. This is especially true in soils with a high clay content and is called **puddling**. Heavy machinery can destroy soil structure by compacting the soil. Raindrop impact on the soil surface can break up the aggregates causing the soil surface to crust when it dries. Soil erosion, whether by wind or water, removes the topsoil that has the most structural development. The negative effects of these latter two factors can be minimized by maintaining an adequate amount of crop residue on the soil surface.

Effects of Texture and Structure

Soil pore size is determined by texture and structure, which, in turn, determine the amount of water and air in the soil. Coarse textured soils, such as sands, have fewer but larger pores that hold less water. Fine textured soils, such as clays, have more pores that are smaller, but hold more total water. Soil structure also determines pore size and soils with well-developed structure have larger pores than those with less developed structure. Soil pore size affects the water holding capacity of a soil, the ease of movement of water into the soil, and the ease with which roots can grow through the soil.

Infiltration rate, also called **intake rate**, is the speed with which water moves into the soil. It is necessary to maintain this rate as high as possible to minimize runoff during periods of heavy rainfall. The rate with which water moves within the soil is called the **percolation rate**, or **permeability rate**.

Organic Matter

Organic matter is the soil component derived from living organisms. These living organisms include microorganisms such as fungi, mycorrhizal fungi, bacteria, and actinomycetes, as well as insects, earthworms, plant roots, plant residues, and others. Although the amount of organic matter shown in Figure 5-1 is 5%, the actual amount of organic matter that a soil contains depends on precipitation, temperature, and cultural practices. As annual precipitation increases, the soil can support more plant life that results in an increase in soil organic matter. As mean annual temperature increases, the rate of decomposition of organic matter increases resulting in less organic matter in the soil. Therefore, soil organic matter generally increases from south to north and from west to east across the North American Plains. Tillage increases the rate of decomposition of organic matter by incorporating it into the soil. The organic matter content of North America's prairie soils has shown a steady decline since tillage began.

Classification of Organic Matter

Litter, more commonly called plant or crop **residues,** is the undecayed leaves, stems, and other plant parts that are still recognizable. Residue on the soil surface is usually not classified as soil organic matter. Incorporated residues, however, are included as part of the organic matter of the soil. **Duff** is partially decomposed organic matter. It is matted and moldy and may not be recognizable. **Humus** is organic matter that is mostly resistant to further decay. It is the stable part of the soil's organic matter and determines mostly the organic matter level of a soil, such as the 5 percent shown in Figure 5-1. The first two types are very unstable and will continue to decompose until they reach the humus stage.

Contributions of Organic Matter

Organic matter is an important source of plant nutrients. Crop residues contain plant nutrients that are released for new crop growth through decomposition by microorganisms. This process is called **mineralization**. In the organic form, these nutrients are not easily leached or eroded from the soil. However, they also are not available to plants in the organic form. Figure 5-3 shows the process of mineralization of plant nutrients.

Mineralization is sometimes called **"chemical recycling."** Nutrients are absorbed from the soil by the roots, and these nutrients are incorporated into the plant through photosynthesis and metabolism. When the plant matures, the crop residue containing the plant nutrients in organic form is returned to the soil. Through microbial decomposition, the nutrients in the crop residue are changed from the **organic** (unavailable) form to the **inorganic** (available) form. The nutrients are absorbed again by the roots of the new crop and the cycle is complete.

Organic matter increases the water **infiltration rate** and **soil water holding capacity**. Soils higher in organic matter will maintain a higher water infiltration rate because soil aggregates will be more stable and raindrop impact is less destructive. Soil organic matter holds more water than the mineral matter. The amount of added organic matter that is retained as humus is very small, about 5 percent or less. Therefore, changes in soil water characteristics occur slowly when large additions of organic matter are made over long periods of time.

As mentioned earlier, organic matter helps to maintain and improve soil structure. Microbial decomposition products are coating and cementing agents for soil aggregates. Soil organisms themselves improve soil structure. For example, earthworm casts are more water stable than soil aggregates, and fungal mycelia bind soil particles together. The organic matter is also a source of nutrients and energy for the microbial population. This may be a cause and effect relationship because the population of microorganisms in the soil is directly related to the amount of nondecomposed organic matter in the soil and the amount of nondecomposed organic matter is directly related to the microbial population. As mentioned before, the soil is a dynamic, living medium!

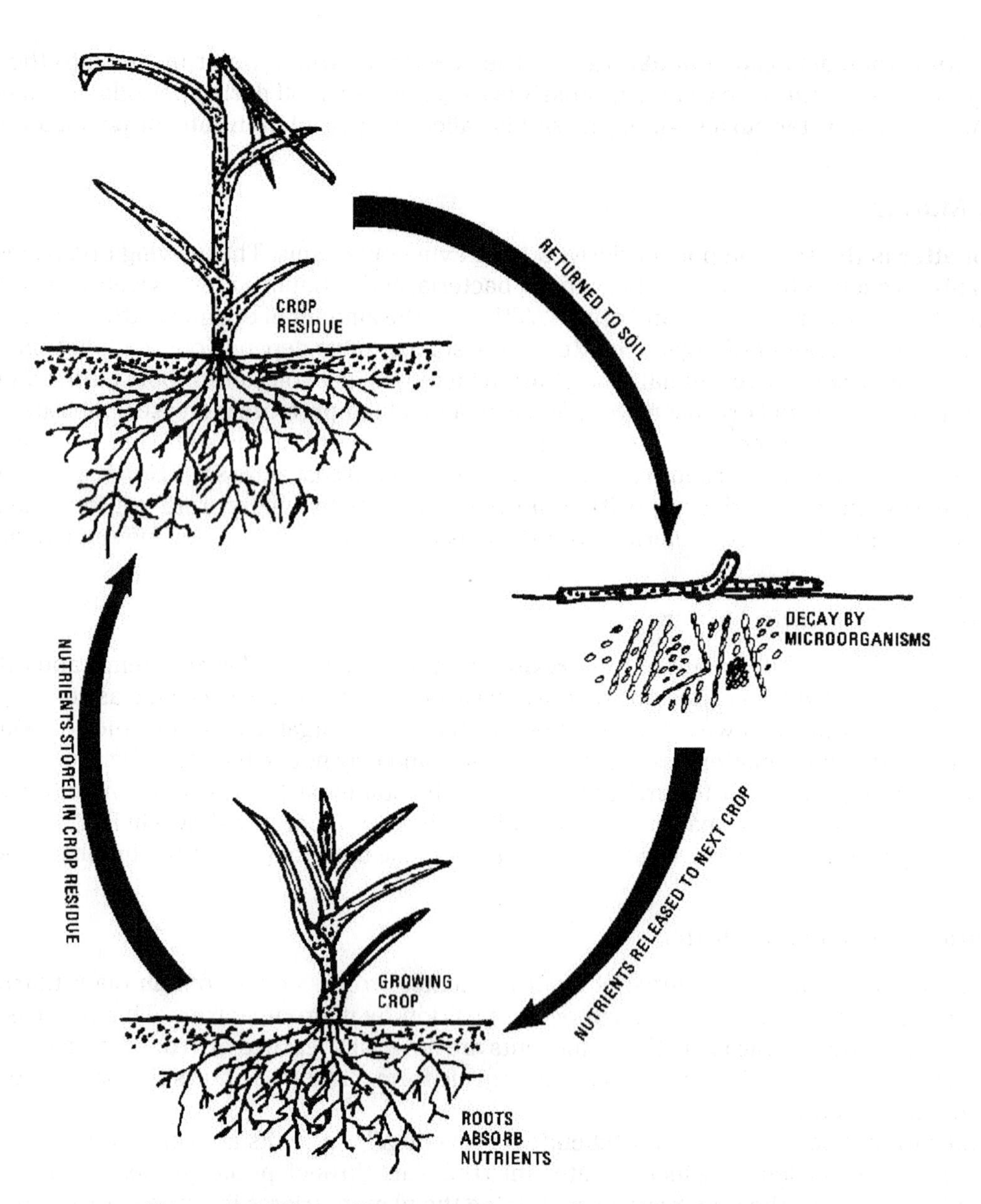

Figure 5-3 Mineralization and recycling of crop nutrients.

Most organic matter, soil organisms, and plant roots are found in the upper foot of the soil because that is the most favorable environment. Organic matter is greatest at the soil surface because of additions from the aboveground plant parts and from decaying roots. Many soil organisms feed on organic matter so they will be concentrated where the food supply is the greatest. Also soil oxygen is greater near the soil surface that most soil organisms require.

Soil Water

Water is the substance used in the greatest quantity by plants and practically all of it comes from the soil. The ability of the soil to store and release water greatly determines its capacity to support crops. The amount of water in the soil is constantly changing as the roots extract it and precipitation and/or irrigations replenish it. Soil water is analogous to a bank account with the plants making withdrawals and precipitation and irrigation making deposits. The producer tries to manage the crop to not overdraw the account.

To fully comprehend the principle of soil water, you must realize that water is held in the soil by capillary forces exerted by the walls of the soil pores (formed by soil particles and aggregates), and by the cohesion of the water itself. Soil water is measured as **soil water potential (SWP)**. Since soil water is held under a tension, SWP is a negative number. SWP is commonly expressed in **bars**. A bar is 750 mm of mercury, or 14.5 psi. A bar is also equal to one million dynes/cm^2. One dyne is the unit of force equal to the weight of one milligram.

The roots must overcome the tension the soil exerts on soil water to absorb it. The amount of water in the soil and the size of the soil pores affect SWP. Water in larger pores is held at a lower tension than water in smaller pores. As a soil dries out, the water will be retained in increasingly smaller pores, thus decreasing SWP. Therefore, not all soil water is equally available to plants.

Definitions of Soil Water

Gravitational water, also called **free water**, is that which remains in the soil only temporarily and moves down through the soil due to the force of gravity. **Available water**, also called **capillary water**, is held in soil pores by capillary forces and can be absorbed by plant roots. **Unavailable water** is held so tightly by the soil particles that plant roots cannot extract it. Some unavailable water is called **hygroscopic** or **bound water**. Figures 5-4 and 5-5 illustrate types of soil water.

At **saturation capacity** all the pores are saturated or completely filled with water. At this point much of the water in the soil is gravitational water and the SWP is essentially zero. When all of the gravitational water has moved down through the soil and the pore space is filled with only the water that the soil can hold against the force of gravity, the soil is at **field capacity**. At this time the SWP is about minus one-third bar. Field capacity is the upper limit of available water. As the soil dries due to evaporation and water uptake by plant roots, the SWP continues to decrease until it reaches -15 bars. At this point the roots can no longer extract soil water and plants begin to wilt. This is called the **wilting point** and water left in the soil is unavailable to plants. Table 5-1 shows the amount of water present in different textured soils at various water levels.

Soil Air

Soil air is a noncontinuous, nonuniform system within the soil. It is higher in carbon dioxide and lower in oxygen than atmospheric air. Respiration by roots and soil microorganisms uses oxygen and releases carbon dioxide. The decomposition of organic matter also increases soil carbon dioxide.

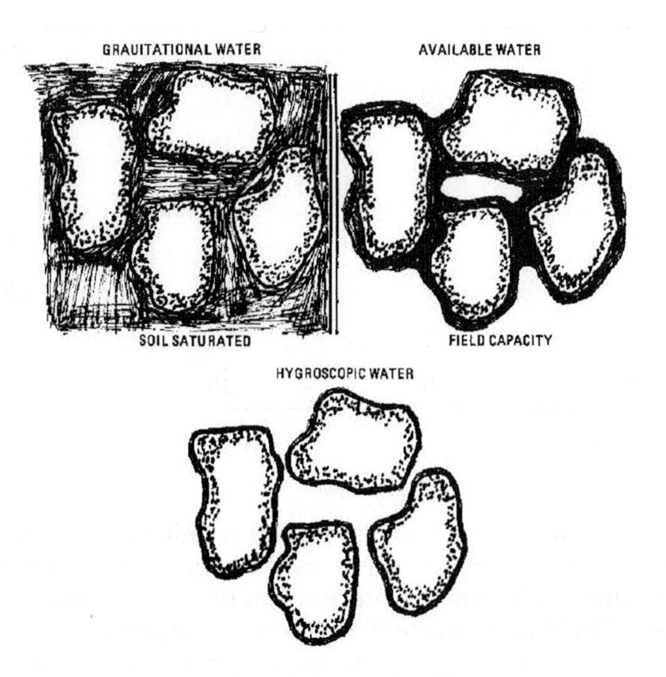

Figure 5-4 Diagram of soil water.

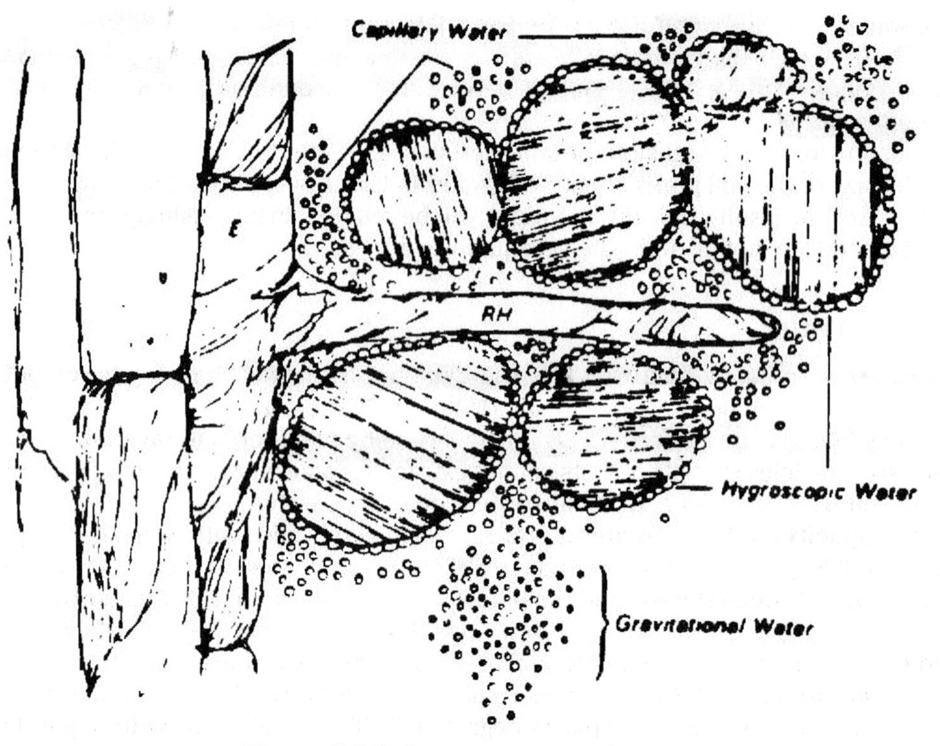

Figure 5-5 Soil water next to plant roots.

Table 5-1 Percent by volume of water in different textured soils.

Texture Class	Permanent Wilting Point %	Available Water %	Field Capacity %	Gravitational Water %	Saturation Capacity %
Sand	2	7	9	19	28
Sandy loam	4	10	14	20	34
Fine sandy loam	6	12	18	22	40
Loam	8	15	23	23	46
Silt loam	10	17	27	24	51
Clay loam	12	20	32	25	57
Clay	14	22	36	26	62
Heavy clay	18	20	38	30	68

The main concern with soil air is maintaining good **soil aeration** to provide an adequate amount of oxygen for plant roots to properly grow. Roots use oxygen for growth and other metabolic processes. The uptake of water and nutrients is an active process that requires energy from respiration with subsequent use of soil oxygen. Under normal conditions there are no problems with soil aeration, but certain conditions can lead to poor aeration and subsequent lack of oxygen. Poorly drained soils contain excessive amounts of gravitational water and will be poorly aerated because most of the pore space is filled with water. Soils with poorly developed structure or with a crusted surface will be poorly aerated because air movement into and within the soil is restricted.

Poor aeration can cause a decline of beneficial microorganisms in the soil and a reduction in the decomposition rate of organic matter. Detrimental microorganisms that can survive low levels of oxygen will increase in numbers and may form substances in the soil that are toxic to roots. The effects of inadequate oxygen are discussed in more detail in the section on respiration in Chapter 9.

SOIL CHEMISTRY

The soil is a very active chemical medium. Soil water contains many dissolved substances which makes it chemically reactive with clay, humus, and plants roots. Although a full study of the chemistry of soils is beyond the scope of this book, there are two chemical properties of soils that are important to discuss because of their influence on plant nutrients found in the soil.

Cation Exchange Capacity

Cation exchange capacity (CEC) is a measure of the ability of clay and humus to react with ions dissolved in the soil water. A cation is a positively charged particle, and an anion is a negatively charged particle. The clay and humus particles in the soil have negatively charged sites that enable them to function as anions. These negative sites will attract cations in the soil solution. CEC represents the total quantity of negative charges available to attract cations and is expressed in terms of milliequivalents of negative charge per 100 grams of dry soil. CEC values can range from as low as 3 to 5 meq/100 g for sandy soils and 20 to 50 meq/100 g for clay soils, to 50 to 100 meq/100 g for soils high in organic matter.

CEC is important in determining the ability of the soil to retain plant nutrients. Nutrients with positive charges (cations) will be held in the soil by the cation exchange sites and subsequently absorbed by the roots. Common cation nutrients are potassium (K^+), calcium (Ca^{++}), and magnesium (Mg^{++}). Soils with low CEC may lose even cation nutrients to water moving down through the soil by a process called **leaching**. Due to their inability to attach to cation exchange sites, anion nutrients are easily leached from any soil. The most commonly leached anion nutrient is nitrate-nitrogen (NO_3^-). Leaching of nitrate-nitrogen is discussed later in this chapter.

Soil pH

Soil pH is a measure of the acidity and alkalinity of the soil. The pH is measured on a scale of 0 to 14 with 7 being neutral. When the pH is below 7, the soil is acid; above 7, it is alkaline. Most field crops prefer soil pH to be near neutral to slightly acidic (6.0 - 7.3).

Unless the pH is extremely low or high, the direct effects on crops are small. However, soil pH does affect many chemical reactions in the soil particularly those involving plant nutrients. The availability of many nutrients is affected by soil pH and this is discussed later with the affected nutrients. Nitrogen-fixing Rhizobia bacteria associated with soybeans and other legumes are harmed when the soil pH drops below 6. Extremely acid or alkaline soils can be detrimental to beneficial soil microorganisms and root growth. Some elements that can be toxic to plants, such as aluminum, can become soluble when the pH is very acid resulting in crop damage.

Soils slowly become more acid with time. Alkaline elements are more likely to be leached below the root zone; and plants remove more alkaline elements than acid elements. Many applied fertilizers are acidic. Soil acidity is corrected by additions of **lime** (calcium carbonate, $CaCO_3$) to the soil.

SOIL FERTILITY

Next to providing adequate water, the biggest concern of most crop producers is providing adequate nutrients to the growing crop. The ability of the soil to store plant nutrients and make them available to plants has already been mentioned, also the mineralization of organic matter. Most cropping systems require the addition of plant nutrients, or fertilizers to maintain crop yields. These fertilizers can come from a variety of sources and the amount required depends on the crop grown and the yield goal.

Soil nutrient levels are determined by carefully gathering soil samples that are then analyzed in a testing laboratory. How often soils need to be tested depends on the type of soil and the crop being grown. Soils with high CEC values retain nutrients better and may need to be tested less often. Soil pH affects nutrient availability as shown in Lab Figure 6-2 on page 261. Soils will need to be sampled more often when crops are grown that use large amounts of nutrients, such as corn and cotton.

Sources of Plant Nutrients

There are many sources of nutrients for field crops. Crop residues release nutrients when they decompose through mineralization. Crop residues such as corn, small grains and sorghum contain about 0.5-1.0% nitrogen and phosphorus by weight. The return of residues to the soil is important in nutrient maintenance. When non-legume forage crops, such as corn silage, are continuously harvested from a field, the requirements for added nutrients increase considerably.

Animal wastes are another important source of nutrients. The amount of nutrients in animal wastes varies with the type of animal, the ration fed, the waste handling system, and whether bedding is included. Other organic sources of nitrogen include sewage sludge, peat moss, and other organic waste materials.

The most widely used sources of soil nutrients are commercial fertilizer. Most commercial fertilizer is inorganic which makes it more readily available to plants. The organic sources just discussed must be changed to inorganic forms through mineralization before they can be absorbed by plants. Commercial fertilizers are the most widely used sources for most nutrients because they are more convenient and more economical than other sources. Also, presently, there is an insufficient supply of organic sources.

Commercial fertilizers can be purchased as dry granules, liquids, or even as a gas. All commercial fertilizers are required by law to show the plant nutrient content or **fertilizer grade**. Fertilizer grade is the percent nitrogen (N), phosphorus as P_2O_5, and potassium as K_2O as determined by chemical analysis. These forms of phosphorus (P_2O_5) and potassium (K_2O) do not occur in any fertilizer materials and are merely relics of antiquated methods of analysis and reporting of fertilizer grade. Phosphorus pentoxide (P_2O_5) contains 42% actual phosphorus by weight and potassium oxide (K_2O) contains 83% actual potassium. There has been some effort within the fertilizer industry to change fertilizer grade to actual nitrogen, phosphorus, and potassium but these efforts have had only limited success. Examples of fertilizer grades will be given as different types of commercial fertilizers are discussed.

Classification of Plant Nutrients

Plant nutrients can be classified as **macronutrients**, or those that are used in large quantities by plants, or as **micronutrients** that are used in only very small amounts by plants. The relative proportions of the various nutrients used by plants, however, are not an indication of their importance. A deficiency of a micronutrient can be just as devastating to a plant as a deficiency of a macronutrient.

Plant Nutrition, pages 260–263, discusses nutrient mobility within the plant and deficiency symptoms. Ability to recognize nutrient deficiency symptoms in plants is essential for understanding and diagnosing problems in a crop field.

Macronutrients

The macronutrients for plants are carbon (C), hydrogen (H), oxygen (O), nitrogen (N), phosphorus (P), potassium (K), calcium (Ca), magnesium (Mg), and sulfur (S). Carbon, hydrogen, and oxygen come primarily from air and water, and are seldom lacking as nutrients. In a drought, water can be inadequate for maintaining cell turgidity, but not as a source of hydrogen and oxygen for photosynthesis. Carbon from carbon dioxide can be deficient as a reactant in photosynthesis when stomata are closed; again, that is due to drought stress. (See Chapter 8). It is possible to increase the rate of photosynthesis by adding carbon dioxide to the air, but this is only possible in a closed environment such as a greenhouse. In field crops these nutrients are always assumed to be in adequate supply. The remaining macronutrients are supplied by the soil.

Nitrogen

Nitrogen (N) is used in large amounts by most crops, and consequently is most likely to be deficient in the soil. All nitrogen, whether from organic sources or from added fertilizers, will eventually be converted to the **nitrate** form by soil microorganisms through a process called **nitrification**. Nitrate-nitrogen (NO_3) is easily leached out of the root zone because it is extremely soluble in soil water and is unable to attach itself to the cation exchange sites as discussed previously. **Leaching** is the movement of a nutrient, or other soluble substance, below the root zone with the gravitational water. Loss of soil nitrogen can also occur during prolonged soil saturation through a process called **denitrification.** Nitrate is converted into gaseous nitrogen (N_2) or other N compounds and lost to the atmosphere. Denitrification occurs because certain soil microorganisms will use the oxygen in the nitrate to fuel respiration during anaerobic conditions. Because of these factors, more nitrogen fertilizer is applied than any other nutrient.

Nitrogen is absorbed by the plant primarily in the nitrate form. Small amounts are also taken up in the ammonium (NH_4^+) and amino (NH_2^+) forms. Natural forms of nitrogen are in organic forms that are subsequently changed to inorganic forms through mineralization and subsequent nitrification.

Nitrate-nitrogen absorbed by the plant is converted to the ammonium form and subsequently incorporated into amino acids, proteins, chlorophyll, nucleic acids, and coenzymes. Adequate N promotes vegetative growth.

A plant that is deficient in N will show **chlorosis**, or a yellowing of the leaves, followed by "firing" or death of leaf tissue beginning at the tips and sides of the leaves and progressing to the base. Nitrogen deficiency will first appear in older leaves at the bottom of the plant and progress up the plant if not corrected. Plants deficient in N will also be stunted and will not flower or produce seed properly.

Many nitrogen fertilizers are sold in the dry granular form. Ammonium nitrate is one of the most common types. Its fertilizer grade is 33-0-0 which means that it contains 33% nitrogen, 0% phosphorus, and 0% potassium. Urea (45-0-0) is another type of dry granular nitrogen fertilizer, also ammonium sulfate (21-0-0) which also contains 16% sulfur as a plant nutrient. Liquid forms of nitrogen fertilizer are non-pressure solutions made by dissolving ammonium nitrate and/or urea in water. The most common solution grades are 28-0-0 and 32-0-0. The nitrogen in liquids is no more available to plants than other forms, but is more convenient than granular forms. Liquids can be easily transferred using pumps.

Gaseous nitrogen fertilizer is anhydrous ammonia (82-0-0) which is ammonia gas liquefied under pressure. It is a liquid in the tank but becomes a gas when injected into the soil. Anhydrous ammonia requires special equipment and must be injected into the soil. Once in the soil, it is quickly dissolved in the soil water in the ammonium (NH_4^+) form. While in the ammonium form it attaches to soil organic matter or clay particles which prevent leaching. However, it is converted to the nitrate form by soil microorganisms through **nitrification** as discussed previously. Anhydrous ammonia is the most economical form of nitrogen and is the nitrogen source used in making many other forms and types of nitrogen fertilizer.

Another important source of nitrogen for plants is legumes. Legumes have the unique ability to form a symbiotic relationship with ***Rhizobia sp.*** bacteria resulting in nitrogen fixation. Nitrogen fixation converts atmospheric nitrogen into organic nitrogen and subsequently makes it available to the host plant. As discussed in Chapter 3, **symbiosis** is a mutually beneficial relationship between two organisms. Here, the host legume plants protect the bacteria that live in nodules on the legume roots and provide them with sugar for food. In return the bacteria provide needed nitrogen to the host plants. The bacteria usually fix more nitrogen than the host plant requires and this additional nitrogen becomes available to succeeding crops through the mineralization cycle. Before inexpensive commercial fertilizers were available, the use of legumes in rotation with other crops was the primary method of maintaining adequate amounts of soil nitrogen. It is still a valuable practice, especially as fertilizer prices continue to rise.

Phosphorus

Phosphorus (P) is required by plants in smaller quantities than N or K. It is important in the storage and transfer of energy as part of ATP. Phosphorus is a constituent of many proteins, coenzymes, and in other compounds. There are high concentrations of phosphorus in meristems, seeds and fruits. A good supply of phosphorus early in the plant's life is essential. About 75% of total phosphorus will be taken up by grain crops by the time they have produced only 25% of their total growth.

Phosphorus is absorbed by plants mostly in the orthophosphate form ($H_2PO_4^-$) and to a lesser extent in the monohydrogen phosphate form (HPO_4^{2-}). These forms, however, are not readily stored in the soil so their supplies are constantly renewed. The amount of available phosphorus is greatly affected by soil pH. Phosphorus is most readily available when soil pH is from 5.5 to 7.0.

Deficiency symptoms of phosphorus include a purple color in the leaves beginning with the older leaves. Plants with severe or prolonged phosphorus deficiency will become severely stunted. Sometimes, even when P is adequate, a purple color will develop in seedlings, especially corn, when the plants are exposed to cold, wet conditions. This is caused by problems with P availability in the soil and uptake by the plant roots due to the existing weather conditions. If phosphorus supply is adequate, however, the plants will recover rapidly when the weather improves and yield will not be affected.

Phosphorus fertilizer is required in many soils, especially weathered and sandy soils, and in **calcareous soils** that contain free lime. Phosphorus is mostly immobile in the soil so it is not subject to leaching like nitrogen. Crop residues are important sources of phosphorus and animal wastes usually contain about one-half as much phosphorus as nitrogen. Other organic sources such as those mentioned before also contain phosphorus. However, as with nitrogen, most of the phosphorus fertilizer comes from commercial fertilizer.

All types of phosphorus fertilizers come from rock phosphate deposits. Rock phosphate itself is unavailable to plants so it must be processed before it can be used as a fertilizer. Superphosphate (0-20-0) is formed by treating rock phosphate with sulfuric acid. It also contains about 24% sulfur. Another important commercial fertilizer is triple superphosphate (0-45-0) which is made by treating superphosphate with phosphoric acid. These two fertilizers are marketed in the dry granular form and are often mixed with other fertilizers to provide complete fertilizer mixtures. Liquid phosphoric acid is also available for use with other liquid fertilizers. If phosphorus fertilizer is needed, banding a concentrated supply of P fertilizer near the seed at seeding is an efficient method of application as discussed later in this chapter.

Potassium

A field crop may take up more **potassium (K)** than any other soil nutrient, but it is less likely to be deficient than nitrogen. Potassium is mostly stable in the soil and is not easily lost. Only a small amount of potassium is removed in the grain or seed and a significant amount is returned to the soil in the residue. However, careful attention must be given to potassium supply in the soil.

Plants absorb potassium as K^+. Potassium is used in many metabolic processes within the plant, although it is not a component of any organic compound. Potassium is essential for photosynthesis, sugar translocation, and enzyme activation. Plants deficient in potassium exhibit marginal **necrosis**, or death of leaf tissue along its edge. Older leaves will first show deficiency symptoms. Stalk or stem strength is also affected by K supply. Plants without adequate potassium will be more likely to lodge late in the season.

With high levels of potassium in the soil, the plant will exhibit "luxury consumption" or excessive uptake of K. Although this may not harm the plant directly, it may interfere with the uptake and utilization of other nutrients, especially magnesium, and lead to a deficiency of those nutrients. Potassium fertilizer is needed in some soils, especially those that have been highly weathered or those formed from parent material low in potassium.

Crop residues are excellent sources of potassium. Most of the potassium absorbed by plants is found in the leaves, stems, and roots, and is returned to the soil through mineralization. Animal wastes and other organic sources usually contain about as much potassium as nitrogen.

As with the previous two nutrients, the most common source of potassium is commercial fertilizer. The most common potassium fertilizer is potassium chloride (0-0-62) which is commonly called muriate of potash. Deposits of this are found in various parts of the world. Florida is the primary supplier in the U.S. Another common source is potassium-magnesium sulfate (0-0-27) which also contains 22% sulfur and 8% magnesium. These are usually sold in the dry form. Liquid forms are also available.

Calcium, Magnesium, and Sulfur

Calcium and magnesium are seldom lacking as nutrients. **Calcium (Ca)** is a major component of the cell wall and is involved in cell division. It is absorbed by plants as Ca^{2+}. **Magnesium (Mg)** is an important component of chlorophyll and aids in P uptake. It is absorbed by plants as Mg^{2+}. Both nutrients are stable in the soil. Although rarely lacking as a nutrient, great quantities of calcium are applied each year as lime to some soils to correct soil acidity, as was discussed previously. Calcium is rarely applied to crops as a nutrient. Crop that do respond to calcium fertilizer are peanut and some vegetables. The quality of these crops is enhanced by timely applications of calcium. Potassium-magnesium sulfate (8% magnesium) is the most common commercial source of magnesium fertilizer. Crop residues are important sources of calcium and magnesium.

Sulfur (S) is required by plants in about the same quantities as phosphorus. It is a part of several amino acids and is crucial in protein synthesis. It is also involved in respiration. Plants absorb it as SO_4^{2-}. Because of its similarity to N in the plant's metabolism, a deficiency of S will look similar to a deficiency of N. However, S deficiency will begin on the younger, upper leaves, not the older, lower leaves as described for N deficiency.

A sulfur deficiency is less likely to occur than deficiencies of N, P, or K. Soil organic matter is an important source of sulfur. Sulfur is mostly stable in organic matter and is released through mineralization. Sulfur can be deficient in some acid and/or sandy soils.

Crop residue, animal waste, and other organic sources are important in maintaining soil sulfur levels. Common commercial sources of sulfur are elemental sulfur (100% S), gypsum (24% S), ammonium sulfate (16% S), and potassium-magnesium sulfate (22% S).

Micronutrients

Micronutrients, also called **minor elements,** are those nutrients that are used only in small quantities by plants. For example, a hectare of corn can require over 224 kg (200 lb/A) of nitrogen, but only 70 to 140 g (1-2 oz/A) of zinc. However, as mentioned before, they are just as important to plants as those used in large quantities and deficiencies are just as harmful. Since the plants need small amounts, micronutrients are not as likely to be deficient in the soil, but deficiencies do occur in some soils. The micronutrients are iron, zinc, manganese, molybdenum, boron, chlorine, copper, and cobalt.

Iron (Fe) and **zinc (Zn)** are the two most common micronutrient deficiencies to occur in U.S. soils. Iron deficiency is common in **calcareous soils** (soils with free lime). Here, the iron is unavailable because of the high pH of these soils. Additions of iron fertilizer to the soil do not always correct the problem because the added iron also becomes unavailable. Repeated foliar applications of an iron sulfate solution can correct the deficiency but is quite expensive. Usually the most economical solution to the problem is to raise only crops that are tolerant to low amounts of iron such as wheat or alfalfa. Zinc deficiencies can occur on some acid sandy soils, calcareous soils, or soils that have been leveled for irrigation or otherwise disturbed. The most common zinc fertilizer is zinc sulfate.

Plants absorb iron as Fe^{2+} or as a complex with organic salts (chelate). Iron is involved in many enzymatic reactions in the plant and in the production of chlorophyll. A deficiency, called **iron chlorosis,** shows up as a yellowing or even whitish coloration between the leaf veins. If severe, stunting will also occur. Plants absorb zinc as Zn^{2+}. Zinc is a part of enzymes and serves to activate enzymes. Deficiency symptoms include a rosette growth pattern due to little internode elongation on the stem. Leaves will be short and chlorotic.

NUTRIENT UPTAKE

Nutrient uptake, especially phosphorus and micronutrients, is enhanced in many crops by the symbiotic relationship between roots and mycorrhiza fungi living on the root surface. The mycorrhiza send filaments, called hyphae, out from the root surface that absorb nutrients and make them available to the root. This benefit is most important when nutrients levels are low.

Nutrients contact the root surface for absorption by one or more of three methods as shown in Figure 5-6. Although some examples are given for each method, any nutrient can contact the root using any of these three methods. All three methods are in constant activity in the soil during nutrient uptake.

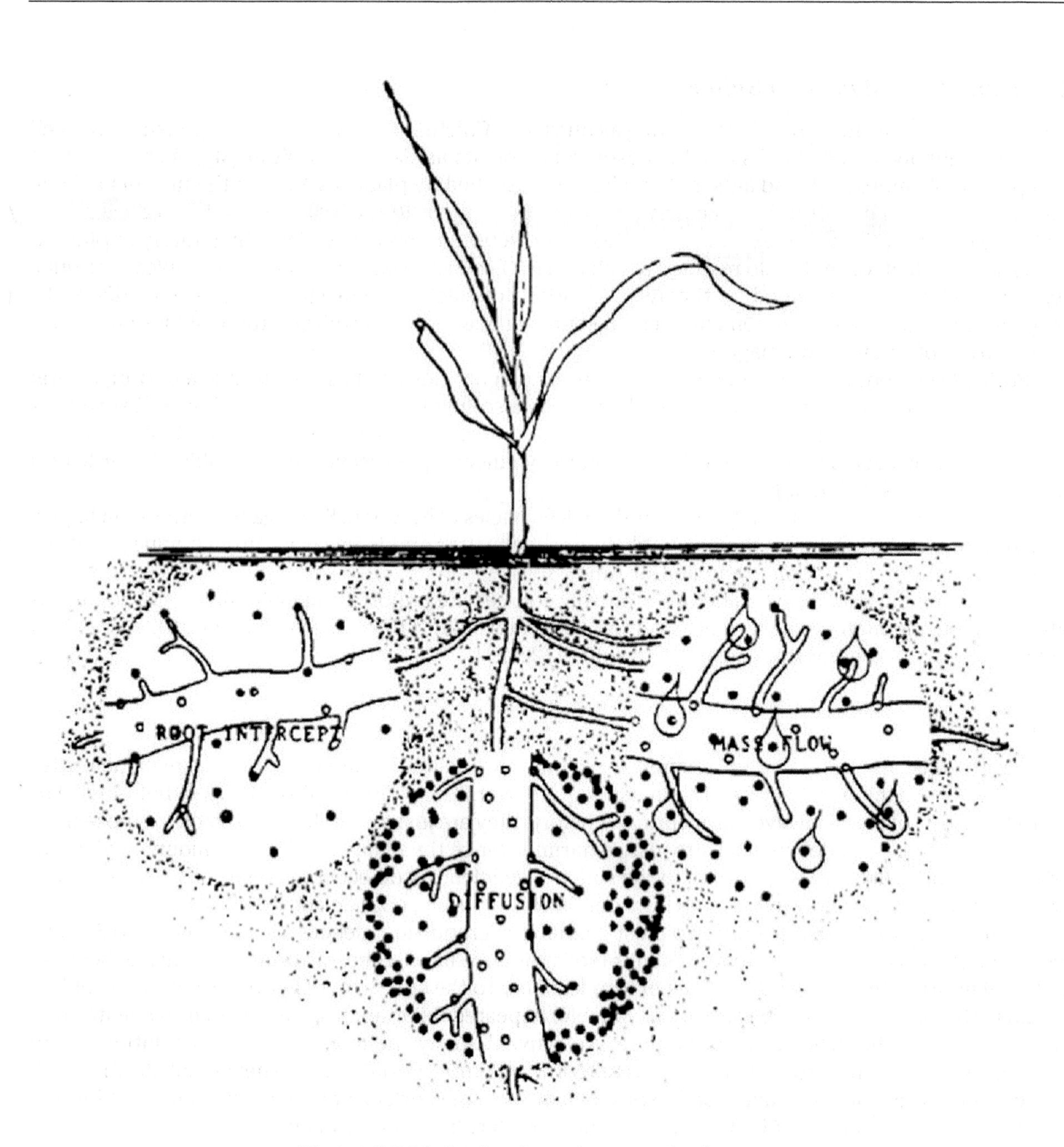

Figure 5-6 Methods of nutrient uptake by roots.

Mass flow accounts for about 80% of the contact between roots and nutrients. As discussed previously, the plant roots exert a tension on the soil water to draw it out of the soil pores. With mass flow, the nutrients are dissolved in the soil water and move with the soil water as it is drawn to the roots. Most of the nitrogen and sulfur are absorbed in this manner.

The second mechanism by which roots contact soil nutrients is **diffusion**. Diffusion is a natural process in which a molecule will move from an area of higher concentration to an area of lower concentration. As the roots absorb a nutrient and remove it from the soil solution, the concentration of that nutrient will become lower. More molecules of the nutrient will then move or diffuse toward the root surface. Much of the phosphorus and potassium, which are attached to soil clays and organic matter, move toward the root by diffusion.

Root interception is the third mechanism that enables roots to contact nutrients. In this method, the roots simply encounter the nutrients as they grow through the soil pores. Most of the calcium, magnesium, and molybdenum are taken up in this way.

The actual mechanism by which nutrients enter the root is complex and not well understood. A nutrient must pass through the cell membranes in the root which involves both physical and chemical processes. It involves a combination of ion exchange with attachment of the nutrient to a carrier that moves it across the membrane.

METHODS OF FERTILIZER APPLICATION

The best time to apply fertilizer to a field depends on the crop, the availability of equipment and labor, and the type of fertilizer. Nitrogen should be applied before grain crops flower; while phosphorus, potassium, and most other nutrients should be applied at, or before, seeding. Fertilizers subject to leaching, such as nitrate-nitrogen, should be applied later than those that are immobile in the soil. There is no best time to apply all the fertilizer nutrients to a crop, but often only one fertilizer application is made to minimize labor and equipment costs.

Preplant

Fertilizers can be applied before planting. **Preplant applications** usually occur during seedbed preparation and frequently involve incorporation of the fertilizer as well. Often the fertilizer is mixed with a herbicide or other chemical to combine field operations in the spring. Preplant incorporation of fertilizer can occur in the fall or spring for spring seeded crops, or in the late summer for fall seeded crops. Most fertilizer can be applied with little danger of loss from the soil unless the fertilizer contains nitrate-nitrogen. If nitrogen is to be applied more than a few weeks before planting it is better to use an ammonium form, such as anhydrous ammonia, to minimize leaching losses.

Seeding

Fertilizer can also be applied at seeding time as a starter. **Starter fertilizer** is banded below and to the side of the seed row. The purpose of starter fertilizer is to provide the seedlings with a readily available and concentrated source of nutrients while the root system is still small. The result is quicker and more uniform emergence especially when the soil is cool and wet. Although starter fertilizer may or may not increase crop yields, it is an effective method of applying fertilizers, especially phosphorus and micronutrients. Fertilizer can also be applied on the soil surface with or without incorporation at planting. Again, a common practice is to apply it with a herbicide. However, this type of application is not a starter because the fertilizer is not readily available to the newly emerged seedling.

Postemergent

Fertilizer can also be applied postemergent, or after the crop has emerged. This is the most efficient way to apply nitrogen since it is applied just before the greatest crop demand and consequently is rapidly absorbed by the crop before losses occur. **Side dressing** is a postemergent application between the rows of row crops. **Top dressing** is used when the rows are too narrow for application between the rows, such as with small grains or alfalfa. Top dressed fertilizer is not incorporated into the soil and the fertilizer is not available to the crop until rain or irrigation moves it into the root zone. During this time some volatilization loss can occur especially if the nitrogen is in the urea form.

Another postemergent method of fertilizer application is **fertigation**, or applying the fertilizer through the irrigation system. Although fertigation is used mainly with sprinkler irrigation systems, any method of irrigation can be used. Fertigation is a very efficient means of applying nitrogen as it is immediately available to the crop and requires little labor and equipment.

TILLAGE AND SEEDBED PREPARATION

Tillage is the mechanical, soil stirring action conducted for nurturing plants. Tillage requires over half the engine power used on farms. To be economically feasible, all tillage operations must return more in crop yield or improved soil conditions than the cost, and, as mentioned before, excessive tillage can destroy soil structure. Proper tillage, however, can aid in weed control, maintain good soil tilth, and optimize production.

Purposes of Tillage

Most tillage is conducted to control weeds. Tillage can kill the growing weeds and bury weed seed too deep for successful emergence. Successful weed control reduces the competition between the weed species and the planted crop for light, water, and nutrients. Tillage often provides a soil surface and/or a seedbed which increases the effectiveness of herbicides.

Another important purpose of tillage is **seedbed preparation**. Proper tillage can improve the soil environment for seed germination. First, it can increase the soil temperature. Tillage incorporates plant residues from the soil surface and a bare soil warms faster than a soil covered with crop residues. A bare soil will lose moisture faster from evaporation that also enables it to warm faster. The biggest disadvantage of leaving the soil surface without a mulch of crop residue is the increased risk of erosion. Although it is cooler, a mulched soil conserves more soil water than a bare soil.

Tillage also improves the soil environment for seed germination by improving soil aeration. Tillage loosens the soil and increases the amount of large air spaces in the soil. It can also eliminate a soil crust that hinders movement of air through the soil. However, as mentioned before, tillage destroys the inherent soil structure and may ultimately lead to less total pore space in the soil. In most soils, tillage operations must be carefully timed and minimized for best results. Sometimes, tillage can increase soil moisture content by destroying a crust on the soil surface and by increasing surface roughness which aid in water infiltration into the soil. However, as previously discussed, a mulched soil will usually absorb and hold more precipitation than a bare soil. Under some conditions the soil may be too wet for optimum germination and tillage may be useful in drying out the soil.

In spite of the disadvantages of incorporating crop residue, tillage is important in the management of these residues. A reduction of residue may be necessary to avoid interference with subsequent cultural practices such as seeding, cultivation, irrigation, or fertilization. Residue management can also reduce the incidence of certain insects or diseases that overwinter or live in the residue.

Tillage can be necessary to incorporate soil amendments such as lime and fertilizer or herbicides. Some fertilizers and herbicides will volatilize and require incorporation to avoid losses. To be most effective in reducing soil acidity, lime should be thoroughly mixed with the soil tillage layer. Although a loose mulch-covered surface provides the most effective erosion control, a bare soil is sometimes roughened by tillage to reduce surface wind velocities and subsequent wind erosion. Tillage is sometimes used to expose the soil to freezing and thawing, and wetting and drying to improve soil tilth. However, this provides only temporary modification of surface soil structure and should be only a part of a good tillage management system.

Obviously, the factors that decide the optimum amount and type of tillage are complex and vary between soils and years. To properly manage the soil, the crop producer must have a keen sense of what the soil requires and match the tillage operation to those needs.

Tillage Implements

Tillage implements can be divided into two major categories: stirring implements that cause major alteration of the soil surface and crop residue, and subsurface implements that minimize surface changes. The tillage implements that incorporate maximum amounts of residue include the moldboard plow and various types of disks. Subsurface implements leave most of the residues on the soil surface and include the sweep plow and the rodweeder. Table 5-2 shows the amount of residue left by different kinds of tillage implements.

Primary tillage is the first tillage operation after the harvest of the preceding crop. Primary tillage is mainly used to incorporate some crop residue and kill any existing weeds. The most common primary tillage implements are the moldboard plow and the offset or tandem disk. As Table 5-2 shows, the moldboard plow will bury essentially all of the residues, while the disk will bury about one-half. Subsurface implements may be used when it is necessary to leave most of the residues on the soil surface for erosion control.

Secondary tillage is any subsequent tillage operation in the preparation of the soil for seeding. Sometimes, secondary tillage is not used and the soil is tilled only once before seeding. The moldboard plow is usually not used for secondary tillage but any other type of implement can be used. Secondary tillage can be used to break up large soil aggregates and firm the soil to remove large air pockets. It can also be used to level and firm the seedbed. However, the usual purpose of secondary tillage is to kill weeds that have sprouted.

Table 5-2 Amount of residue incorporated by tillage and seeding implements.

Machine	Residue Incorporated (%)
Moldboard plow	100
Oneway disk	40
Tandem or Offset disk	
18-22 inch disks	40
24-26 inch disks	50
Anhydrous applicator	20
Chisel plow	25
Mulch treader	20-25
Sweep plow, 30" or wider blades	10-15
Rod weeder	5-10
Grain drill, double disk openers	20
Planter, double disk openers	10
Slot planter	0
Till planter	20

It is possible to seed a crop with very little tillage at all. **Conservation tillage** is a system that minimizes disturbing the soil and residue. Sometimes it is called **no tillage** or **minimum tillage**. This practice takes specialized seeding equipment, such as the slot planter or till planter, to cut through the crop residue, but can provide savings in fuel, labor, and equipment. Weed control is more difficult with little or no tillage, and sometimes the increased cost of herbicides offsets the savings in fuel and labor. Conservation tillage may not work well on poorly drained soils due to difficulties with getting the soil dry and warm in the spring. However, this practice provides maximum soil conservation and is becoming more popular each year.

Tillage Problems

Any tillage will destroy soil structure because the implement disturbs the soil as discussed previously. **Soil compaction** is caused by heavy tractors, harvesters, and tillage equipment that press or compact the soil pores resulting in smaller pore size. Soil compaction results in a reduction in soil aeration and difficulty in root growth through the compacted layer. Compacted soil is usually found just below the tillage zone and is sometimes called a tillage pan.

Soil compaction can be reduced by minimizing the number of tillage operations, avoiding tillage when the soil is wet, and running tractor and machinery tires over the same track with each tillage or harvesting operation. Compacted layers can be broken up with deep tillage by a chisel plow but care must be taken to avoid the conditions that caused the compaction or a tillage pan may develop even deeper in the soil.

REVIEW QUESTIONS

1. What is soil? Why is it important?
2. How does the soil function in plant growth?
3. What is the difference between mineral matter and organic matter?
4. What is the relationship of the volume of air and water in the soil?
5. What are the differences between soil texture and soil structure? Which one is easily changed?
6. How do texture and structure affect soil infiltration rate and percolation rate? Why are these important?
7. What factors affect soil organic matter content?
8. What is mineralization? Why is it important to crops?
9. What effects does organic matter have on soil?
10. What is soil water potential? How does it affect the movement and storage of soil water? Why is it expressed as a negative number?
11. Is all water equally available to plants? Is all available water equally available? Why?
12. Why is soil aeration important? What affects it?
13. What are primary, secondary, and micronutrients? Is a deficiency of one more damaging than another? Why?
14. How is a nutrient leached? What nutrient is subject to leaching?
15. What is fertilizer grade? How is it used by a grower?
16. Why are commercial inorganic fertilizers used in greater quantities on crops? Are they more effective?
17. When is the best time to apply nitrogen, phosphorus, and potassium? What are the advantages and disadvantages of preplant and postemergent applications?
18. Why is tillage used on crops? Is it absolutely necessary?
19. Why is residue management important? Should all residues be left on the soil surface?
20. What causes soil compaction? How does it harm the plant?

CHAPTER SIX

Seeds and Seeding

INSTRUCTIONAL OBJECTIVES

Upon completion of this chapter, you should be able to:

1. Discuss the importance of seed size, shape, and color. Give an example of each.
2. Discuss the importance of seed composition. List seeds that are either high or low in oil, protein, fiber, and starch.
3. Define fruit, true seed, and caryopsis. List two examples of each.
4. List the three major components of seeds. Discuss the function and importance of each.
5. Define radicle, cotyledon, endosperm, plumule, epicotyl, hypocotyl, and coleoptile. Discuss how each functions in germination and seedling emergence.
6. Discuss the differences between dicot and monocot seeds.
7. Define seed viability. Discuss four factors that affect it during storage.
8. Discuss the three stages of germination.
9. Define epigeal and hypogeal emergence. List the embryonic structures responsible for emergence for each. List at least three examples of crops for each.
10. Discuss the advantages of hypogeal emergence.
11. Discuss the four factors that affect germination.
12. List four causes of germination failure. Give an example of each.
13. Define seed dormancy, physical dormancy, and physiological dormancy. Discuss the methods used to break dormancy.
14. Discuss the difference between dormant seed and quiescent seed.
15. Discuss the factors that affect seeding rate, seeding depth, and seeding date for spring and fall seeded crops.

INTRODUCTION

With grain crops, the roots, stems, leaves, and flowers are all grown to produce seed. All of the management decisions and cultural practices, such as tillage, fertilizers, and variety selection, are designed to maximize seed production. A large, vigorous, healthy plant will produce the most seed, and a large segment of crop science is involved in the improvement and production of high quality seed.

Nature was most ingenious in the design of seeds! They are miniature plants naturally packaged and ready for shipment, storage or planting. Each seed contains an **embryo** that is a miniature plant complete with stem, leaves, and root. It contains a **food supply** to nourish the live embryo during storage. The embryo is in a somewhat suspended animation, or quiescence, so that the food supply will not be rapidly consumed. Finally, the seed is encased in a **seedcoat** that protects it from the outside world. However, when conditions are right, the embryo springs to life and rapidly develops into a full-fledged plant as we know it!

Seed Shape, Size, and Color

Seeds vary greatly in shape, size, and color, which enable us to identify them. It can also affect the handling characteristics during storage and seeding. **Seed shape** is the form of the seed. Seeds can be egg-shaped as in wheat and rye, triangular as in buckwheat, round as in soybean, slender as in oats and rice, flat as in corn, and curved or coiled as in alfalfa. The shape of seeds decides the type of planting equipment mechanism that must be used.

The planter must move the seed and meter it out uniformly for best results. Many planters use plates for this and the type of plate must be matched to the seed being planted. Round seeds will move much easier than flat or triangular seeds. Some perennial grasses such as bluestem have glumes and awns attached which makes them fuzzy and causes them to stick together. With these types of seeds, it is necessary to have mechanical agitators in the seed box of the planter to ease the movement of the seed to the metering device.

You don't have to look at very many different types of seeds before you realize that they vary tremendously in size between crops and even between varieties of the same crop. One of the largest seeds is the coconut that can weigh as much as 10 kgs (22 lbs). Most crop seeds are medium in size. Soybeans average about 5,500 seeds per kilogram (2,500/lb), and grain sorghum averages about 33,000 seeds/kg (15,000/lb). Some crop seeds, however, are quite small. Tobacco averages about eleven million seeds/kg (5 million/lb)!

Seed size does not determine the size of the plant that will be produced. Tobacco seeds, which average about five million seeds/pound produce mature plants as large as soybean seeds that average about 2,500 seeds/pound. Another good example is the Giant Sequoia of Northern California. Its seeds are quite small, and the full-grown tree is about 400 million times larger than the seed from which it began! Seed size, however, does affect the type of planting mechanism needed to meter the seed. As mentioned before, the seeds must be individually metered during seeding to achieve a uniform stand of plants in the field. Incorrectly adjusted planters can drop two seeds at once if set too large, or they can crack the seeds if set too small.

Seed size also decides seeding depth. Larger seeds have more food reserves than smaller seeds and may also have larger embryos. Conversely, they have the energy to emerge from deeper in the soil before the food supply is exhausted. For example, alfalfa seed is only planted about 0.6 to 1.2 cm (1/4-1/2 in) deep while corn is planted about 4 to 5 cm (1.5-2 in) deep. It is usually easier to achieve satisfactory germination and emergence with larger seeds because the soil is much more likely to be moist at the deeper depths.

Seed color is also highly variable between crops and even between varieties of the same crop. Seeds can be just about any color and some are even multicolored. Here are just a few examples:

RED:	grain sorghum, kidney beans, cranberries
ORANGE:	sorghum
PURPLE:	red clover
YELLOW:	corn, soybeans, oats, barley
BROWN:	wheat, sorghum
WHITE:	field beans
GREEN:	field peas
BLACK:	beans
SPOTTED:	castor, pinto bean
STRIPED:	sunflower
VARIEGATED:	alfalfa

Seed color does not affect planting or production, but it can affect marketability. For instance, a variety of oats was developed that was higher in food value than other oats, but it never really became popular because it had a black hull which made it appear unpalatable to livestock and humans.

IMPORTANCE OF SEEDS

As mentioned before, the primary purpose of field crops is seed production. The plants themselves use seeds for the perpetuation and propagation of the species. Early people learned this when they began saving seed to plant back the next year as was discussed in Chapter 1. Unless some sort of disaster strikes, the plant will always produce more seeds than itself. For example, one corn seed produces 1,000 to 2,000 seeds, one wheat seed produces 50 to 80 seeds, one sorghum seed produces 1,000 to 2,000 seeds, and one soybean seed produces 50 to 150 seeds. As you can see, plants vary greatly in their ability to increase themselves, or perpetuate the species.

Crops also differ in how they can propagate themselves through seed production. Pure lines of self pollinated crops produce progeny or offspring exactly like the parents. For instance, a pure variety of wheat, oats, or soybeans will produce that variety unless physically contaminated. On the other hand, cross pollinated crops produce progeny that may be similar to, or widely different from, the parents. For example, hybrid corn and grain sorghum seed must be purchased every year to maintain genetic purity because these crops are cross pollinated, and seed fields of most perennial grasses and legumes must be isolated to prevent contamination with foreign pollen from other varieties. The subject of genetic purity is discussed in more detail in Chapter 10.

Seeds are also important as a source of food and fiber for people and animals. Most of our food comes from grain crops either directly or indirectly through animals. Seeds of grain crops are sources of starches as in corn and small grains, oil and protein as in soybean, and fiber as in cotton as well as beverages, flavorings, and spices. Table 6-1 shows the chemical composition of several crop seeds. **Ether extract** is a measure of oil content, **ash** measures mineral content, and **nitrogen free extract** measures carbohydrates, primarily starches.

As shown in Table 6-1, seeds vary considerably in their composition. Seeds that are high in protein are usually also high in oil. Seeds that are high in oil and proteins are usually low in carbohydrates (starches), and vice versa. The reason for this is that seeds usually use either starches or oils as the primary food source for the embryo. Seeds that still have the hulls attached, such as oats, are higher in

Table 6-1 Chemical composition of seeds. Source: National Academy of Sciences, National Research Council, Publication 505, Composition of Cereal Grains and Forages.

	Dry Matter (%)	Crude Protein (%)	Ether Extract (%)	Crude Fiber (%)	Ash (%)	Nitrogen Free Extract (%)
Barley	90.7	14.2	2.1	6.2	3.3	74.2
Corn	89.3	10.9	4.6	2.6	1.5	80.4
Oats	90.3	14.4	4.7	11.8	3.8	65.3
Peanut	--	30.4	47.7	2.5	2.3	11.7
Rice	88.6	9.2	1.4	2.7	1.8	84.9
Rye	--	14.7	1.8	2.5	2.0	79.0
Sorghum	88.7	12.9	3.6	2.5	2.0	79.0
Soybean	--	37.9	18.0	5.0	1.6	24.5
Wheat	88.9	14.2	1.7	2.3	2.0	79.8

fiber. These differences in seed composition enable people and animals to obtain a balance of necessary nutrients through the consumption of several different sources of food. Severe dietetic deficiencies can develop when only one source of food is used, resulting in many nutritional problems being experienced in many parts of the world today.

SEED STRUCTURE AND FUNCTION

It is important to understand the botanical difference between a seed and a fruit. Botanically, a **fruit** is the mature ovary that contains one or more **ovules** surrounded by an **ovary wall**, and a **seed** is classified as a mature **ovule** containing an **embryo**. A **true seed** contains an embryo and food source within a seedcoat. Other seeds are really dry, one-seeded fruits called **caryopses** (singular: **caryopsis**) because the seedcoat has been fused to the ovary wall. They are not true seeds because the ovary wall is present. Most grain crops produce caryopses including corn, sorghum, wheat, oats, rye, barley, millet, and rice. Examples of crops with true seeds are soybean, field bean, peas, and alfalfa. With these crops, the ovary wall is the pod that is discarded at maturity leaving only the mature ovules or seeds.

From an agricultural or agronomic standpoint, however, *seed* refers to the unit being planted, despite whether it is a true seed or not. Therefore, we will not always make the distinction between seeds and caryopses when discussing seeding. However, it is important that you understand the differences so that you don't get confused when studying the botany of crop plants.

Seed Components

As mentioned before, a seed is composed of a seedcoat, an embryo, and a food supply for the embryo. The seedcoat protects the seed from damage from insects or diseases, or from mechanical damage during harvesting, storage, or seeding. Seeds that have cracked or broken seedcoats are much more likely to be damaged by any of the problems above. The food supply is not only the source of energy for the embryo during storage and germination. It also serves as a source of energy for the young seedling during emergence and for up to four to six weeks after emergence. The food supply is contained in the endosperm in grasses, and in the cotyledons in legumes. The endosperm is composed mostly of starch and other carbohydrates. The endosperm of wheat and other grains is used for flour. The cotyledons are part of the embryo and are discussed later.

The Embryo

The **embryo**, as mentioned before, is a miniature plant consisting of many distinct and important parts. Figure 6-1 shows the parts of a corn see and Figure 6-2 shows the parts of a bean seed. These two seeds are typical of the grass and legume families, respectively. Grasses belong to the *Monocotyledonae* subclass, **monocots** for short and legumes belong to the *Dicotyledonae* subclass, or **dicots** for short. These two subclasses of the botanical classification system have distinct differences in the characteristics of their seeds that will become apparent as we study the different parts of the embryo.

The **radicle** is the embryonic root. It is the first structure to emerge from the seed during germination. It then develops into the primary root system. During germination, the most critical need of the embryo is water, as the uptake of water is responsible for most of the increase in cell size within the embryo. Thus, it is not surprising that the first thing done by the developing seedling is to send out its root to reach additional supplies of water. The radicle in some advanced monocots, such as grasses, is covered by a protective sheath called a **coleorhiza**.

The **cotyledon** is a specialized seed leaf of the embryo. The number of cotyledons in the embryo determines whether a plant is classified as a monocot (one cotyledon), or a dicot (two cotyledons). In dicots, the two cotyledons serve as the food supply for the embryo and young seedling until the leaf area and its photosynthetic activity are sufficient to maintain the plant. If the cotyledons are damaged or broken off, the young seedling grows slowly. In advanced monocots, such as grasses, the single cotyledon is modified into two specialized structures. The **scutellum** absorbs food from the endosperm and transfers it to other parts of the embryo. The **coleoptile** is a protective cap over the embryonic shoot bud or **plumule**.

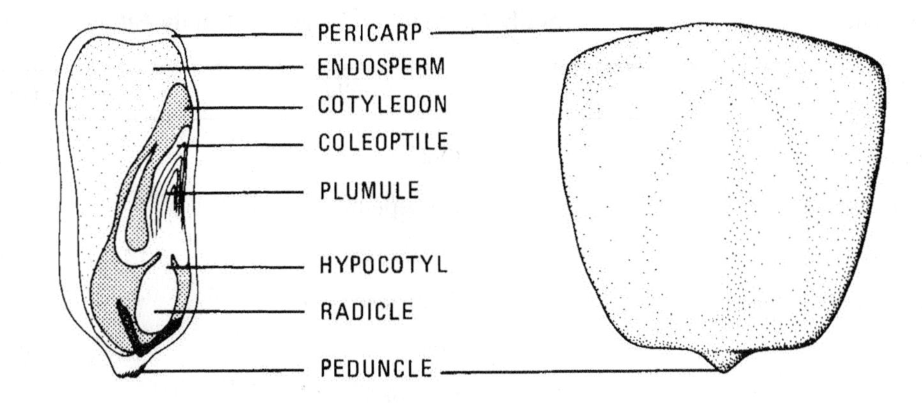

Figure 6-1 Parts of a corn seed (caryopsis).

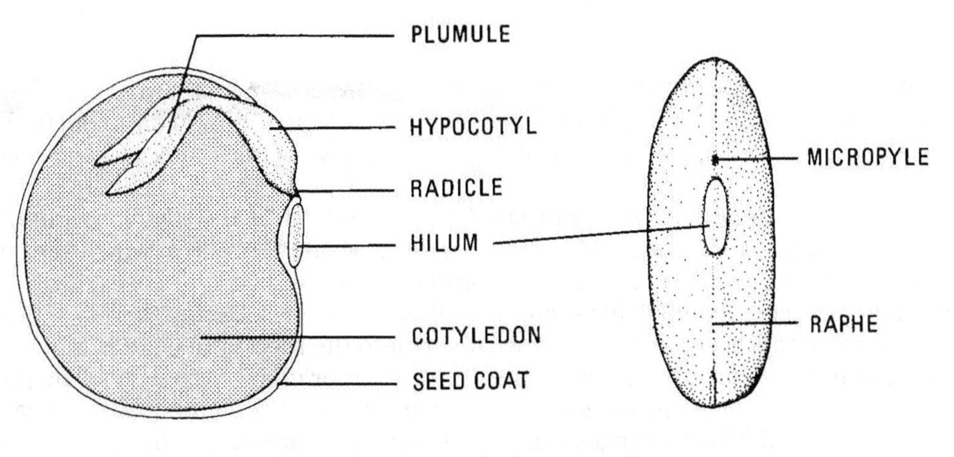

Figure 6-2 Parts of a bean seed.

The **epicotyl** is the embryonic stem above the cotyledons and below the plumule. The **hypocotyl** is the embryonic stem below the cotyledons and above the radicle. These embryonic stems can elongate and function in seedling emergence. The **mesocotyl** is a specialized embryonic stem found in advanced monocots. It elongates during emergence thus becoming the first internode in grass seedlings.

The **plumule** is the embryonic shoot bud. It is located above the cotyledons and above the epicotyl, and contains a compacted layer of **meristematic cells**. Meristematic cells are capable of dividing. In dicots, such as legumes, the plumule consists of the **apical meristem** and usually two embryonic leaves. In grasses, the plumule consists of the **intercalary meristem** and two to three embryonic leaves. Apical and intercalary meristems are discussed in Chapter 8.

Since we have been contrasting grasses and legumes, it might be helpful to summarize their differences. This is done in Table 6-2.

SEED VIABILITY AND GERMINATION

Viability is the capacity of a seed to germinate. While in storage, most seeds have very low rates of metabolism. A viable seed is one that can resume growth when conditions are favorable. Viability is usually expressed as germination percentage, or the number of normal, viable seedlings produced from 100 seeds. However, viability involves more than just the ability to germinate. Good viability also includes a rapid rate of germination, followed by normal vigorous seedling growth.

Table 6-2 Differences between grass (Family *Poaceae*) and legume (Family *Fabaceae*) seeds.

	Grass Family	Legume Family
Cotyledon	One	Two
Food Supply Location	Endosperm	Cotyledons
Food Supply Type	Mostly starch	Protein and oil
Function of Cotyledons	Food absorption	Food storage
Coleoptile	Present	Absent
Coleorhiza	Present	Absent
Hypocotyl	Inactive	Active
Epicotyl	(Mesocotyl)	Present

Germination

Germination is the process by which a seed embryo develops into a seedling and subsequently a plant. The process of germination involves several stages that are distinct and can be identified by rate of water uptake and visible changes in the seed. Figure 6-3 shows water uptake by the seed during germination.

Stage I, called the **activation stage,** is primarily characterized by the rapid absorption of water by the seed. During this stage, the seedcoat softens, swells, and often ruptures, which is easy to observe. Water absorption during this stage is a passive, physical process. In other words, it requires no metabolic energy and can occur in nonviable seeds as well as viable ones. During this stage the seed is drier than the surrounding soil, and the soil water will move into the seed until equilibrium is reached.

The increase in the water content of the seed causes other changes to occur. The protein synthesis system is activated. Some enzymes produced during seed formation are reactivated, and new enzymes are synthesized. These changes cause the respiration rate within the embryo to increase dramatically. Stage I may be completed within a matter of minutes or hours depending on the type of seed, the size of the seed, and water availability in the soil.

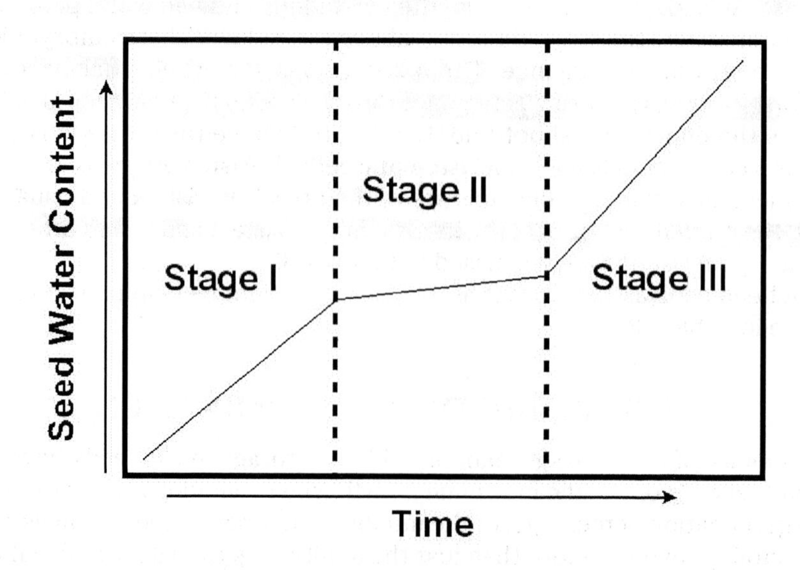

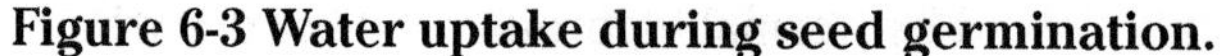

Figure 6-3 Water uptake during seed germination.

Stage II is the **digestion and translocation stage** of germination. During this stage, food reserves are broken down and translocated to the growing points of the embryo, namely the radicle and plumule. Embryo metabolism is now very rapid and specific enzymes are synthesized for each specific compound and end product. Complex carbohydrates are changed to starches, which are further reduced to sugars. Proteins are broken down to amino acids and other nitrogenous compounds. Fats and oils are changed to fatty acids, which are further reduced to sugars. During this stage, water uptake continues at a steady rate although not as rapidly as during the first stage. There are no visible changes in the seed during this stage, as it involves primarily metabolic processes.

Stage III involves **cell division and elongation** within the embryo and is clearly visible to the viewer. Cell division begins in the growing points located in the radicle and plumule. The radicle emerges first from the base of the embryonic axis and water uptake increases again as shown in Figure 6-3. Shortly after that, cell elongation begins in the embryonic stem and emergence begins. Development of the young seedling is a further extension of Stage III.

Seedling Emergence

Once the radicle has emerged from the seed, it develops into the primary root, and, in monocots, some **seminal roots** develop rapidly from the seed. This initial root system anchors the young seedling in the soil.

Subsequent development of the seedling depends on the type of emergence that occurs. **Hypogeal emergence** involves elongation of the epicotyl or mesocotyl that leaves the cotyledon(s) in the soil. **Epigeal emergence** involves elongation of the hypocotyl that pushes the cotyledon(s) above the soil surface. Figure 6-4 shows hypogeal emergence of corn, and Figure 6-5 shows epigeal emergence of bean. Notice the difference in the location of the cotyledon(s) after emergence.

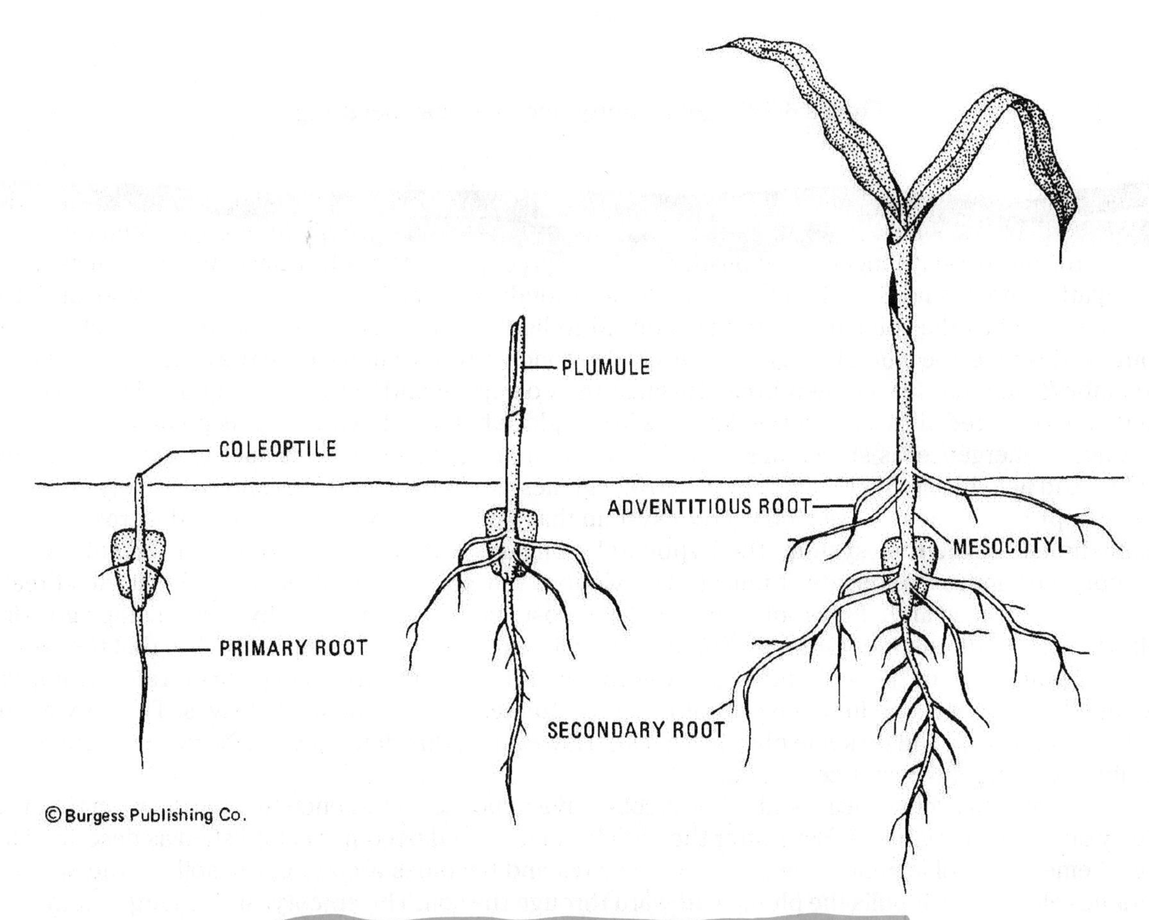

Figure 6-4 Hypogeal emergence of a corn seedling.

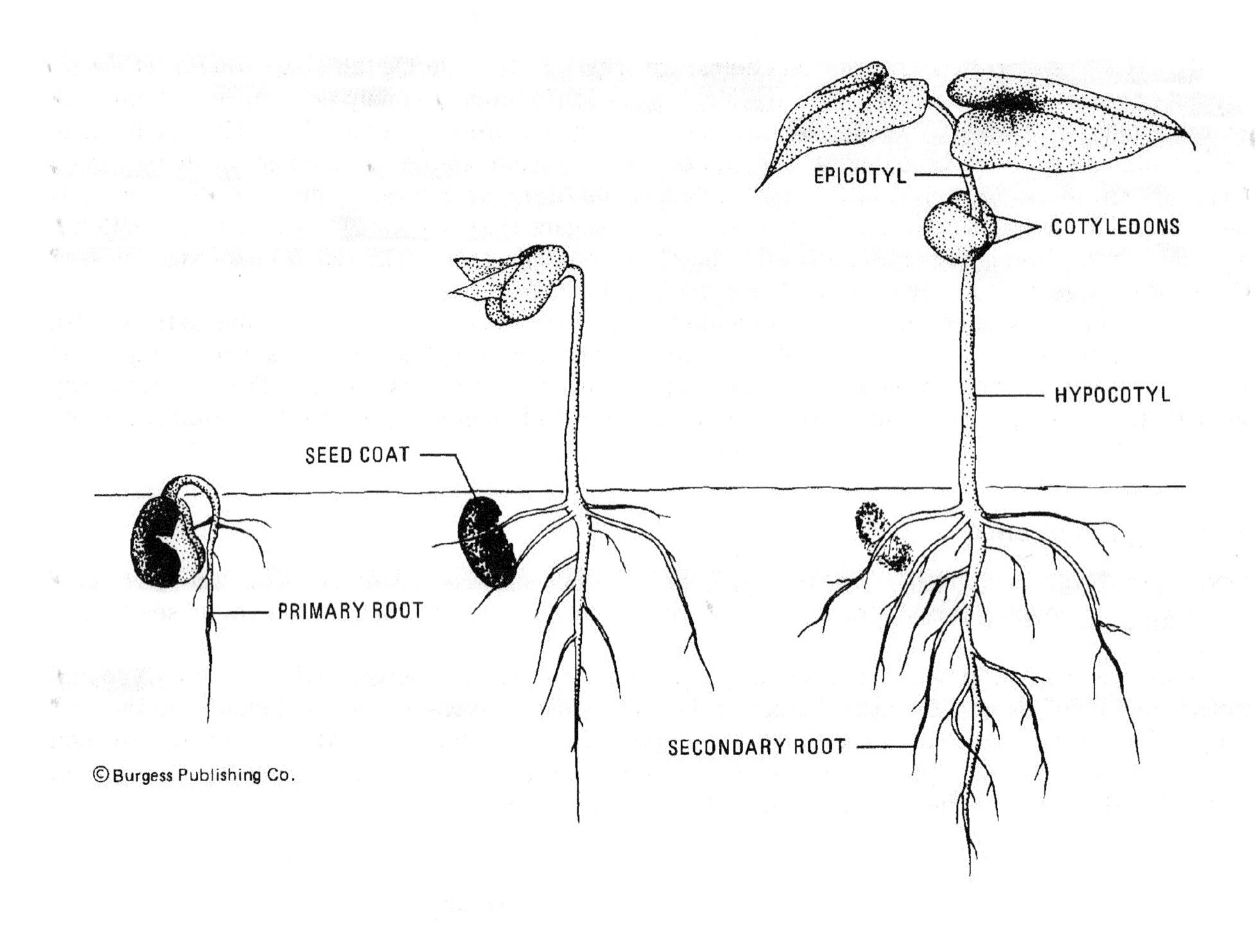

Figure 6-5 Epigeal emergence of a bean seedling.

Hypogeal emergence is common in monocots, and all grasses have hypogeal emergence. Crops with hypogeal emergence include corn, wheat, sorghum, rye, oats, and barley. In hypogeal emergence of grasses, the mesocotyl elongates and pushes the coleoptile toward the soil surface. While the mesocotyl is elongating there is no growth of the plumule or secondary roots. The mesocotyl elongates until the coleoptile reaches the soil surface and is exposed to light, which signals the mesocotyl to stop elongating. At this time, the plumule begins growth and secondary roots emerge from the base of the plumule. The embryonic foliar leaves then break through the coleoptile and emergence is complete. Since all elongation occurred above the cotyledon, it is left in place in the soil where it was planted.

Epigeal emergence, as shown in Figure 6-5, occurs in many legume crops such as soybean, alfalfa, clovers, and field beans. In epigeal emergence of legumes, the radicle emerges and eventually develops into the taproot. There is no secondary root system that subsequently develops as in the grasses. After establishment of the root system, the hypocotyl elongates and becomes arched in the soil. As the hypocotyl continues to elongate, it pulls the cotyledons upward toward the soil surface. It then breaks the soil surface in an arch. In response to light, the most exposed part of the hypocotyl stops growing while the underside continues growth. This results in a straightening of the hypocotyl so that the cotyledons are pulled out of the soil and above the hypocotyl. After the hypocotyl has straightened, the plumule develops into the first leaves that are **unifoliolate**, or a pair of single leaves. The cotyledons produce chlorophyll and do some photosynthesis. However, at this time they are more important to the seedling as a continuing source of food.

Some legumes such as peas, and other dicots, have hypogeal emergence that differs from that previously described for grasses. Here, after the radicle emerges and becomes established as described for epigeal emergence of legumes, the epicotyl elongates and becomes arched in the soil. As the epicotyl continues elongation, it pulls the plumule upward through the soil. The epicotyl arch emerges at the soil surface, and, in response to light, it straightens as described previously. After the epicotyl has straightened, the plumule develops and the first leaves open and enlarge. The cotyledons and seedcoat remain in the soil.

It usually takes less energy for hypogeal emergence than for epigeal emergence. It is physically easier to push a slender coleoptile or epicotyl through the soil than to push an arched hypocotyl through the soil while pulling the cotyledons along, and finally straightening at the surface. Because of this, it is usually possible to seed crops with hypogeal emergence deeper in the soil than similar sized seeds that have epigeal emergence. This can be a real advantage as the soil is usually drier near the surface. Another advantage of the emergence of grasses is that the growing point is left protected near the soil surface which makes it much less susceptible to late spring freezes or other weather hazards as discussed in Chapter 12.

FACTORS AFFECTING THE VIABILITY OF STORED SEEDS

As mentioned before, seed viability is the ability of the embryo to germinate and emerge as a seedling, and become established as a plant. With grain crops that have been grown primarily for feed or food, seed viability may not be as important. Although it is important to maintain the quality of the grain, some embryos can die without greatly affecting the food quality of the grain. Many factors that affect seed viability also affect grain quality, but it is more difficult to maintain high seed viability during storage than it is to maintain good grain quality.

As we discuss the four major factors that affect seed viability, note the interaction between the factors. Interactions occur throughout the study of crop science, as you have already seen. Understanding these interactions enables you to substitute factors. When the ideal of one factor cannot be attained, its effects can be at least partially overcome by changing another factor.

Table 6-3 Viability of cotton seed stored 15 years at 1° C (33° F).

Seed Moisture (%)	Germination (%)
14	None
13	Poor
11	Fair
9	Good
7	Excellent

Seed Moisture

The moisture content of the seed during storage is the most important factor affecting seed viability. The best method to maintain viability in crop seeds is to thoroughly air dry them to 8-10% moisture when they mature, followed by storage under dry conditions. A moisture content of 4-6% may be desirable for prolonged storage of several years. In contrast, market grain that will not be used for seed can be safely stored at 12-14% with little loss of quality. For seeds not adversely affected by low moisture, each 1% decrease in seed moisture from 14% to 5% doubles the life of the seed. A higher seed moisture is possible if the storage temperature is lowered.

Increasing seed moisture can cause various problems. At 8-9% seed moisture, storage insects may become active, and at 12-14%, fungi (molds) may become active. At 18-20% seed moisture, heating may occur as the respiration rate of the seeds is high enough to release significant quantities of heat. At 40-60%, germination occurs. Table 6-3 shows the viability of cotton seed stored for 15 years at various moisture levels and illustrates the importance of maintaining proper seed moisture.

Temperature of the Storage Area

As mentioned before, lowering the temperature can at least partly offset the effect of seed moisture. Cool temperatures of 2-7° C (35-45° F) are the most satisfactory for prolonged seed storage. It is best to keep the storage temperature above freezing. Seeds may be stored at temperatures below freezing, but the seed must be in equilibrium with 70% relative humidity or lower, or the free water in the seeds may freeze and cause injury. Each decrease in temperature of 5° C (9° F) from 44 to 0° C (112 to 32° F) doubles seed life.

Humidity of the Storage Area

A range of 20-40% relative humidity is satisfactory for seed storage, although the optimum is about 20-25%. Humidity affects the moisture content of the seed. Since seeds are relatively dry, they can absorb moisture from the air if the humidity is higher.

In commercial seed storage, environmental factors are carefully controlled for optimum seed longevity, but it is usually not possible to store seed under ideal conditions after purchase. When seed will be planted within a few months, storage conditions are not critical if the seed is kept dry. When storing seed for longer periods, such as seed leftover from the previous year, adequate viability can be maintained if the seed is kept cool and dry. Cool, dry basements are satisfactory, but most storm cellars and damp basements are too humid unless a dehumidifier is used. Storage in buildings such as machine sheds is not good if the temperature will get quite hot in the summer, and quite cold in the winter.

Seed Age

As you would expect, the viability of seed decreases with increasing age, but there are crop and variety differences. The seeds of cereal grains and legumes may remain viable for 20-30 years under proper storage, while perennial grass seeds remain viable only 5-6 years under ideal storage conditions. Generally, the more vigorous the germination and seedling emergence of a particular crop or variety, the greater the seed longevity.

Most crop seeds maintain satisfactory viability for one to two years under farm storage conditions. However, crop seed stored on the farm for one year or more should be tested for germination before planting, even if stored under good conditions. The price of a germination test is inexpensive when compared to the loss incurred from planting seed low in viability and vigor.

ENVIRONMENTAL CONDITIONS AFFECTING GERMINATION

In order for a seed to germinate, it must receive adequate water and oxygen, be exposed to a proper temperature, and sometimes be exposed to light. Although adequate water is the most likely factor to be limiting in the soil, it is important to understand that all factors are equally important in the germination process. A deficiency of any factor will prevent germination from proceeding.

Water

Seeds must absorb sufficient water for rapid germination, and, as mentioned before, the uptake of water begins almost immediately when the seed is placed into moist soil. Water, which is so vital to all organisms, does many important functions during germination. Water softens the seedcoat so the radicle and plumule can emerge. Water aids in changing insoluble food to a soluble or usable form for the embryo, and is the solvent in the translocation of the solubilized food to the embryo. Water also eases the entrance of oxygen into the seed by softening the seedcoat and by increasing membrane permeability to oxygen.

The amount of water absorbed by a seed varies with the type of crop. Corn will absorb 40% of its dry weight in water, and wheat absorbs 60% of its dry weight in water. Seeds high in protein and oil, such as legumes will absorb up to 90% of their dry weight in water. The amount needed for germination to begin is less. Corn must take up 35% of its dry weight before germination begins, and wheat requires 45%. Soybean, a legume, requires about 60% of its dry weight in water to begin germination. When seeds absorb these amounts of water, the seedcoat has sufficiently softened, and adequate food reserves have been mobilized, to allow the respiration rate of the embryo to increase rapidly. The amount of water required for germination is a direct function of the protein content of the seed. Seeds higher in protein require more water for germination than those lower in protein.

Oxygen

Most dry seeds will not absorb oxygen until after they have absorbed enough water to begin germination. An adequate supply of oxygen for germination is primarily dependent on soil aeration, which is discussed in detail in Chapter 5. Seeds placed in a properly prepared seedbed will not usually have problems getting enough oxygen for germination.

Adequate oxygen is necessary for respiration within the embryo during germination. The respiration rate of germinating seeds is several hundred times greater than dry seeds in storage. One-half of the dry weight of seeds is commonly used in the respiration of germination. During germination, 63 kg/ha (1 bu/A) of wheat, which is the common seeding rate, uses the oxygen in 25.5 m^3 (900 ft^3) of air and produces the energy equivalent to plowing that area 15 cm (6 in) deep!

Lack of sufficient oxygen for germination under field conditions seldom occurs, but may result if the soil is completely saturated with water that displaces the soil air. It can also occur if the soil is severely compacted or crusted which presents a physical barrier to the movement of oxygen into the soil. The seeds of some hydrophytes can germinate under conditions of little or no oxygen. For example, rice will germinate on the soil surface under six inches of water. Some weed seeds also can germinate under water or in mud.

Temperature

Proper temperature is necessary to enable the embryo to attain an adequate level of respiration for germination. The temperature must be high enough for the biochemical processes of food solubilization, and cell division and growth to occur. Respiration is discussed in more detail in Chapter 9. Field crop seeds will germinate in a temperature range from 1°C to 49° C (33°F to 120°F) depending on the crop species. Table 6-4 shows the temperature requirements for germination of several crops.

The minimum, maximum, and optimum temperatures at which crop seeds will germinate are called the **cardinal temperatures** of germination. Cardinal temperatures also exist for other stages of growth of each crop as discussed in Chapter 12. The **minimum temperature** is the temperature below which germination will not occur. The **maximum temperature** is the temperature above which germination will not occur. The **optimum temperature** is the temperature range where germination goes most rapidly.

Table 6-4 Temperature requirements for germination of several crops.

Crop	Minimum		Optimum		Maximum	
	°C	°F	°C	°F	°C	°F
Alfalfa	5	41	20	68	44	111
Corn	10	50	20-30	68-86	48	119
Cotton	16	50	20-30	68-86	--	--
Rice	21	70	22-30	72-86	--	--
Sorghum	21	70	22-30	72-86	48	119
Soybean	16	60	20-30	68-86	--	---
Wheat	1	33	15-20	59-68	40	104

Cool-season plants, such as wheat, have lower cardinal temperatures for germination than warm-season plants such as corn or sorghum. Knowledge of the minimum temperatures for germination is important in deciding the time limits for seeding a crop. For instance, corn can be seeded earlier in the spring than sorghum as it can tolerate soil temperatures up to 11° C (20° F) colder.

Most crop seeds germinate best when subjected to alternating temperatures than when exposed to constant temperatures. This is logical since seeds are adapted to conditions where the soil temperature usually rises during the day and drops at night. Scientists have recognized this fact by developing a cold test for germination. This test is used with corn and other crops to measure seed viability under the conditions of cool soil temperatures and high soil moisture that would occur in the field in a cool, wet spring. The test is conducted using unsterilized field soil. After the seeds are planted, the soil is watered to field capacity. The seeds and soil are maintained at 10° C (50° F) for six to ten days, and then the temperature is raised to 21° C (70° F) for three to five days and the emerged seedlings are counted. Placing seeds under minimum conditions for germination is a good measure of their vigor.

Light

The seeds of some plant species require a brief exposure to light to germinate. This sensitivity to light occurs after the seed has imbibed water and is otherwise ready to germinate. Light requirement for germination is usually found in small-seeded species. Small seeds are very limited in their food reserves. By sensing light, the seed knows it is close enough to the soil surface to successfully emerge.

Most crop seeds do not require light for germination, partly because people naturally selected against the trait over the centuries without realizing it. Seeds that were planted that required light usually would not germinate, so no seed was harvested from them, and the requirement gradually disappeared. Also many crop seeds are large and probably never had the trait. Most weed seeds require light for germination and the absence of light in buried weed seeds allow them to remain viable for years. Tillage can kill existing weeds but also brings new weed seeds to the surface where they may germinate if soil moisture is favorable. Some preemergent herbicides, or those applied at seeding, take advantage of the fact that crop seeds do not require light while weed seeds do. The herbicide is placed on the surface, or in the top 3 to 6 mm (1/8 to 1/4 inch) of the soil where weeds germinate. The crop, meanwhile, germinates below the herbicide layer and is not affected by it.

CAUSES OF GERMINATION FAILURE

There are many reasons why seeds fail to germinate. Some have already been discussed. Nonviable seed will not germinate and seedlings low in vigor may not emerge. If an essential environmental factor, namely water, oxygen, temperature, or light, is missing, germination will not proceed. However, sometimes seeds will not germinate or emerge even if all these factors are adequate. It is important to understand these other causes of germination failure and the steps necessary to prevent or overcome them to maximize crop production.

Seed Dormancy

Dormant seed is seed that contains a viable embryo and does not germinate when exposed to favorable environmental conditions. Sometimes people speak of any dry seed as dormant or in a dormant state, since its metabolism is extremely low, and it shows no visible signs of life. To prevent this confusion, seed that will germinate when exposed to favorable environmental conditions is called **quiescent seed.**

Seed dormancy is a natural phenomenon that allows seeds of many crops and weeds to remain viable in the soil over a period of years. Dormancy prevents all the seeds from germinating in a single year thus protecting the species from a disaster that could cause extinction. With dormancy, only some seeds germinate each year to propagate and perpetuate the species. Practically all weed species exhibit seed dormancy that makes their eradication virtually impossible. It is not uncommon for weed seeds to remain dormant but viable in the soil for ten to fifteen years or longer. Most crop seeds, however, are not dormant, probably because farmers, in the cultivation of crops over the years, selected for nondormancy.

Physical Dormancy

Physical dormancy is usually caused by a hard seedcoat and such seeds are called **hard seeds**. Usually, the seedcoat is impermeable to water or oxygen, thus eliminating an essential factor for germination. In other seeds, the seedcoat is resistant to embryo expansion, which also effectively prevents water uptake by the cells.

The most common crops with hard seedcoats are small-seeded legumes such as alfalfa, sweetclover, and red clover. Plump seeds of some cereals also exhibit physical dormancy for a brief period after harvest. In cereals, such as wheat, rye, and barley, the dormancy is usually gone within 20 to 30 days after harvest and is no problem at the time the seed is planted. Normal or slightly shriveled seeds of cereals usually do not have hard seedcoats.

In crop seeds with prolonged physical dormancy, such small-seeded legumes, germination can be improved through scarification. **Scarification** is a seed treatment in which the seedcoat is clipped, scratched, or cut to increase seedcoat permeability. The most common type of scarification is abrasion of the seedcoat by mechanical means. Abrasion is commonly used with legumes. Other methods of scarification include exposing the seeds to hot water, acid, infrared light, or microwave radiation. Scarification must be done very carefully to avoid injury to the embryo.

Physiological Dormancy

Physiological dormancy is caused by a biochemical block within the seed or in the seedcoat that prevents germination. Physiological dormancy is usually caused by a metabolic inhibitor within the embryo itself, but the inhibitor can also be present in the seedcoat or the endosperm. Sometimes, however, the dormancy is caused by an immature embryo that must finish development before germination can occur.

In many cases of physiological dormancy, an after-ripening period is needed to dissipate or decompose the inhibitor, or to allow complete development of the embryo. This after-ripening period is usually only a few weeks or months. When seeds have prolonged physiological dormancy, germination can be improved by stratification. **Stratification** is the process of exposing the dormant seeds to alternating hot and cold, or wet and dry conditions, or a combination of both. Some types of peanut and many perennial forage grasses have physiological dormancy.

Both scarification and stratification subject seeds to many normal weathering factors which occur in the soil that naturally break dormancy. The soil solution is usually mildly acidic and can etch the seedcoat. Seedcoat abrasion can be caused by normal freezing and thawing, and the action of plant roots, soil organisms, and tillage. The soil goes through cycles of hot and cold temperatures, and wetting and drying. Eventually, even the most dormant seed will germinate if left in the soil long enough.

Poor Seedbed Preparation

Another cause of germination failure is poor seedbed preparation that results in an unfavorable environment for germination. Sometimes, the seedbed is too loose, or insufficiently packed. Seeds need firm contact with the soil to quickly absorb the water necessary for germination. The soil above the seed must be firm to insure normal seedling growth and emergence. For instance, the coleoptile of grasses, and the hypocotyl or epicotyl of legumes is light sensitive. Loose soil over the seed that permits even diffuse light to penetrate may cause embryonic leaves to develop beneath the soil surface that prevents normal emergence. Finally, loose soil dries out quickly and may result in inadequate water for the seed and seedling. On the other hand, the soil can be too compact. Soil compaction presents a physical obstruction to the developing seedling. It becomes too difficult for the radicle to penetrate the soil and for the embryonic shoot to move toward the surface. Soil compaction also increases the susceptibility of the soil surface to crusting which can become an impenetrable barrier to the emerging seedling. Soil crusting is also more of a problem with a bare soil than with a mulched soil as discussed in Chapter 5.

Diseases and Insects

Germination failure can also be caused by diseases and insects which attack the seed or seedling. The most common seedling disease is **damping off**. This disease usually results in the death of the germinating seed or the emerging seedling. Damping off is mostly associated with high soil moisture, low soil and air temperatures, high relative humidity, poor soil aeration, and poor surface or internal soil drainage. The disease is caused by several types of soil fungi.

Damage from damping off can occur either preemergent or postemergent. When preemergent, the seed or seedling decays before emergence. After emergence, the stem usually develops a rot near the soil surface that either kills the seedling or severely lowers its vigor. The disease can also infect the roots, causing them to die. Losses from damping off and other seedling diseases can be severe enough to result in the reseeding of the crop or another crop. The most effective method of preventing damping off and other seedling diseases is seed treatment with a fungicide. Most commercially available seed of corn, sorghum, and other crops is treated before sale. Other seeds may be treated if the producer is concerned about disease.

Soil insects can attack the entire seed, the embryo, or the seedling. The most common seed or seedling pests are: wireworms, grubworms, cutworms, seed corn maggot, and seed corn beetle. Losses from these pests can be quite extensive, just as with diseases. Often, it is necessary to treat the soil with an insecticide to prevent total crop loss. Insects which attack the seed can also be controlled by applications of insecticide to the seed, usually with a fungicide as mentioned before. Besides insects, birds such as pheasants, and rodents such as field mice or ground squirrels can also eat seeds and seedlings. Damage from these pests, however, is usually quite localized and rarely results in total crop loss.

SEEDING OF CROPS

There are several factors that must be considered when seeding a crop, including the seeding rate, row spacing, seeding depth, and time of seeding.

Seeding Rate and Row Spacing

Seeding rate and row spacing determine the amount of space each plant will have in the field and the pattern of the plants. Final plant population in the field is decided by seeding rate and percent seed germination in the field.

As will be discussed in Chapter 8, proper plant population and row spacing result in an efficient leaf canopy for light interception. The factors that decide the actual plant spacing for a given field involve primarily the size of the plant and expected water supply.

Water supply is the most limiting factor to crop yield in the field. Water use by the crop must be matched to expected water supply during the growing season. Water supply includes the water stored in the soil at seeding and the water expected to be received from rainfall and/or irrigation from seeding to maturity.

There are differences between crops in the efficiency with which they use water. The most important factor, however, is the size of the plant, or how much it grows during its life span. Larger plants produce more leaf area and more leaf area means more water loss from transpiration. Therefore, plant populations are lower for larger plants than smaller plants. For example, under similar conditions corn would have a population of 25,000 plants per acre and wheat would have almost a million plants!

Seeding rates will be higher when more water is expected to be available. The same crop variety will be seeded at a higher population in an irrigated field when compared to a dryland field. When stored soil water at seeding is lower, seeding rates should be adjusted downward.

Seeding Depth

The proper depth to place seed in the soil is decided mostly by the size of the seed and soil conditions. Germinating seeds need adequate supplies of water and oxygen, and a proper temperature in the soil. Good soil structure also aids in proper germination.

Larger seeds can be planted deeper in the soil than smaller seeds. The germinating seed uses stored food reserves to provide the energy needed for growth and development until the seedling emerges from the soil and begins photosynthesis. Larger seeds have more food reserves and can therefore emerge from deeper in the soil without running out of food.

There may be varietal differences in germination and emergence that can affect seeding depth. Varieties with shorter stalks or stems may not emerge from as deep in the soil as taller varieties. This is especially true with some short varieties of wheat. Stem length depends on internode elongation and sometimes the seedling will have less elongation as well which results in less ability to reach the soil surface.

There is a relationship between soil temperature, soil moisture, and seeding depth. Wetter soil is usually cooler as it takes more heat energy to warm water than soil. The recommended seeding depth for a particular crop or variety is usually expressed as a range and seeding should occur in the shallow end of the range in cooler, wetter soils to ensure warmer temperatures for germination. In warmer, drier soils seed is placed in the deeper end of the range to ensure adequate water for germination.

Seeding Date

Optimum date of seeding is decided primarily by the environmental conditions needed for germination. However, damage by pests and diseases and the ability to compete with weeds are also factors that are affected by seeding date. There are different factors involved when considering Spring or Fall seeded crops.

Spring-seeded Crops

As discussed previously, seeding date of spring-seeded crops is decided by minimum soil temperature for germination. Often, a crop should be seeded as early in the spring as allowed by environmental conditions and acceptance of risk.

Early seeding allows use of longer maturing varieties or hybrids that have more yield potential. Every day after the growing season begins that there is not an emerged crop growing in the field is a day in which potential photosynthesis is lost. Earlier seeding closes the leaf canopy sooner resulting in more photosynthesis for growth and yield and providing improved weed control.

Earlier seeding also allows for more crop growth before the start of feeding by some pests. A larger plant is more able to withstand the feeding of pests. For example, earlier seeded corn will have a larger root system by the time the corn rootworm larvae become active and will suffer less damage. However, with other pests, earlier seeding is not an advantage. Earlier seeded corn will attract the European corn borer moth during egg laying.

Earlier seeding may result in more problems with seed and seedling diseases, especially if the weather turns cool and wet shortly after seeding. If the soil temperature drops below the minimum for germination and emergence, the seedling will be subjected to much stress that makes it more susceptible to disease. Seed treatment with a fungicide can help but will not always prevent diseases.

Fall-seeded Crops

With fall-seeded crops, such as winter wheat, soil temperature is usually near the optimum and proper seeding date is decided by other factors. Winter wheat should be seeded to provide adequate growth before winter dormancy, to minimize water use, and to maximize protection against pests and diseases.

If winter wheat is seeded too early, it will grow excessively before winter dormancy and will use too much soil water. On the other hand, delaying seeding too much will not give the young plants time to adequately develop before dormancy that may result in increased winter killing. Delayed seeding also may result in inadequate plant cover for erosion protection during the winter.

Disease control also is affected by seeding date. Plants that grow too luxuriously in the fall or do not adequately develop before dormancy are more susceptible to disease. Control of pests, such as Hessian fly, is aided by seeding after the adult fly has laid her eggs.

States that commonly grow winter wheat have developed specific recommendations for seeding dates. Producers are more likely to seed too early, however, than too late. Often wheat is seeded early so it is finished before harvest of spring seeded crops begins. If water is adequate, earlier seeding is usually not too much of a problem. Figure 6-6 shows the recommended seeding dates for winter wheat.

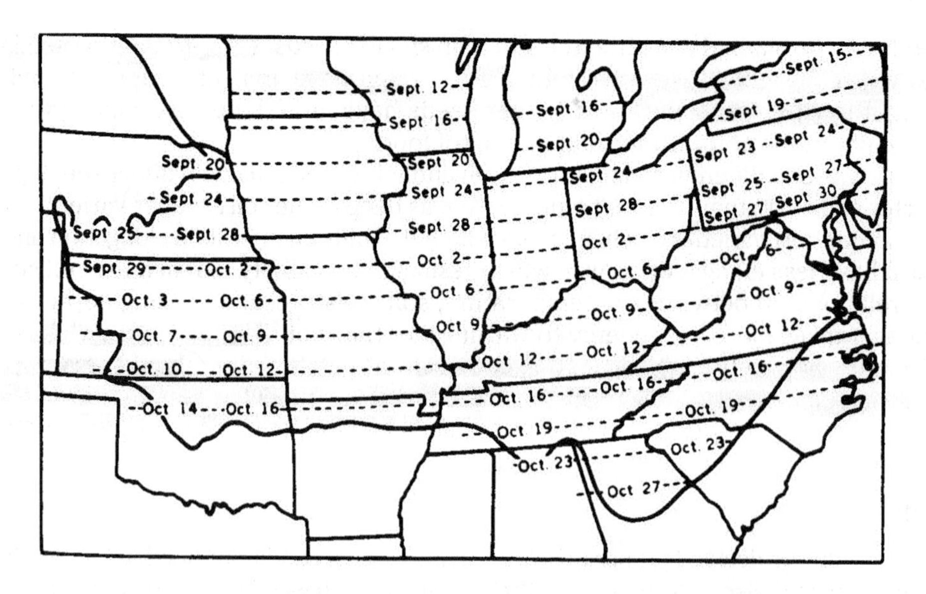

Figure 6-6 Recommended seeding dates for winter wheat.

REVIEW QUESTIONS

1. In what ways are seed size, shape and color important to a crop producer?
2. Which seeds are highest or lowest in protein, oil, or starch? How do these components relate to each other?
3. What is the difference between a seed and a fruit? Why is a caryopsis a fruit?
4. How do the different structures in the embryo function in seedling emergence? How are these functions affected by the type of emergence?
5. What occurs during the three stages of germination? What role does water play during these stages?
6. What are the advantages and disadvantages of epigeal and hypogeal emergence?
7. How do the factors that affect the viability of stored seeds interact with each other? What are the best conditions for storing seeds for long periods of time?
8. What soil factors besides moisture, temperature, and aeration can affect germination?
9. Why is light usually not needed to germinate crop seeds?
10. What is the difference between dormant seed and quiescent seed? What crops exhibit dormancy and quiescence?
11. What are the causes of physiological and physical dormancy? How do these differences affect the methods used to break each type of dormancy?
12. What soil conditions increase the risks from seedling diseases and insects? How can the producer protect against these pests?

CHAPTER SEVEN

Crop Roots

INSTRUCTIONAL OBJECTIVES

Upon completion of this chapter, you should be able to:

1. Define root tip, root cap, apical meristem, elongation zone, zone of maturation, and differentiation zone.
2. Define root hair. Explain its importance.
3. Define epidermis, cortex, endodermis, pericycle, vascular system, and stele. Discuss the function of each.
4. Discuss the differences in the structure of monocot and dicot roots.
5. Define adventitious root, seminal root, and brace root.
6. Define taproot and fibrous root. Discuss the differences in the root systems that develop from these root types.
7. Discuss the difference between a branch root and an adventitious root.
8. List five functions of roots.
9. Discuss the distribution of roots in the soil. List two factors that can affect it.
10. Discuss the functions of the xylem and phloem.
11. Compare the surface area of roots to the plant shoot.
12. Compare the extent of root growth and distribution in various crops.
13. List six factors that can affect root growth.
14. Define tropism. List four tropic responses of roots.
15. Discuss the effect of soil moisture, soil fertility, soil temperature, and soil aeration on root growth.
16. Compare the concentrations of nitrogen, oxygen, and carbon dioxide in soil air to the concentrations in the atmosphere. Discuss the reasons for the differences.

INTRODUCTION

Roots are one of the three main organs of the crop plant, but since they are mostly inconspicuous, they often do not receive the important attention they deserve. Much less effort has been directed toward the study of the growth and physiology of roots than to the stems and leaves above the ground. This is partially because root studies in the soil are difficult and awkward to conduct. To remove all or even a part of the root system for observation or measurement destroys a part of the root system and alters the soil environment.

The physiological functions of roots are most commonly studied in water or sand cultures in the greenhouse or growth chamber. Although these studies are carefully conducted, the results must be adapted to soil and field conditions. However, crop producers and researchers are becoming increasingly aware of the importance of the root system and the root environment as a primary factor controlling crop growth and subsequent yield.

ROOT STRUCTURE AND GROWTH

In any explanation of root growth, the structure most commonly shown and discussed is the root tip (Figure 7-1). The root tip and the structure of the root tip are highly emphasized because root growth originates in this area and nearly all of the absorption activity occurs at the root tip. Although this region probably represents less than one percent of the total crop root mass, there are hundreds of thousands of root tips in the total root system and the metabolic activity that occurs in this region determines the growth and development of the entire plant.

To accomplish these functions, the root structure develops in the following manner. An **apical meristem** is at the tip of the root. In this **meristematic region,** cells divide rapidly with new cells being produced toward the tip and back of the root. Below and surrounding this meristematic region is the **root cap** that is continually regenerated by new cells from the meristematic region. The root cap protects the root tip as it grows through the soil. The cells of the root cap are continually sloughed off as the root tip is pushed through the soil. This sloughing of root cap cells deposits a microscopically thin gelatinous coating on the adjacent soil particles which eases root growth through the soil. It is estimated that this sloughing of root cap cells plus root secretions may use 20 to 30 percent of the plant food energy produced in photosynthesis.

Other cells produced in the meristematic region develop into root tissue. The root grows some by cell division, but the primary region of root extension, that is, increase in length, occurs directly above the meristematic region in the **elongation zone**. In this region, cells enlarge, primarily longitudinally or lengthwise, to push the root tip through the soil. This region varies in length from about 1 mm in timothy to 10 mm in field corn. Although some absorption of water and nutrients occurs in this region, the primary function is root elongation, as roots increase in length only in this region.

From the zone of elongation, the root gradually changes into a **differentiation zone** where the cells begin to develop into different tissues such as xylem and phloem. In this region, many epidermal cells produce lateral extensions called **root hairs**. The root hairs do not form until the cells in that part of the root have ceased elongation. Any further elongation would shear off the fragile root hairs.

The number of root hairs is enormous with as many as 200 root hairs per square millimeter of root. One estimate showed that a four month old rye plant had 14 billion root hairs with a surface area of 400 m^2 (4300 ft^2). These root hairs grow rapidly and reach full size in a few hours. It is estimated that root hairs increase the absorption surface area of roots 20 to 30 times. Crop plants could not absorb adequate amounts of water and nutrients for rapid growth if root hairs did not exist.

In the **maturation zone,** cells become more specialized in function and group together to form tissues as shown in Figure 7-1, 7-2, and 7-3. The outermost row of cells develops into the root **epidermis** that protects the root. The region directly behind the epidermis forms the **cortex** that is the food storage area of the root. Another cell layer that is one cell thick becomes the **endodermis**, a tissue found only in roots. Within the endodermis, the cells develop into the **pericycle** and the **vascular system**. The pericycle is the layer of meristematic cells from which branch roots develop while the vascular system is composed of the **xylem** and **phloem** tissue. Collectively, the pericycle and the vascular system are called the **stele**. Xylem and phloem transport water, nutrients, and sugar as discussed later in this chapter.

Cross sections of mature monocot and dicot roots are shown in Figures 7-2 and 7-3. Notice that the stele is much larger in a monocot root than in a dicot root. Another noticeable difference is the arrangement of the vascular system. Monocot roots have a scattered vascular system around the perimeter of the stele, while dicot roots have it arranged in a cross in the center of the stele.

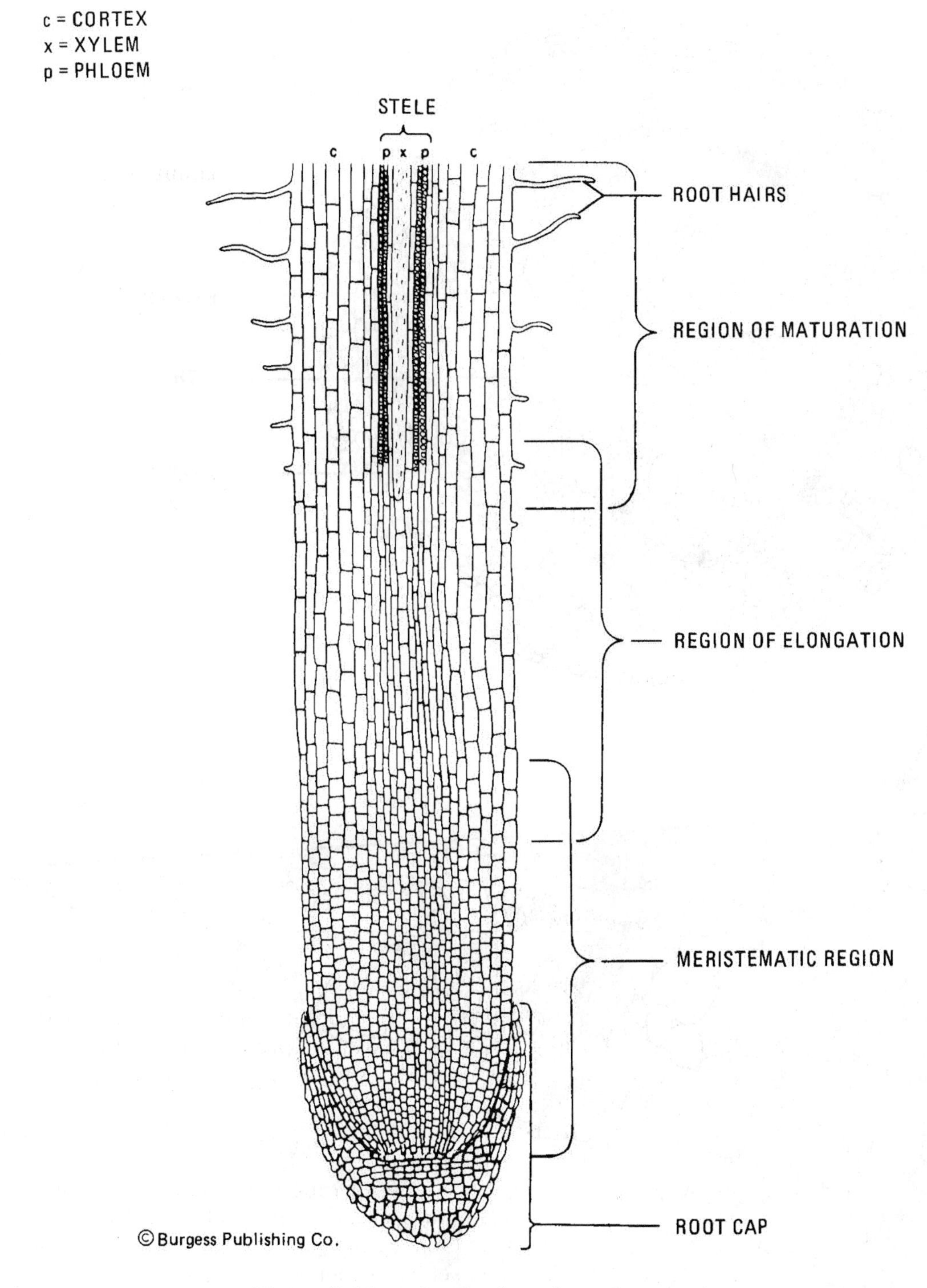

Figure 7-1 Longitudinal section of a root tip.

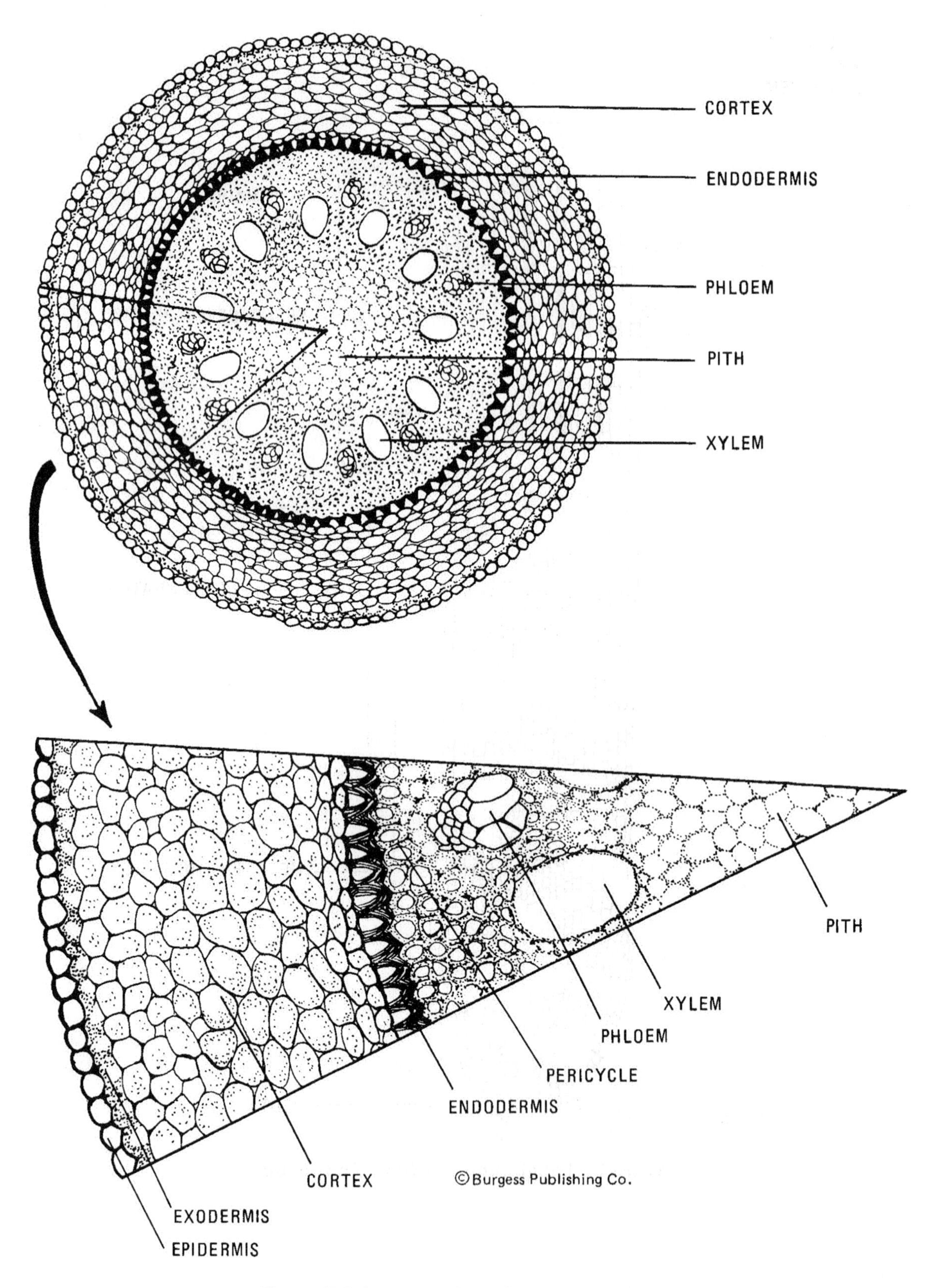

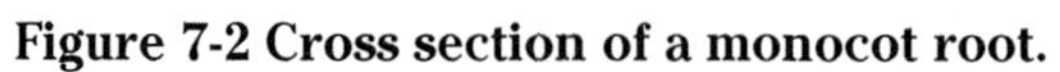

Figure 7-2 Cross section of a monocot root.

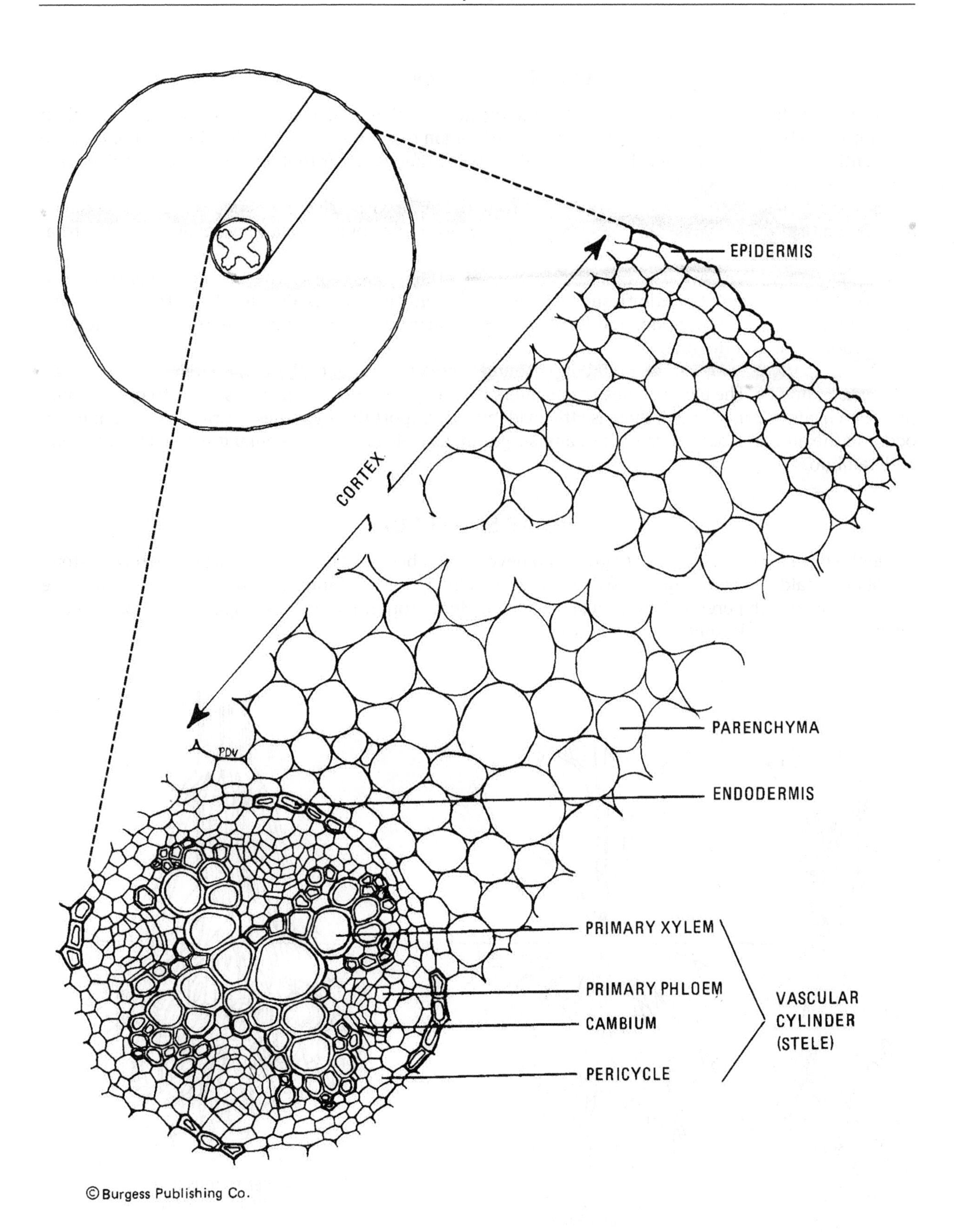

Figure 7-3 Cross section of a dicot root.

ORIGIN OF ROOTS

The **radicle** is the first structure to emerge from the germinating seed as discussed in Chapter 6. It then develops into the primary root and is the first plant organ to exhibit complete development and tissue differentiation. The greatest need of the seedling at this time is water that is necessary for cell expansion and elongation for growth.

Seminal roots are seed roots that originate from the embryo. Seminal roots occur in grass seedlings and are important during seedling establishment. However, they only form a very small part of the total root system of these plants.

Adventitious roots are roots that originate from other than root tissue. Adventitious roots can develop from any meristematic tissue. In grasses, adventitious roots develop from the intercalary meristem above the stem node, and form the secondary root system that eventually becomes the major root system of the plant.

Brace roots, or prop roots, are above ground adventitious roots that originate at the nodes as shown in Figure 7-4. The name of these roots infers that they develop as a plant response to provide support for a heavy stem, although there is little evidence to support this hypothesis. Brace roots commonly occur in the annual grass crops of corn and sorghum, as well as such perennial grasses as sugar cane and bamboo.

ROOT SYSTEMS

When the primary root continues to grow and develop and becomes the central part of the root system, the plant is said to have a **taproot system**, as shown in Figure 7-4. Taproots can grow very deep in the soil, especially with perennial crops such as alfalfa. Most root crops, such as sugarbeet or carrot, have taproots, also most legumes, and other dicots.

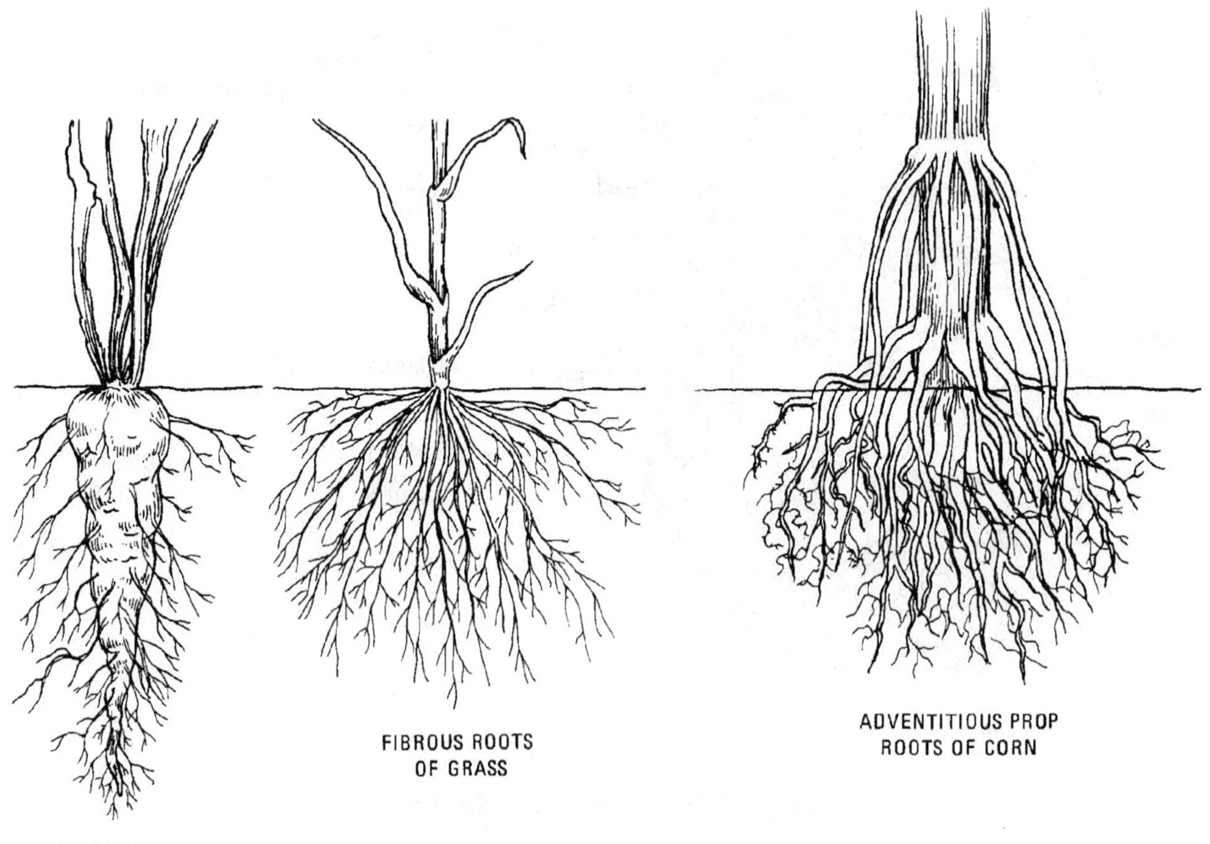

Figure 7-4 Types of root systems.

A **fibrous root system** is one in which there is no main central root from which all other roots originate as shown in Figure 7-4. Instead, there are many roots originating from the plant, all of which are about the same size. In most plants with fibrous roots, the primary root did not continue to develop and most of the roots are secondary roots that are adventitious. All grasses have a fibrous root system. Fibrous roots do not usually grow as deep as taproots, but they usually are more thoroughly distributed in the soil. Taproot systems can absorb water and nutrients from deeper in the soil, but fibrous root systems can absorb more water and nutrients within the root zone. Thus, each system has its advantages.

The degree of branching of roots within the root system varies with the crop, but fibrous roots usually branch less that taproots. Branch roots that develop are designated as follows:

Secondary roots:	Branch from the primary root or stem tissue.
Tertiary roots:	Branch from secondary roots.
Quaternary roots:	Branch from tertiary roots.
Quinary roots:	Branch from quaternary roots.

It is possible to have more than five subdivisions of roots on the plant. Do not confuse secondary roots that branch from the primary root or stem tissue with the *secondary root system* found in grasses and other monocots.

FUNCTIONS OF ROOTS

Although the specific types of roots and pattern of growth vary from crop to crop, the major functions of roots are the same for all plants. Roots primarily function to absorb water and nutrients from the soil and to support and anchor the plant. However, roots also have other important functions.

Absorption of Water and Nutrients

Roots are the plant's contact with the soil and one of their major functions is to absorb the water and nutrients necessary for plant growth. Practically all of the absorption occurs at the root tip, primarily through the root hairs. Because of this, roots must continually grow into new soil areas to contact soil water and plant nutrients. When the soil is moistened from rain or irrigation, new roots must grow into the moistened zone. The old roots presently there cannot absorb the water but instead send out new branches.

Roots do not grow toward water and nutrients, nor is there any substantial movement of water and nutrients to the roots. However, as roots grow, they encounter these materials necessary for growth. As discussed in Chapter 5, water is held in very small pores in the soil. Nutrients are dissolved in the soil water in very low concentrations or are adsorbed (held) on the surface of the soil and organic matter colloids. Roots absorb these nutrients from the soil and the colloids as they grow through the soil. Chapter 5 also discussed mechanisms for nutrient uptake by roots including the importance of mycorrhizal fungi in aiding nutrient uptake.

Since water and nutrients are absorbed from the areas where they occur, the zone of greatest root activity and greatest absorption is usually in the upper layers of the soil. A useful generalization, although not always accurate, is that about 40 percent of the roots and subsequent absorption occur in the upper one-fourth of the root zone, as shown in Figure 7-5. Most crops have an effective rooting depth of two meters (6 ft), so about 70% of the roots are found in the top meter (3 ft) of the soil. Although only 30% of the absorption occurs below one meter, this extra supply of water and nutrients is very important to the plant, especially during a drought.

The management of the water in the soil and the depth of fertilizer placement can affect the zone of maximum root activity. For example, 90% of the total roots of irrigated corn are frequently found in the upper meter of soil because of the constant supply of water in the top of the soil. Conversely, a study with cotton showed that although 55 percent of the root dry weight occurred in the upper 15 cm (6 in) of soil, only 30 percent of the total water absorption occurred in that region as shown in Table 7-1.

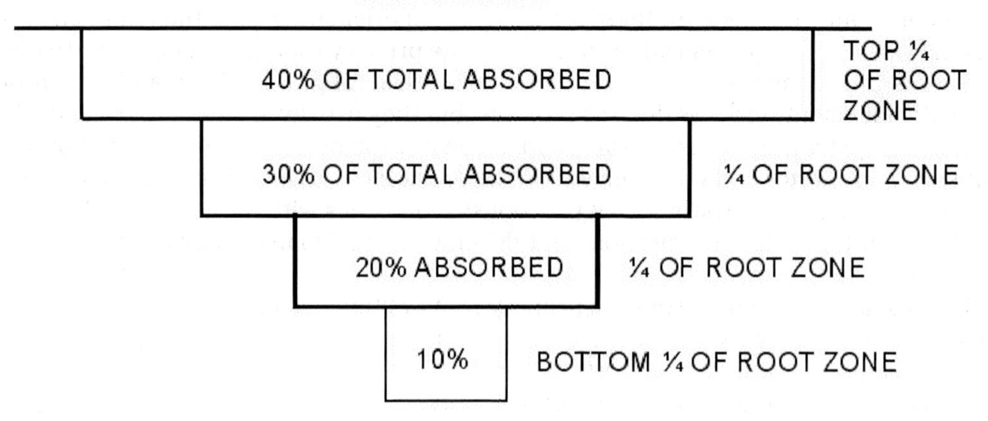

Figure 7-5 Absorption pattern of water and nutrients in non-irrigated crops.

Table 7-1 Relative proportion of root mass and water absorption by cotton in a sandy loam soil.

Soil Depth (cm)	Soil Depth (in)	Root Yield (%)	Water Uptake (%)
0-15	0-6	55	30
15-30	6-12	20	28
30-45	12-18	10	22
45-60	18-24	7	16
60-75	24-30	8	4

Transport of Water and Nutrients

A third function of the total root system is the transport of the absorbed water and nutrients from the root tip, through small branch roots, to larger roots, and finally to the stem for transport to the leaves. Roots are differentiated in specific transport tissue as discussed previously. The **xylem** moves water and nutrients through the roots and stem to the leaves. The **phloem** moves plant food (the products of photosynthesis) from leaves through the stem and roots to the root sites of growth and absorption. If these transport tissues (vascular system) are damaged or do not function because of disease or injury, the plant will wilt and may eventually die.

Storage

Storage is another important function of roots. Most of the plant food translocated to the roots is used for root growth, cell maintenance, and as an energy source for nutrient and water uptake. Biennial crops, such as sugarbeet, carrot, and turnip, store large quantities of plant food as carbohydrates in their large, fleshy taproots during the first year of growth. Figure 7-4 shows a taproot used for storage. These root crops are usually harvested at the end of the first year of growth. If allowed to complete their life cycle, the second year of reproductive growth would begin from the stored root reserves. Regeneration of aboveground growth from stored root reserves is a regular occurrence with perennial forage crops that have been cut or grazed. Even annual crops which have been defoliated due to frost, hail, insect feeding, or disease regenerate growth from root reserves.

Other Functions

Some crop roots have special functions or uses. For example, legume roots have a symbiotic relationship with *Rhizobia sp.* bacteria and fix large quantities of nitrogen for the current crop and future crop use as discussed in Chapter 5. The roots of some plants, such as sweet potato and asparagus, are used for plant propagation as discussed in Chapter 10.

EXTENT OF ROOT GROWTH

The root system of crop plants totals at least one-third to one-fourth of the total dry weight produced. Some researchers estimate that the below-ground dry weight production is equal to the above ground portion but is not measured in root studies since many small branch roots and root hairs are lost during removal from the soil. The root surface area is 20 to 30 times greater than the leaf and stem surface area, and if total root hair area could be included, the surface area is probably 100 times greater. The total root system represents a large surface area in intimate contact with the soil environment.

Knowledge of the root growth pattern and area of penetration at various growth stages is essential in planning fertilizer placement, and deciding when and how much irrigation water to apply. When it is desirable to band fertilizer to increase nutrient availability, such as phosphorus fertilizer in calcareous soils, it is also important to place the fertilizer where root growth occurs. Information about root growth patterns is also useful in deciding crop adaptation to soil depth, cultivation depth, drainage till placement, and selection of the optimum cropping system.

Root growth is both vertical and horizontal, and is influenced primarily by crop species, soil moisture, soil temperature, soil aeration, and soil depth. Crops with a taproot system usually penetrate deeper but have less lateral spread than crops with a fibrous root system. Crops grown under light and frequent irrigations will have a higher percentage of their roots in the upper root zone. Nonirrigated crop roots may penetrate deeper into the soil as moisture in the top soil layer is exhausted as discussed previously. Optimum soil temperature and oxygen level for root growth vary with the type of crop. A brief summary of the root distribution of several crops is shown in Table 7-2.

Fall-seeded small grains penetrate 30 to 40% deeper into the soil than spring seeded small grains. Although this is partially a genetic influence, it is also partly due to the longer growing season available to winter annuals. Sorghum and corn appear to have the same general pattern of root growth but

Table 7-2 Typical root distribution of crops.

Crop	Type of Root System	Depth of Maximum Uptake (cm)	Depth of Maximum Uptake (in)	Maximum Depth of Roots (cm)	Maximum Depth of Roots (in)	Lateral Spread (cm)	Lateral Spread (in)
W. Wheat	Fibrous	115	45	210	82	20	8
Sp. Oats	Fibrous	75	30	120	48	20	8
Corn	Fibrous	105	41	180	71	105	41
Sorghum	Fibrous	105	41	180	71	90	36
Rice	Fibrous	3-5	1-2	15	6	15	6
Cotton	Tap	30	12	60	24	97	38
Sugarbeet	Tap	105	41	180	71	30	12
Soybean	Tap	30	12	60	24	60	24
Bromegrass	Fibrous	88	35	90	36	60	24
Alfalfa	Tap	120	48	600	240	60	24

sorghum produces smaller diameter roots and the total number of roots in a given soil volume is greater. This undoubtedly contributes to the greater drought tolerance of sorghum compared to corn, and the ability of sorghum to produce equivalent yields with less applied nitrogen fertilizer.

FACTORS AFFECTING ROOT GROWTH

Root growth is directly influenced by the root/soil environment and indirectly by the factors which influence the growth of the entire plant. The initial energy for seedling root growth is derived from the seed. However, the major energy source for expanded and continued root growth is the sugar produced by photosynthesis in the leaves and translocated to the roots.

Tropism

Root growth responds to certain environmental conditions and these responses have been labeled **tropisms**. The downward growth of roots is called positive **geotropism**, or the growth of roots toward the pull of gravity. A crop plant growing in a pot in the greenhouse will exhibit upright stem growth (negative geotropism) and downward root growth (positive geotropism). If the pot is placed on its side, the stem will adjust its growth habit and grow up, and the roots will reorient themselves and grow down. This is caused by a redistribution of a growth hormone, probably auxin, due to gravity. When the pot is placed on its side, the growth is stimulated on the underside of the stem and the upper side of the root causing the reorientation of growth.

Roots also grow away from light, which is called negative **phototropism**. This growth habit is in response to growth hormone distribution due to light effects. Roots grow more rapidly and proliferate more in favorable temperature environments. This is called positive **thermotropism**. Roots also grow more rapidly in a favorable moisture environment. This is called positive **hydrotropism**. Roots do not sense the location of a favorable temperature and/or moisture environment and grow to it. Instead, roots respond to the favorable environment by increasing growth rate.

Genetics

The kind of root system, taproot or fibrous, is determined genetically. A taproot system develops when the primary root becomes the central root with small branch roots. The large central root may exhibit considerable branching as with soybeans and alfalfa, or only a slight amount as with sugarbeet. A fibrous root system develops when the primary root is supplemented by many adventitious roots. All annual and perennial grasses have a fibrous root system.

Besides the species differences in rooting patterns and root surface area, there are also great varietal and hybrid differences within species. Plant breeders have developed corn hybrids with rapid root growth to overcome corn rootworm feeding. Varieties selected for drought tolerance are highly branched and develop a large root volume.

Soil Moisture Level

Crop roots can grow in soils within the range of -1/3 to -15 bars soil water potential as discussed in Chapter 5. At less than -15 bars, the soil is too dry for root growth. In water saturated soil, soil aeration is too low for root growth. Poorly drained soils with a high water table or which usually contain excessive amounts of soil water will restrict root growth. Conversely, excessively drained soils may also limit root growth because soil moisture is lacking.

Crop roots will not grow into or through dry soil to reach moist soil. Since there is very little lateral movement of soil water, roots must continually grow into new moist soil areas to absorb the soil water needed for cell elongation in the root tip. Once the soil moisture in a given soil volume has been depleted, it must be replenished by precipitation or irrigation before renewed root growth occurs. The amount of precipitation that occurs greatly influences the rooting patterns as shown in Table 7-3.

As Table 7-3 shows, the rooting depth decreases and the lateral spread of roots increases as precipitation decreases. With less precipitation, the plant relies almost totally on the precipitation that falls during the growing season since less is stored in the soil before planting. Roots spread farther laterally to absorb all of the limited water supply. They do not penetrate as deeply because the soil does not get moistened very deep.

In irrigated crop production, the lateral spread and depth of root penetration are decided by the volume of soil that is wetted. With drip irrigation, only a very limited soil volume may be moistened, and light and frequent irrigation by sprinklers will moisten only the upper soil volume. However, with furrow irrigation, the entire rooting zone may be moist and suitable for root growth. Whatever the method of irrigation, if water is supplied in amounts that are adequate to meet the crop's needs, acceptable crop yields can be attained. However, there is a danger in moistening only a small volume of soil. If the irrigation system should fail, the crop may be lost because the plant has an insufficient root system.

Physical Conditions of the Soil

Root growth can be limited by such physical barriers as impervious rock, gravel subsoil, heavy clay pan, or a compacted soil layer from tillage or heavy traffic. Roots can penetrate physical barriers if the barrier strength is less than the root pressure. Root pressures vary with the crop species, but can reach maximums of 9 to 13 bars. If roots cannot penetrate a physical barrier, growth will be horizontal at the face of the barrier.

Deep tillage to fracture heavy clay pans or tillage pans have been effective in improving root penetration of cotton on sandy soils in California, and sugarcane in the Cauca Valley of Colombia, South America. However, this technique has not had lasting effects on most agricultural soils in the Midwest.

Soil Fertility and pH

Root growth is more extensive and branched in a fertile soil than in an infertile soil. Roots will proliferate extensively around a lump of animal manure or fertilizer band in the soil. Roots will not, however, grow directly into the zone of high nutrient concentration, and often will show some inhibited growth and deformities at the point of first contact with the concentrated supply. This is due to the **"salt effect"** of the concentrated fertilizer. The zone of concentrated nutrients affects the level of chemical salts in the soil water near the fertilizer band, and roots cannot grow into it.

Both high and low soil pH decrease root growth. At pH levels above 7.5, the availability of phosphorus and the micronutrients iron, manganese, zinc, copper, and cobalt, are reduced. An adequate level of available phosphorus in the soil is essential for proper crop root growth. High soil pH can be caused by an accumulation of sodium salts that may be toxic and/or interfere with water absorption by roots. At low pH levels, the increased solubility of iron and aluminum may cause these elements to reach toxic levels or reduce the availability of other nutrients. The relationship of soil pH and nutrient availability was discussed in more detail in Chapter 5.

Soil (Root) Temperatures

Optimum soil temperatures for root growth vary between crops as shown in Table 7-4, but are usually lower than the optimum temperatures for stem and leaf growth since the soil is usually cooler than the air above it. Also, variations in the root environment temperature are usually much less than air temperature during the growing season.

Table 7-3 Influence of annual precipitation on root growth of winter wheat.

Precipitation		Root Depth		Lateral Spread		Plant Height	
(cm)	(in)	(cm)	(in)	(cm)	(in)	(cm)	(in)
65-80	26-32	150	60	30	12	100	40
50-60	20-24	120	48	50	20	90	36
40-50	16-20	60	24	60	24	65	26

Table 7-4 Optimum root and shoot temperatures of various crops.

Crop	Roots		Shoot	
	°C	°F	°C	°F
Alfalfa	20-28	68-82	20-30	68-68
Barley	13-16	43-55	15-20	60-69
Corn	25-30	77-86	25-30	77-86
Cotton	28-30	82-86	28-30	82-86
Oats	15-20	60-69	15-25	60-77
Rice	26-29	80-85	26-29	80-85
Wheat	18-20	64-68	18-22	64-72

In a favorable soil environment, root growth will increase as temperature increases up to the optimum. Cooler temperatures reduce the rate of biochemical reactions and membrane permeability and increase the viscosity of the cell fluids. In addition, at cooler temperatures, the rate of mineralization of plant nutrients from the organic (unavailable) to the inorganic form (available) is reduced. Although there is considerable species variation at cooler soil temperatures, roots are larger in diameter, and less branched than at warmer temperatures.

Soil temperatures are influenced by the soil moisture content and wetter soils are usually cooler than drier soils. This is due to the fact that water has a specific heat about five times higher than the other soil components as discussed in Chapter 5. Soil color also influences soil temperatures. Soils inherently darker in color or darker due to organic matter content absorb more heat. A bare soil will be warmer than a soil covered with crop residue. Any factors that affect soil temperature also influence root growth.

Soil Aeration

About 20-25% of the volume of a typical soil is air as discussed in Chapter 5, and about 20% of the soil air is oxygen. Oxygen is essential for root respiration that provides the energy needed for continual cell growth and maintenance, and nutrient uptake. In some circumstances, energy is required for water absorption. Under **anaerobic** conditions (without adequate oxygen), roots and soil organisms may produce toxic substances that inhibit growth.

Crops vary in the optimum level of oxygen for satisfactory root growth. For example, rice and buckwheat can tolerate low soil oxygen levels while corn and field peas require higher oxygen levels than most other crops. However, the root growth of most crops is retarded when the soil oxygen content is less than 10%. Root elongation of cotton and soybean roots is reduced at soil oxygen levels less than 10% but exhibits the same rate of growth in the soil oxygen range from 10-21%.

Soil air is about ten times higher in carbon dioxide and slightly lower in oxygen than atmospheric air as shown in Table 7-5, although the composition will vary with soil type, organic matter level, crops grown, and atmospheric conditions. The roots and soil organisms use oxygen and produce carbon dioxide. Many soil pores are not contiguous with the atmosphere that reduces the exchange of gases between the soil and the atmosphere. Coarse textured soils have larger pores and are better aerated than fine textured soils.

For most crops, root growth is not seriously hindered until oxygen levels in the soil drop below 10%. Hydrophytes can tolerate oxygen levels lower than 10%. There are contradictory reports concerning the carbon dioxide levels that inhibit root growth. Generally, it appears that the usual range of carbon dioxide seldom reduces root growth, but, under conditions of compacted or wet soils, carbon dioxide levels can reach inhibitory levels.

Table 7-5 Relative composition of atmospheric air and soil air at 15 cm (6 in).

	Nitrogen (%)	Oxygen (%)	Carbon dioxide (%)
Soil Air	79.2	20.6	0.25
Atmosphere	70.0	21.0	0.03

REVIEW QUESTIONS

1. Where do water and nutrient absorption occur in the root? How does this affect the growth and distribution of roots?
2. What is the importance of the root cap and root hairs?
3. What is the origin of the primary root? What is a secondary root?
4. In which tissue do branch roots originate?
5. What are the differences between a taproot system and a fibrous root system?
6. In which tissue does carbohydrate storage occur in roots?
7. How does a tropic response occur in a plant?
8. How do the amount and distribution of soil water affect root growth rate and distribution?
9. Why are optimum temperatures for growth of roots lower than for the shoot?
10. Why is the oxygen content of soil air lower than the atmosphere? Why is carbon dioxide content higher?

CHAPTER EIGHT

Crop Stems and Leaves

INSTRUCTIONAL OBJECTIVES

Upon completion of this chapter, you should be able to:

1. Define stem, leaf, shoot, culm, stalk, woody stem, herbaceous stem, node, and internode.
2. Discuss the differences in structure between monocot and dicot stems.
3. Define modified stem, stolon, crown, tiller, spur, rhizome, tuber, bulb, and corm. Give an example of each.
4. Name the structures from which stems originate.
5. Define apical meristem and intercalary meristem. Discuss the differences in growth pattern between them.
6. Define apical dominance. Discuss its effect on stem growth.
7. List six functions of stems. Give an example of each.
8. List three factors that affect stem growth. Discuss how each factor affects it.
9. Define stem lodging. List three factors that can cause it.
10. Define leaf blade, auricle, ligule, collar, stipule, petiole, leaf axil, and sessile.
11. Define leaf margin, leaf venation, parallel venation, net venation, opposite leaf arrangement, alternate leaf arrangement, and whorled leaf arrangement. Give an example of each.
12. Define leaf canopy. List four factors that affect it.
13. Define open canopy and closed canopy. Discuss differences in light interception between them.
14. Define leaf area index, leaf area distribution, leaf angle, and leaf duration. Discuss how each affects light interception in the leaf canopy.
15. Define leaf epidermis, cutin, cuticle, pubescence, and glabrous.
16. Define guard cell, stomata, leaf mesophyll, palisade cell, spongy cell, leaf vein, and bundle sheath cell.
17. Name the structures from which leaves originate.
18. List four functions of leaves. Give an example of each.
19. Define transpiration. Discuss its importance to plants.
20. List three factors that affect stomatal opening. Give an example of how each affects it.

21. Define diffusion. Discuss its importance in transpiration.
22. List five factors that affect the rate of transpiration. Give an example of how each affects it.

INTRODUCTION

The **stem** is the central axis of a plant with appendages. These appendages include leaves, branches, and flowers. All stems have the important functions of supporting the plant so that the leaves are properly displayed to intercept light for photosynthesis, and contain the vascular system that provides the lifelines for the various plant parts. Some stems also have other functions as you will see later in this chapter.

Leaves are flat appendages attached to the stems that are usually arranged to present the maximum surface area to intercept light. The primary function of the leaves of crop plants is to conduct photosynthesis that allows the plant to grow. Leaves are the workhorses of the plant, and many cultural practices of crop production are designed to maximize the efficiency of the leaves.

CROP STEMS

Before discussing stems in detail, it is important that you be familiar with the nomenclature involved. **Shoot** is another word for stem, but often it is used to include the stem and leaves. The hollow stems of some grasses, such as small grains, are called **culms**. Solid grass stems, such as those found in corn and sorghum, are called **stalks**. The term, stem, is usually used with legumes such as soybean and alfalfa.

Stems can be classified as **woody** or **herbaceous**. The differences between woody and herbaceous stems were discussed in Chapter 2. Woody stems are found only in perennial plants. In perennial dicots, woody stems can be further classified as **hardwood,** such as in oak, or **softwood**, as in elm. Examples of woody monocots are palm and bamboo. Examples of plants with herbaceous perennial stems are alfalfa and bromegrass. All annual crops have herbaceous stems.

Stem Structure

Stems are composed of nodes and internodes. **Nodes** are areas of compressed tissue that originate other plant parts, such as leaves, branches, and flowers. Sometimes, nodes contain meristematic tissue for growth as discussed later in this chapter. **Internodes** are the areas between the nodes that contain elongated cells. The height of a plant is mostly decided by the length of the internodes. The presence or absence of nodes is the main identification factor for separating stems from other plant parts, especially roots.

Internal Stem Structure

Figures 8-1 and 8-2 show cross sections of monocot and dicot stems. Monocot stems, such as grasses, have scattered systems of vascular bundles; while dicots, such as legumes, have their vascular systems arranged in a circle. The **vascular bundles** contain the **xylem** that conducts water and nutrients from the roots to other plant parts, and the **phloem** that conducts sugars from photosynthesis and other plant originated foods to the roots and other plant parts. Dicot stems also have a cambium that is located between the xylem and phloem. The cambium contains meristematic tissue that produces lateral stem growth, particularly in perennial woody plants. Monocot stems do not have a cambium.

Modified Stems

Although most stems are normal with leaves and flowers attached, there are some types of stems that have been modified to serve specific functions for the plant, such as propagation, and may not look like stems at all. These modified stems can be found either aboveground or belowground.

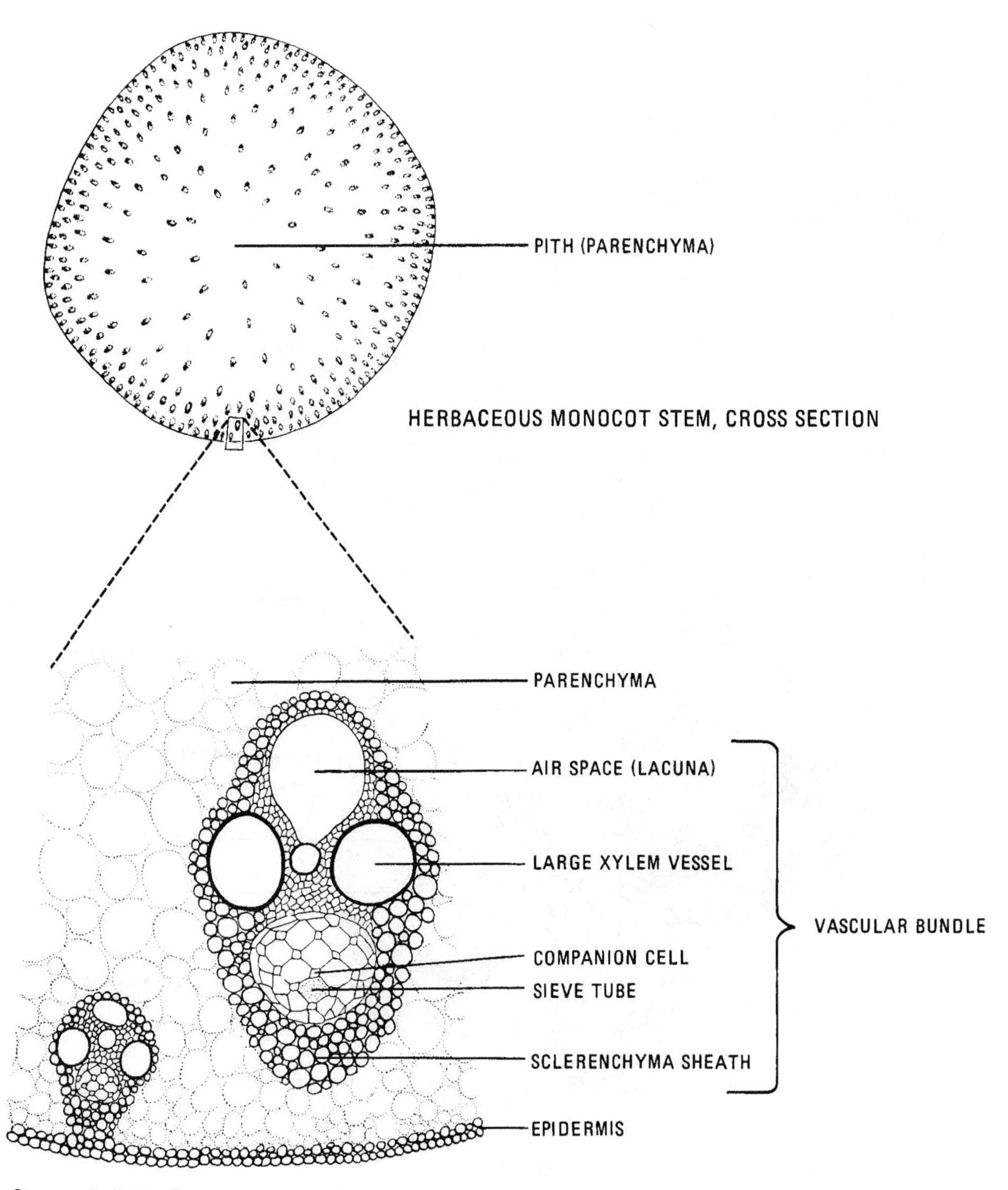

Figure 8-1 Cross section of a monocot stem.

Figure 8-3 gives some examples of aboveground modified stems. **Stolons** are aboveground horizontal stems that function in asexual reproduction. Each node of a stolon has the potential to produce a new plant complete with roots, stems, and flowers. Plants that spread by stolons form dense populations of the species that tend to crowd out other plants. Examples of plants with stolons are white clover, buffalograss, and strawberry. Stolons are found only on perennial plants.

A **crown** is a modified stem that is composed of compressed stem tissue and is usually located near the soil surface. A compressed stem contains many nodes with no internode regions between. Each

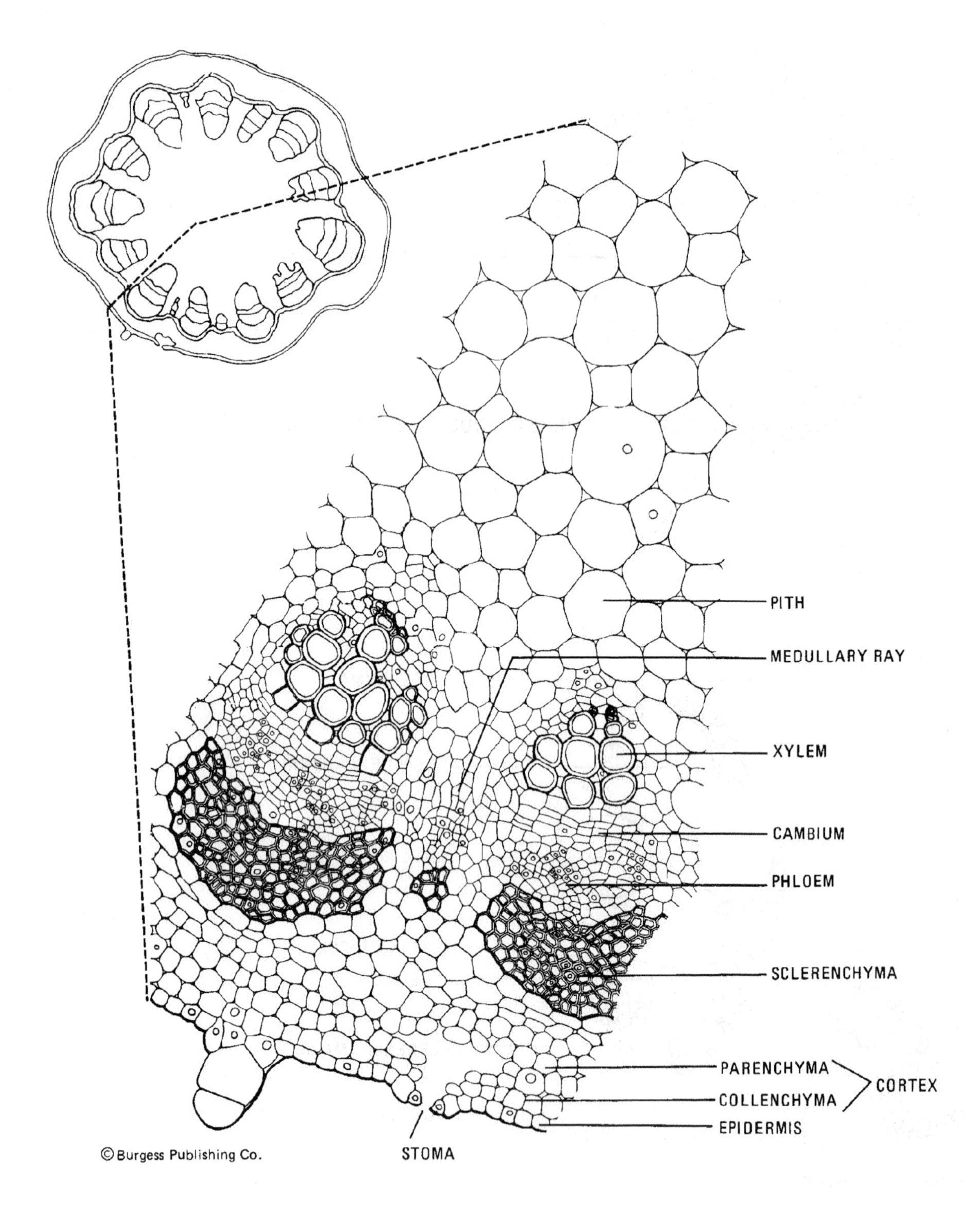

Figure 8-2 Cross section of a dicot stem.

node in a crown has an **axillary bud** which can produce a stem. In grasses, a stem from a crown is called a **tiller**. Most turf and forage grasses, also forage legumes such as alfalfa or clover, contain crowns that can initiate new growth after mowing or grazing. This makes them very adaptable to frequent removal of top growth. Crowns are found in annual, biennial, and perennial plants. **Spurs** are compressed woody stems of woody perennials. The flowers and subsequent fruit of many fruit trees, such as apples and pears, are borne on spurs.

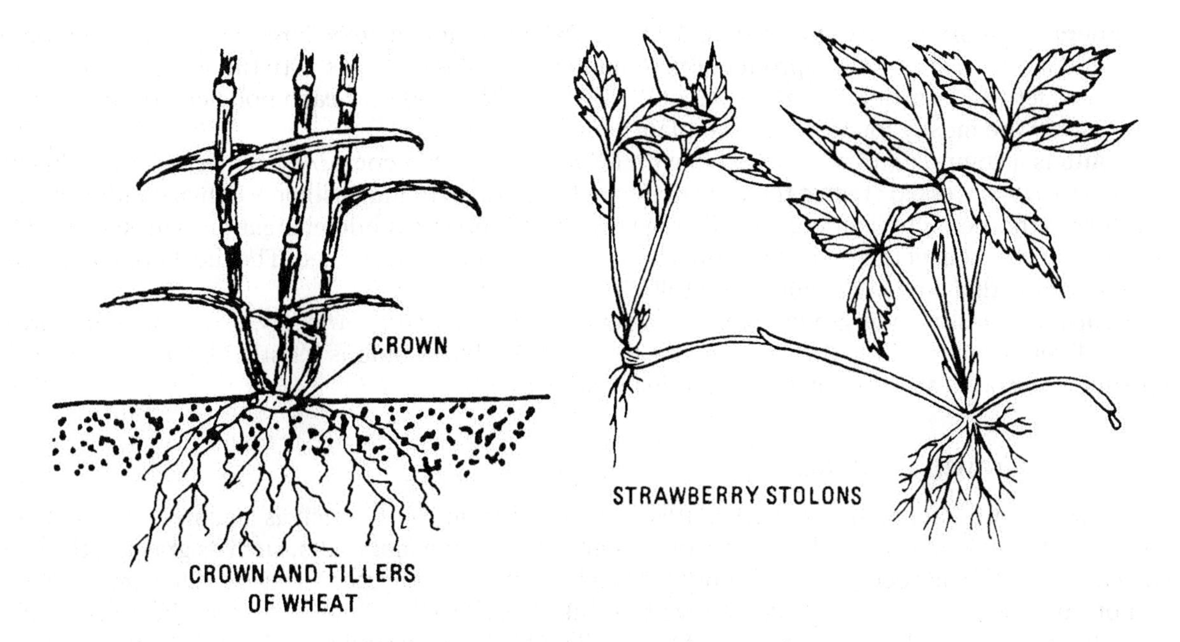

Figure 8-3 Aboveground modified stems.

Figure 8-4 shows some common types of belowground modified stems. **Rhizomes** are modified horizontal stems found on some perennial plants. They are similar to stolons in that they function in the asexual reproduction of the plant. Each node can produce a new plant; and plants with rhizomes also form dense mats or populations and crowd out other plants. Grasses that spread by rhizomes or stolons are called **sod-forming grasses**, while grasses that contain only crowns are called **bunch grasses**. Many turf and forage grasses such as bluegrass and bluestem, as well as many weeds such as field bindweed and common milkweed, spread by rhizomes.

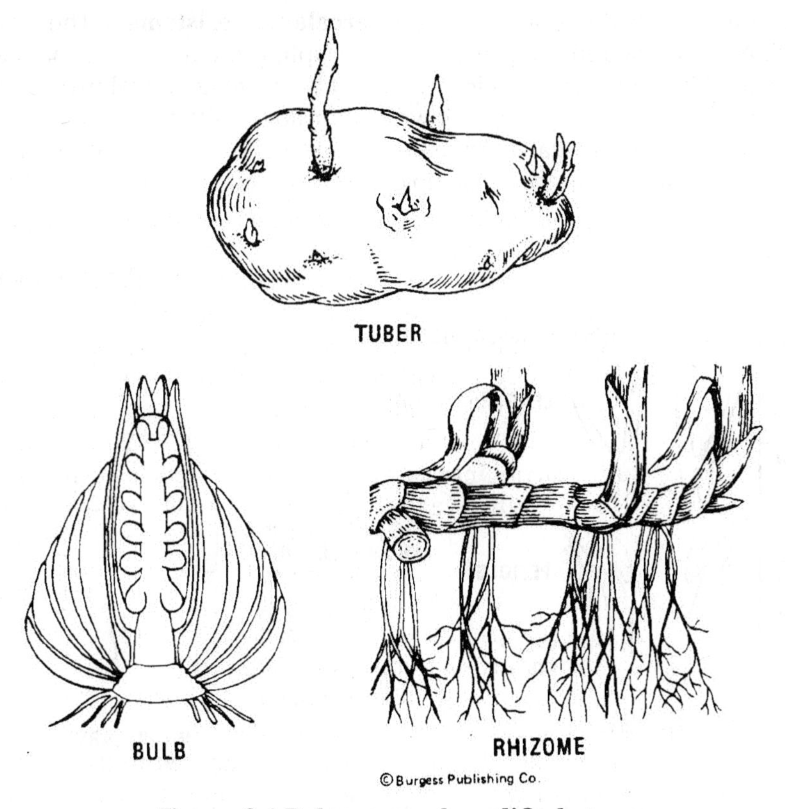

Figure 8-4 Belowground modified stems.

Tubers are short, greatly enlarged underground stems that can store large quantities of carbohydrates and function in asexual reproduction. Common examples of plants with tubers are the common potato and Jerusalem artichoke. The eyes of a tuber are the nodes and each node can produce a new plant. Tubers are mostly found on annual plants.

A **bulb** is a compressed modified stem with leaves. It is mostly composed of fleshy leaves that are attached to a compressed stem at the base of the bulb. Bulbs are found in some biennial plants. The first year of growth, the bulb functions as a storage organ with no internode elongation. The second year, the compressed stem of the bulb sends up a flower stalk that produces seed. The most common example of a plant with a bulb is the onion and tulip.

A **corm** is a fleshy underground stem with very few nodes, and a few scaly, inconspicuous leaves. A corm differs from a bulb in that it is mostly stem tissue while a bulb is mostly leaf tissue. Examples of plants with corms are timothy grass, gladiolus, and crocus.

Origin and Growth of Stems

As discussed in Chapter 5, the stem originates in the embryo. In dicots, such as legumes, the stem originates from the epicotyl and/or hypocotyl of the embryo. In some monocots, such as grasses, the stem originates from the mesocotyl of the embryo. These structures elongate during seedling emergence to push or pull the shoot of the seedling above the soil. These structures also contain the meristematic tissue that produces subsequent growth of the stem and other aboveground plant parts. In an actively growing shoot, an active meristem is called the **growing point.**

The growing point in legumes and most other dicots occurs in an **apical meristem,** which is at the tip of the stem. Figure 8-5 shows an apical meristem. Growth is caused by cell division at the tip of the stem followed by cell elongation within the internodes below. As the cells elongate they begin to differentiate into the various tissues of the stem. When differentiation is complete, the tissue is said to be mature. Little, if any, cell elongation occurs within the nodes and the height of the plant, or the length of the stem, is mostly decided by the length of the internodes. Leaves originate from the nodes, and nodes may also contain **axillary buds** that can originate new stems (branches) or flowers. Axillary buds contain meristematic tissue that is controlled by the apical meristem.

A grass stem has many growing points called **intercalary meristems** at the top of each node as shown in Figure 8-5. As discussed in the previous paragraph, growth occurs from cell division in the meristem and cell elongation in the internodes. Grasses also have an apical meristem at the tip of the stem. This apical growing point develops into flowers after floral initiation as discussed in Chapter 9.

The main differences between apical meristems and intercalary meristems are the location and the direction of growth. Apical meristems are at the tip of the stems and growth is from the top down. Inter-

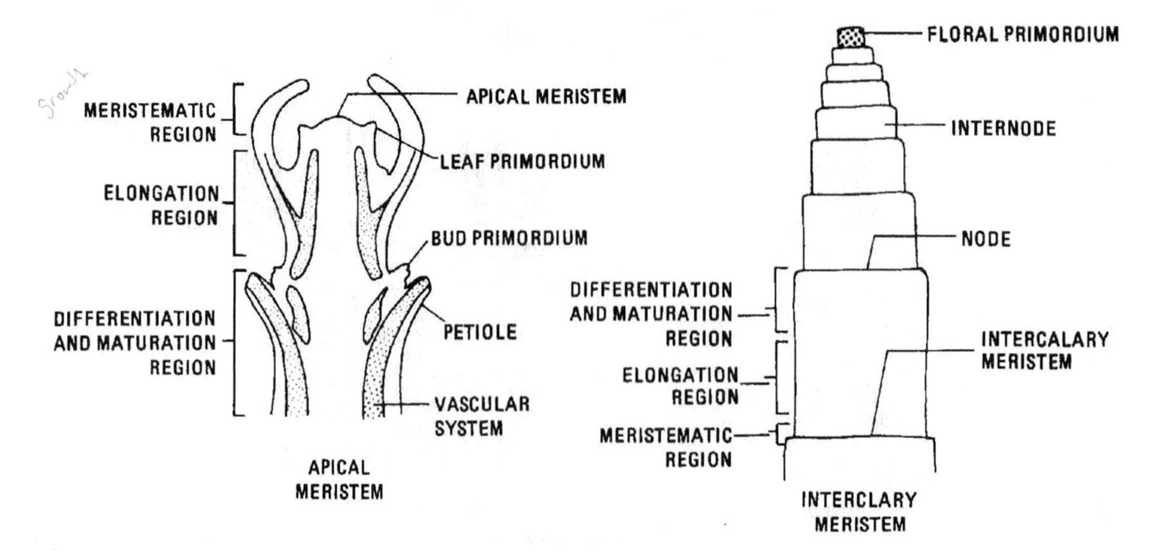

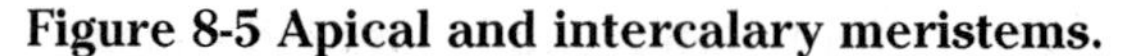
Figure 8-5 Apical and intercalary meristems.

calary meristems are at each node and growth is from the bottom up. Grasses grow in a telescoping fashion from the bottom which makes them very adaptable to mowing and grazing as not all of the active meristems are removed.

Most axillary buds will not become active unless the apical meristem is damaged or removed; a process called **apical dominance**. Apical dominance is found in all higher plants and enables plants to continue growth and reproduction after pests or weather damage, mowing, or grazing. If the apical growing point is damaged, the plant activates one or more axillary buds that grow branches to resume growth. If the plant has a crown, its buds will also be controlled by the apical growing point, and they become active if the top growth is removed. Growth of these meristems is regulated by the production of **auxin,** a growth hormone, by the apical meristem. Auxin suppresses cell division in the lower meristems. When the apical growing point is damaged or removed, auxin production stops and the inhibition of the axillary buds is removed.

Functions of Stems

Stems serve many functions for the plant, some of which have already been mentioned. All these functions are equally important and the plant is severely damaged if one of these functions is hindered due to disease or injury.

Support and Display the Leaves

Since leaves are attached to the stem, they are totally dependent on the stem to display them in a way that most efficiently intercepts light for photosynthesis. The development and arrangement of the leaves on the stem determine the **leaf canopy**. Leaf canopy is discussed in more detail later in this chapter. Stem development determines the type of leaf canopy that will develop and varies between crops. Crops with erect stems will develop a different type of canopy than those with prostrate or horizontal stems. Internode elongation, which determines the distance between leaves also affects the leaf canopy and may differ between cultivars of the same crop species.

Support the Flowers and Fruits

Flowers must be supported and displayed so that proper pollination and fertilization may occur. This is especially important for cross pollinated crops. The fruits, or grain, must be supported so that proper development can occur and harvest can be accomplished.

The flowers and fruits are supported by modified or adapted stems. The ear of a corn plant is supported by a modified branch called a **shank**. Groups of flowers at the top of the stem, such as in small grains and sorghum, are supported by a modified stem called a **peduncle**. Axillary flowers, such as found in soybean, are supported by modified branches called **pedicels**. The development of these modified stems can affect the harvestability of the crop. For instance, proper peduncle elongation in small grains and sorghum allows the grain to dry rapidly at maturity, and allows harvest with a minimum amount of leaves that can interfere with threshing. However, too much peduncle elongation can cause lodging.

Transport Water and Nutrients

As discussed earlier, the stem contains vascular bundles of xylem and phloem. The xylem carries water and soil nutrients from the roots to other plant parts, and the phloem carries sugars and other products from the leaves to the roots and other plant parts. These are shown in Figures 8-1 and 8-2. The arrangement of this conductive tissue within the stem is quite different for monocots and dicots but the transport function is essentially the same.

Storage of Plant Food

In some plants, such as sugarcane and sweet sorghum, the major food storage organ is the stem. In other crops it may be the grain or seed, or the roots. However, some food is stored in most stems.

The type of food reserve stored in the stem can vary with the age of the plant, and may affect its digestibility when used for forage. For instance, alfalfa is cut when one-tenth of its flowers have opened. As the alfalfa plant nears maturity, the carbohydrates in its stem are converted from starch to less digestible forms. Harvesting at one-tenth bloom results in maximum productivity and maximum feeding value because the stems are easily digested and most leaves are still attached. Pasture and range plants are also managed so that grazing occurs at times of optimal forage utilization by the animals.

Plant Propagation

Stems and pieces of stems can function in the asexual reproduction of some plant species. The role of some modified stems in producing new plants has already been discussed. Asexual reproduction is discussed in more detail in Chapter 9. In nature, parts of stems may be broken off and carried to new locations by animals, birds, water, etc. Although most crops do not use stems for propagation, this phenomenon is important with some weeds, and people help spread these species by carrying plant parts on machinery. Sugarcane is a crop commonly propagated by stem cuttings, as are many ornamental plants. Potato is an example of a crop that uses modified stems (tubers) for propagation.

Photosynthesis

Photosynthesis in stems is important in some fleshy plants that have few leaves or no leaves such as cacti. However, it is unimportant in crop plants. Most of the photosynthesis in crop plants occurs in the leaves. This is shown by the fact that crops that have been defoliated by hail or other disasters suffer considerable loss of yield. The cell structure of stem tissue is not conducive to photosynthetic activity.

Factors Affecting Stem Growth

As discussed previously, stems grow by cell division in the apical or intercalary meristem, and by cell elongation in the internodes. There are many internal and external factors that influence stem growth, many of which can be either directly or indirectly controlled by the crop producer. Many of these factors can cause the stem to become weak resulting in stem lodging that is discussed later in this chapter.

Low light intensity on the stem stimulates the production of **auxin** that causes rapid cell elongation resulting in increased internode length or **etiolation**. Excess auxin also decreases the production of lignin in the stem cell walls which makes them weaker. Together, these factors combine to produce tall, spindly plants with weak stems that are very susceptible to lodging.

Low light intensity may be caused by prolonged periods of cloudy weather, but most often it is caused by shading which occurs in the field. Shading can be caused by high plant populations that cause mutual shading of adjacent crop plants. It can also be caused by an excessive supply of nitrogen or high soil water contents that stimulate rapid growth of the crop. Excess nitrogen not only increases shading by stimulating rapid and excessive growth, but it also directly affects stems by reducing cell wall lignin that weakens the stem.

Some factors reduce stem growth. Drought slows the rate of photosynthesis and affects the metabolism of the entire plant. Plants exposed to severe drought are stunted due to decreased internode elongation. Many nutrient deficiencies and diseases can also cause stunting. Anything that affects the general health and rate of growth of the plant also affects stem growth, particularly internode length.

Stem Lodging

Stem lodging is the failure of an ordinarily erect stem to remain erect. Lodging causes disorientation of the leaf canopy that decreases the efficiency of light interception, and ultimately photosynthesis. Lodging also increases mechanical harvest problems and usually increases harvest losses.

As discussed before, lodging can be caused by shading, excessive soil nitrogen, and excessive soil water. Usually a combination of these factors is responsible. Lodging commonly develops in low areas of a field where soil fertility is high, excessive rainfall accumulates due to runoff, and good germination and stand development occur due to the increased water supply. Shading can also occur from weed competition that has the same effect as a high plant population. Other factors can also affect lodging. Diseases that directly attack the stem tissue weaken the stem. Common examples are stalk rot of corn, charcoal rot of sorghum, and brown stem rot of soybeans. Some insects feed on stem tissue and increase lodging. European corn borer and wheat stem sawfly are examples.

Lodging problems can be overcome by carefully controlling the factors that cause it. Proper seeding rates, avoidance of excessive application of fertilizers, especially nitrogen, and management practices that reduce weeds, insects, and diseases; can all reduce lodging. In addition, crop breeders are continually developing new varieties that can resist lodging.

CROP LEAVES

As mentioned before, **leaves** are the workhorses of the plant. They are designed to produce the sugars needed for the growth and maintenance of the rest of the plant, and for our growth and maintenance as well. All animal life is directly or indirectly dependent on the organic products produced in leaves through photosynthesis.

Leaf Structure

Leaves come in many shapes and sizes, and these characteristics are valuable in identification of plant species. Basic knowledge of these characteristics can enable you to find out at a glance whether the plant you are looking at is a monocot or dicot; and even whether it is a grass, legume, or a member of another plant family.

External Structure of Leaves

Leaves are always attached to the stem at a node and contain a **blade,** which is the flattened portion designed to intercept light and conduct photosynthesis. The leaf blade may or may not be attached directly to the stem. Figures 8-6 and 8-7 show the structures of grass and dicot leaves.

In grasses, the blade is attached to the **leaf sheath** that is wrapped around the stem. The region between the blade and the sheath is called the **collar** that provides the support necessary to hold the blade away from the stem. The collar may or not contain **auricles**, which are claw-like appendages on the edges of the collar, and a **ligule,** which is located on the inside of the collar next to the stem. Ligules occur in various sizes and shapes and can be composed of long or short hairs, or a membrane. Auricles can be long or short. The presence or absence of ligules or auricles, and their size and shapes when present are good identifying characteristics among different species of plants.

As Figure 8-7 shows, the typical dicot leaf is also composed of several specialized structures that aid in identification. The **blade** may or may not be attached to a **petiole** or short stalk composed mostly of vascular tissue. When the petiole is missing and the blade is attached directly to the stem, the leaf is said to be **sessile**. The angle formed by the petiole and stem is called the **leaf axil**. The dicot leaf may or may not have **stipules** attached to the node at the base of the leaf. Stipules are small leaf-like bracts or modified leaves that can vary in size and shape.

Figure 8-8 shows some typical leaf margins and venation. The **leaf margin** is the edge of the leaf blade and can be smooth, toothed, or lobed which also aids in identification. **Leaf venation** is the pattern of the veins or vascular bundles in the leaf. Monocots have **parallel venation**, in which the veins run parallel the length of the leaf and are not branched. Dicots have **net venation** in which the veins are highly branched to produce a net effect. Net venation can be either **pinnate** in which a central vein, called a midrib, runs the length of the leaf and all other veins branch from it; or it can be **palmate** in which several major veins originate from the base of the leaf blade and connect to other veins.

Another valuable identification tool is **leaf arrangement** as shown in Figure 8-9. **Opposite leaf arrangement** means there are two leaves at each node on opposite sides of the stem. **Alternate leaf arrangement** means there is one leaf at each node on alternating sides of the stem. **Whorled leaf arrangement** means there are three or more leaves at each node. Grasses always have alternate leaf arrangement while dicots can have any of the three.

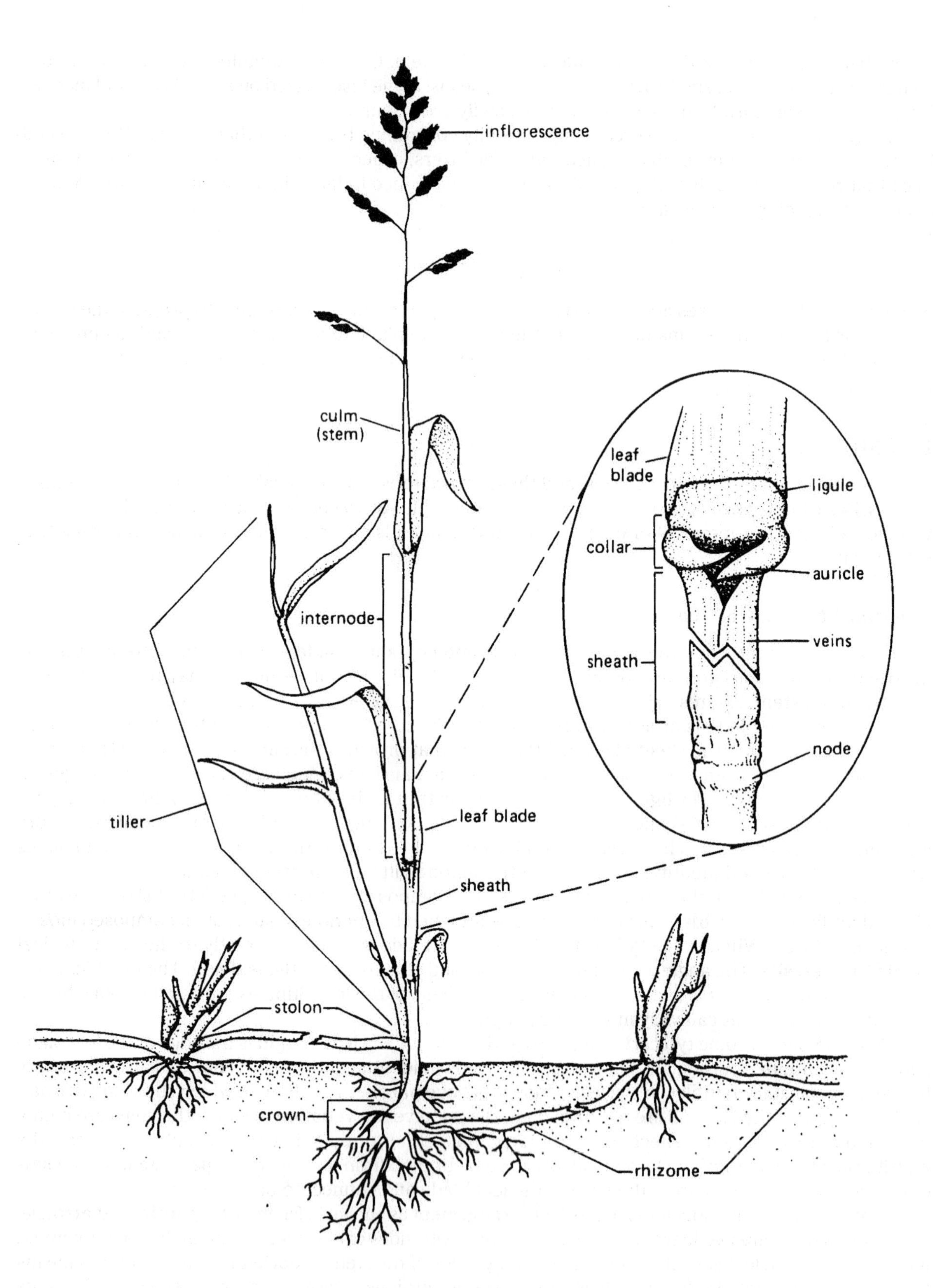

(Figure 8-6 Structures of a grass plant.

(Moore, *Crop Science: A Laboratory Manual*)

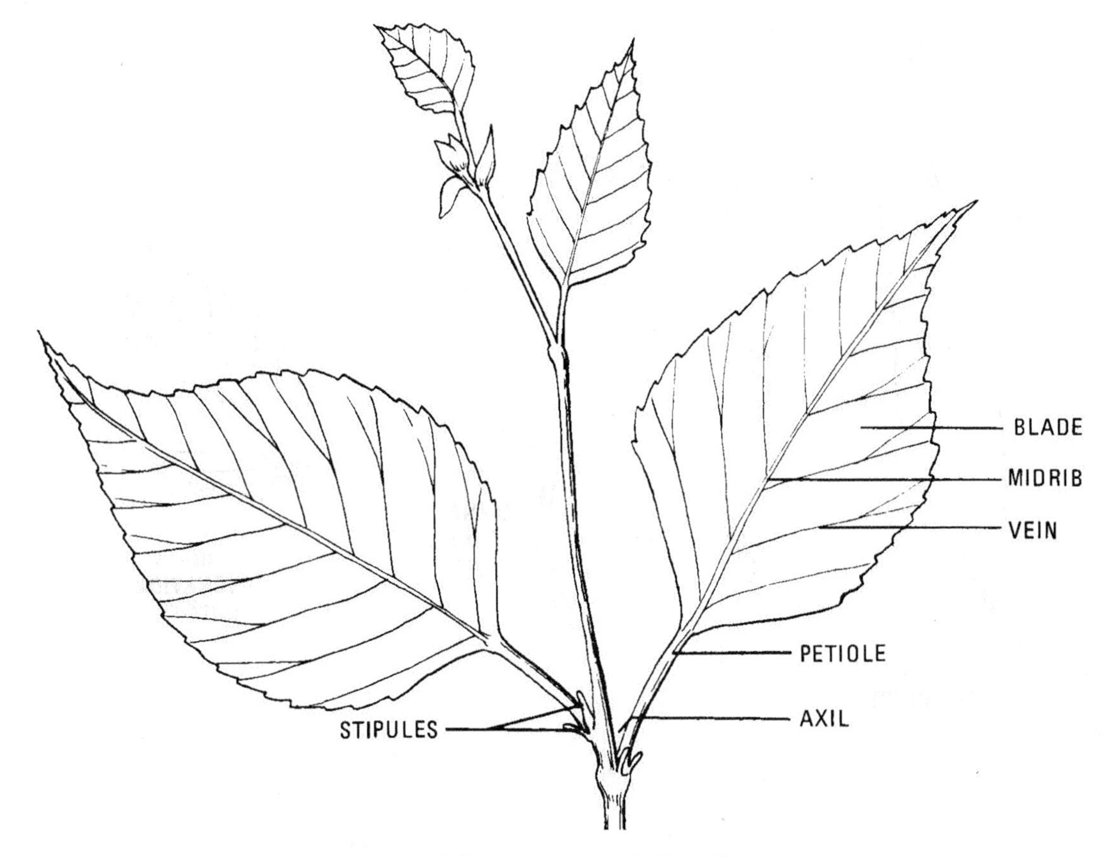

Figure 8-7 Structures of dicot leaves.

Leaf Canopy

Leaf canopy is the pattern of all the leaves of adjoining plants in a field. It is affected by many external features of leaves that we have been discussing, including leaf size, shape, angle of attachment, and arrangement on the stem. Other factors are row spacing and plant population. These factors greatly influence the efficiency of light interception by the leaves in the field. Maximum efficiency occurs when practically all of the incoming sunlight is intercepted and the light is allowed to penetrate deeply into the canopy so that many leaves are active in photosynthesis. Light penetration is important only after the canopy has closed. **Canopy closure** occurs when the leaves of adjoining plants grow together and completely shade the soil.

Leaf size affects **leaf area**, which is also affected by plant population. Leaf area is commonly reported as **leaf area index** (LAI), which is leaf area divided by soil area. Leaf area indexes of 2.5 to 4.0 are common for most crops. This means that there are from 2.5 to 4.0 hectares of leaves in every hectare of cropland. Leaf area index is very important. If it is too low, not all the light will be intercepted; if too high, lower leaves will be shaded too much and will die. Leaf area index is most directly controlled by plant population thus illustrating the importance of proper seeding rates for crops.

Leaf area distribution is the relative proportion of leaf area along the stem. Most crop plants have more leaf area in the middle part of the stem and less at the top and bottom ends. This type of leaf area distribution is very efficient for maximizing light interception within the canopy. Another efficient type of leaf area distribution will increase leaf area from the top of the plant to the bottom. "Christmas Trees" have this type of leaf area distribution. Plants that have most of their leaf area at the top of the stem do not allow much penetration into the canopy.

Leaf angle affects the amount of light intercepted by the upper leaves. Leaves positioned at right angles to the sun intercept the most light and allow less light to get to the lower leaves. Leaves that are more upright allow more light penetration into the canopy thus increasing efficiency by allowing more leaves to photosynthesize. Leaf angle can be controlled genetically and considerable progress has been

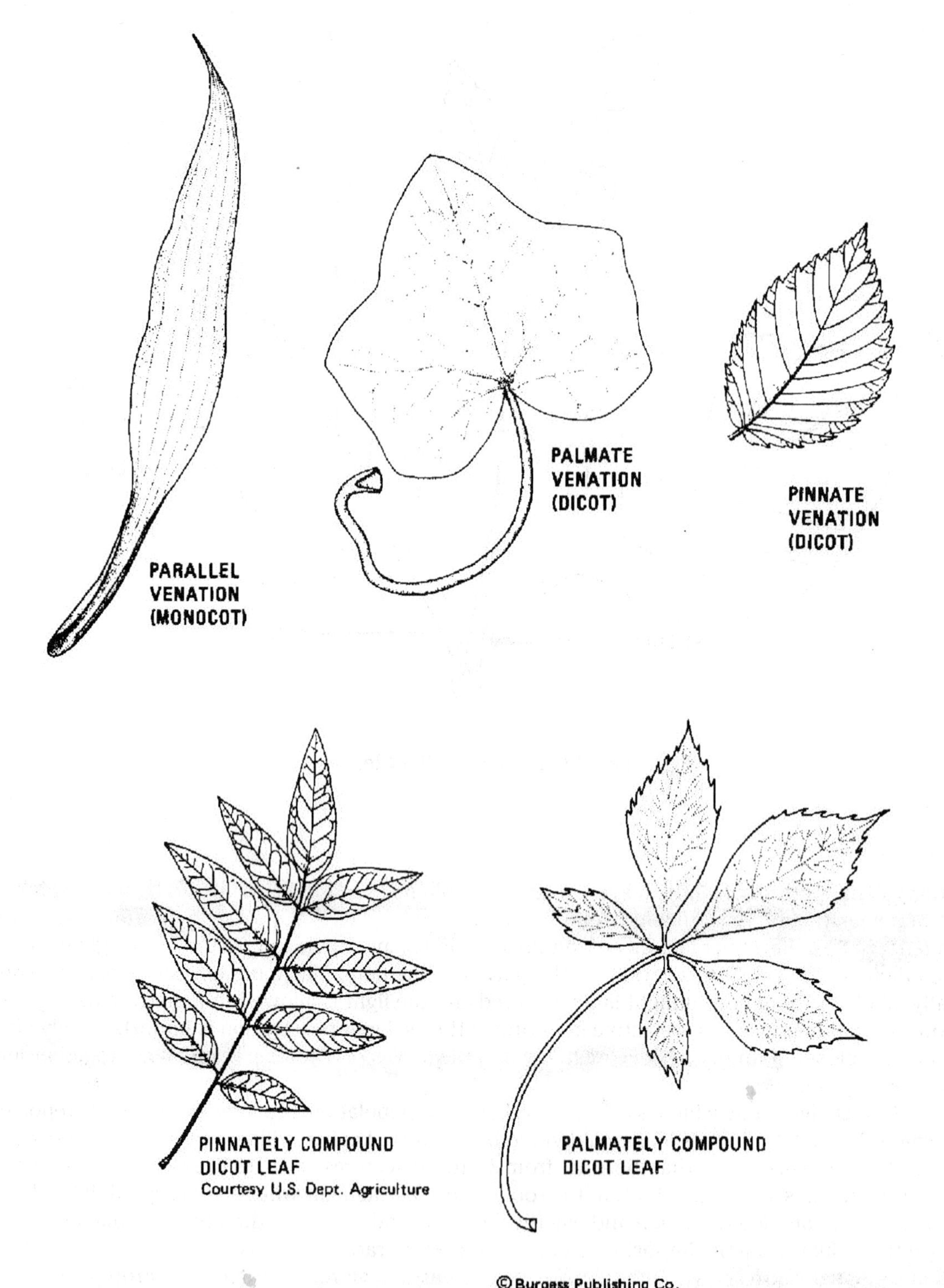

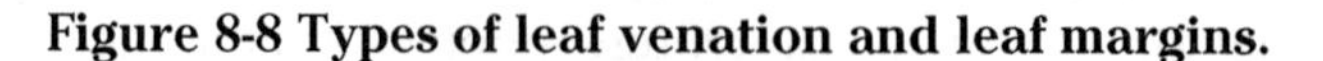

Figure 8-8 Types of leaf venation and leaf margins.

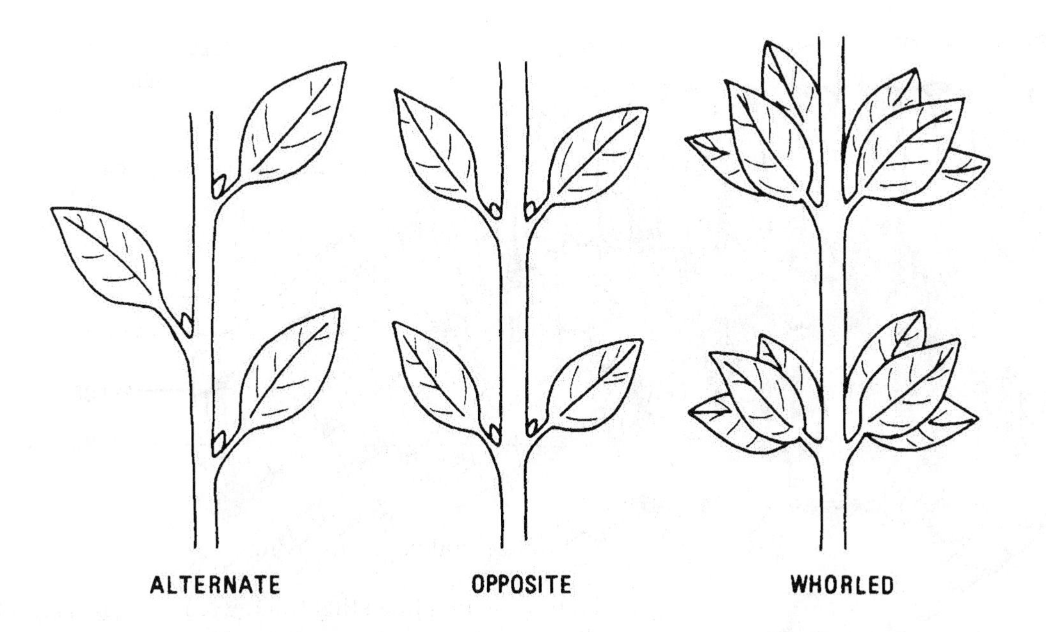

Figure 8-9 Types of leaf arrangements on stems.

made in some crops, particularly corn, to increase canopy efficiency by changing leaf angle. Some leaves are **phototropic**, which means they position themselves at right angles to the sun and move as the sun moves across the sky. An example of a crop with phototropic leaves is soybean. Here, other factors, such as leaf shape and size, must be considered in trying to increase efficiency of light interception. Light penetration in the soybean canopy can be increased by breeding plants with narrower leaves.

Leaf duration, or the time that leaves conduct photosynthesis, is another factor that influences the efficiency of light utilization in the canopy. Crop breeders have made considerable progress in developing plants that do not readily lose their lower leaves. Other factors, such as nutrient availability can also affect leaf life. Deficiencies of many nutrients, including nitrogen and potassium, can cause premature death of lower leaves.

Internal Structure of Leaves

The leaf sheath and petiole, when present, are mostly composed of vascular tissue that connects the leaf blade to the vascular system in the stem. This vascular system, as mentioned before, contains the xylem, which conducts water and plant nutrients to the leaf, and phloem that conducts photosynthate, or the products of photosynthesis from the leaf. The other tissues in these structures are designed to orient the leaf blade for maximum interception of light.

The leaf blade contains many specialized structures that are designed to help the process of photosynthesis. These are shown in Figure 8-10.

The **leaf epidermis** is a single layer of transparent cells located on the upper and lower surfaces of the leaf blade. They are covered with a waxy layer of **cutin**, called a **cuticle**, which prevents the loss of water through the epidermal cells. The epidermis may also have hairs, called **pubescence**, which can provide protection for the leaf surface by reflecting light and reducing wind velocity at the leaf surface. Leaves that are not pubescent are called **glabrous**.

The epidermis also contains green specialized cells called **guard cells**, which form special pores or openings into the leaf called **stomata** (singular, stoma or stomate) which are shown in Figures 8-10 and 8-11. Stomata function in the exchange of gases into and out of the leaf. The number of stomata varies with the species of crop. There are about 5,200 stomata/cm^2 in the upper epidermis of a corn leaf, and about 6,800/cm^2 in the lower epidermis. In alfalfa, there are about 17,000 stomata/cm^2 in the upper epidermis, and about 14,000/cm^2 in the lower epidermis. The plant can control the opening and closing of the stomata. This controls the movement of carbon dioxide into the leaf for photosynthesis, and the loss of water vapor from the leaf in a process called **transpiration**. This is discussed in more detail later in this chapter.

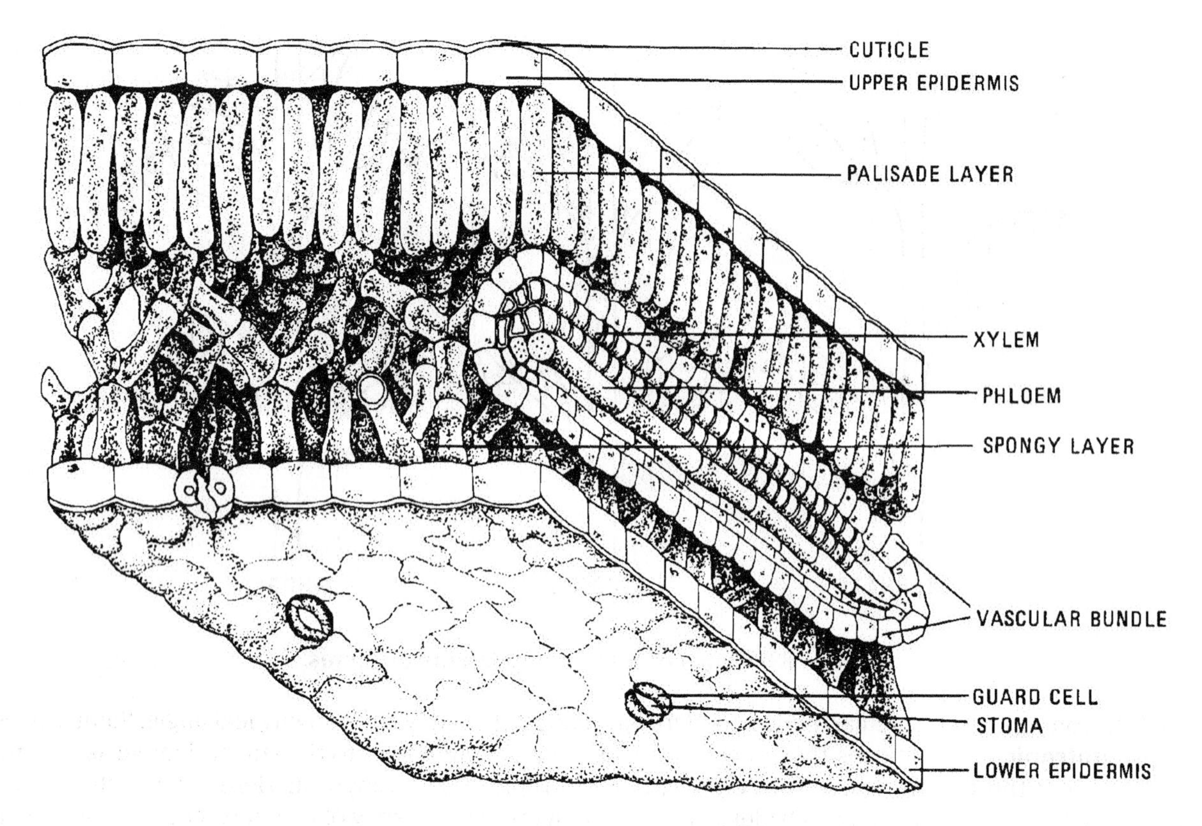

Figure 8-10 Cross section of a leaf blade.

The interior of the leaf blade contains mesophyll tissue and vascular bundles. The **leaf mesophyll** in many plants consists of two kinds of cells. **Palisade mesophyll** cells are cylindrical in shape and are located just below the upper epidermis. **Spongy mesophyll** cells are rounded and are located between the palisade cells and the lower epidermis. These mesophyll cells contain much chlorophyll and are loosely arranged to help movement of carbon dioxide within the interior of the leaf. Some plants, particularly corn, have only spongy cells in their mesophyll. Practically all the photosynthesis in a plant occurs in the mesophyll of the leaves.

The **vascular bundles** of the leaf, also called **veins**, are quite numerous to help the removal of the sugar produced in photosynthesis. Besides the xylem and phloem, the veins in the leaf may contain **bundle sheath cells**. These specialized cells are very active in photosynthesis. In some plants, such as corn, much of the total photosynthesis occurs in the bundle sheath cells. Photosynthesis in bundle sheath cells is very efficient as it is easy to move photosynthate directly into the phloem for transport to other parts of the plant. Photosynthesis is discussed in more detail in Chapter 9.

Origin and Growth of Leaves

Although the first leaves of the plant originated in the embryo as discussed in Chapter 5, all subsequent leaves originate in the **meristematic region** or **growing point** of the stem. The growing point produces a **leaf primordium** at each node as shown in Figure 8-5. This tissue contains an active meristem that then develops into the leaf. Cell division occurs along the edge of the leaf including the tip and sides. The newly produced cells then elongate and differentiate into the various tissues discussed previously. Once the leaf has fully developed, the meristem ceases to function. It is interesting that the developing leaf begins photosynthesis as soon as it is exposed to light, thus manufacturing much of the carbohydrate needed to build itself.

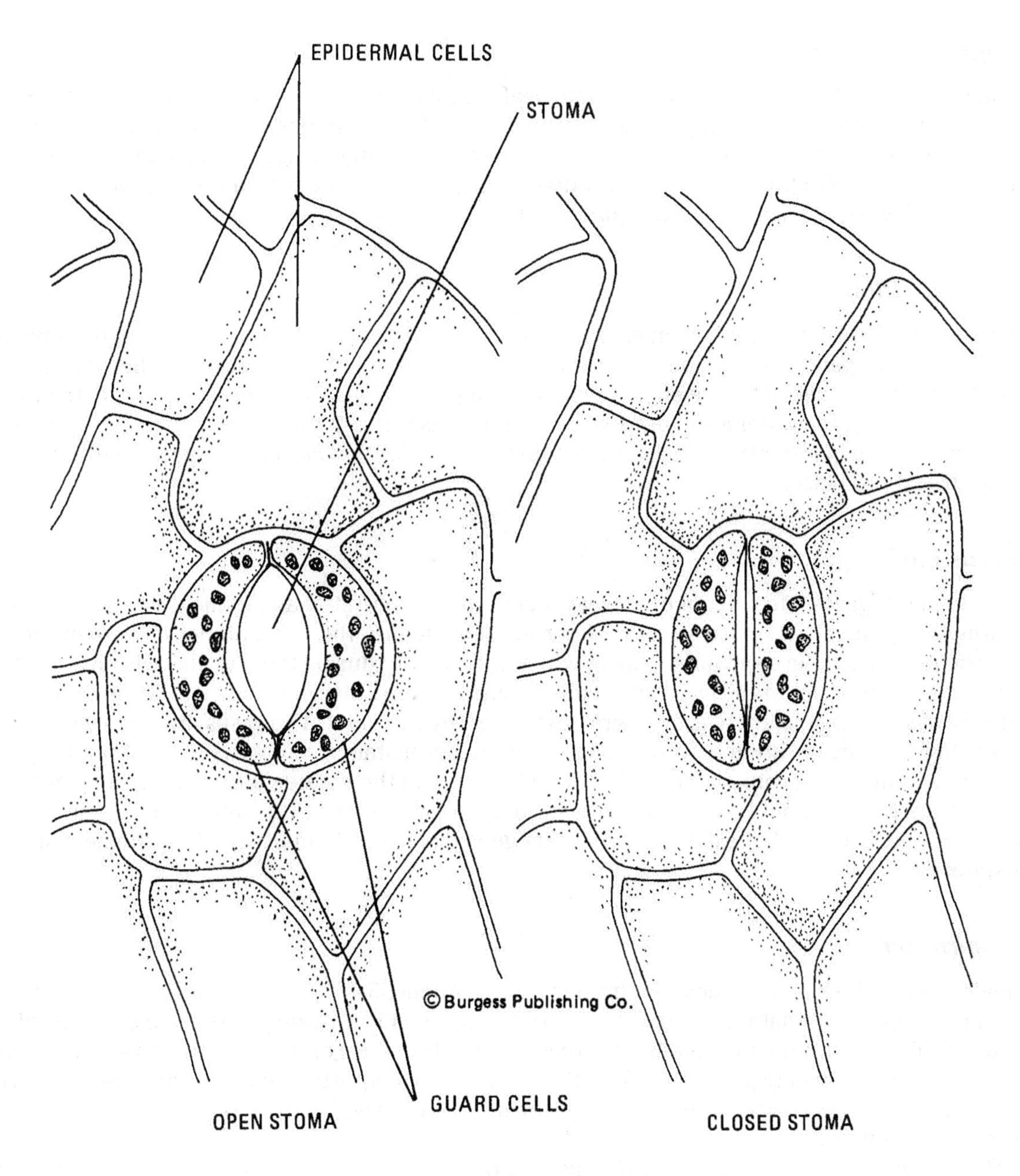

Figure 8-11 Leaf stomata, open and closed.

Functions of Leaves

Many functions of leaves have already been mentioned, but their importance is such that they need to be repeated. As with stems, and other parts of the plant, all these functions are equally important and the loss of any of them can severely disrupt plant growth.

Photosynthesis

As you already know, most of the photosynthesis occurs in the leaves. The absorption and conversion of radiant energy into chemical energy are the basis for growth and maintenance of all higher plants. Photosynthesis is discussed in more detail in Chapter 9.

Storage of Plant Food

Although most photosynthate produced in the leaf is rapidly moved out to other parts of the plant, the leaf is used to store some carbohydrates, proteins, and other plant foods for its own maintenance. The amount stored varies with the species and is important in many forage crops, such as alfalfa, and in green, leafy vegetables for human consumption. Often, however, the total amount stored in the leaves is quite small when compared to other plant parts.

Translocate Plant Food

As mentioned before, the leaf has an extensive system of vascular tissue that is used to move photosynthate to other parts of the plant. Rapid transport of photosynthate is necessary because the rate of photosynthesis will decline if the sugars produced in photosynthesis begin to accumulate in the leaf. In addition, most of the water absorbed by the plant moves through the leaves due to transpiration. Efficient translocation of water throughout the leaf prevents wilting that can disrupt photosynthesis by causing the stomata to close.

Gaseous Exchange

The exchange of gases into and out of the leaf occurs through the stomata. While the leaf is actively photosynthesizing, there is a net movement of carbon dioxide into the leaf, and net movement of oxygen out of the leaf. While the stomata are open, which is primarily during the day, there is also a net movement of water vapor out of the leaf that is called transpiration. Gaseous exchange through the stomata is strictly passive; that is, there is no energy expended by the plant. The gases are simply moving from areas of higher concentration to lower concentration. For instance, while the plant is photosynthesizing, there will be less carbon dioxide within the leaf than in the outside air because the plant is using it. Because of this, carbon dioxide will move from the area of higher concentration, the air, to the area of lower concentration, the leaf. Gaseous exchange is discussed in more detail in the next section on transpiration.

Transpiration

Transpiration is the loss of water vapor from the plant. About 95–97% of the total transpiration in a plant occurs through the stomata. Only about 3–5% of the water is lost through the cuticle of the epidermis because of its design and waxy coating. There is also another form of transpiration that occurs through the special pores of woody plants called **lenticels**, but it is unimportant in crop plants. Since practically all the transpiration in crop plants is through the stomata, we will confine our discussion to only that type of transpiration.

The amount of water lost by stomatal transpiration is quite impressive. Many herbaceous plants transpire several times their own volume in a single day. A single corn plant will transpire up to 1.9 l (2 qts) of water per day, and 3.7 million l/ha (400,000 gal/ac) during the growing season. Since water is the most limiting factor for successful crop production in most of the world, reducing transpiration is a subject of keen interest. Unfortunately, transpiration must occur so that carbon dioxide can enter the leaf for photosynthesis.

Mechanism of Opening and Closing Stomata

Stomata are opened and closed by the guard cells which form them, and the turgidity of the guard cells controls the stomatal opening. When a cell is fully turgid, it is holding as much water as it can. Increasing the turgor of the guard cells causes the stomata to open, and decreasing the turgor causes them to close.

There are three principal factors that influence stomatal opening. As mentioned before, guard cells contain chlorophyll and conduct photosynthesis. When exposed to light, the guard cells of most plants increase their turgidity and open the stomata. Photosynthesis increases the concentration of sugar and decreases the concentration of carbon dioxide in the guard cells which increases their **osmotic pres-**

sure. The increased osmotic pressure causes water in the plant to enter the guard cells which increases turgidity, and the stomata open. The ability to close stomata during the night when carbon dioxide is not needed for photosynthesis enables the plant to conserve much water. A few plants, which are mostly xerophytes, open their stomata during the night when the relative humidity is higher, and close them during the day. These plants have special metabolic pathways that enable them to store carbon dioxide for use during the day. However, this type of stomatal closure does not allow the levels of production that we expect in most crop plants. Pineapple is a crop that uses this type of stomatal closure.

Temperature can also affect stomatal opening and closure. Extremely high or low temperatures can affect photosynthesis in the guard cells that causes the stomata to close. Since these extreme temperatures also cause photosynthesis in the leaf to decline, this again allows the plant to conserve water.

The third factor that influences stomatal opening is the water status of the plant. If the plant experiences a water deficit the guard cells will lose turgidity causing the stomata to close, and the plant conserves water. The guard cells are very sensitive to the water status of the plant and will begin to close the stomata long before the leaves show any visible signs of wilting. It is important to understand that the plant does not simply open and close its stomata. Stomata can be fully open, fully closed, or anything between. The ability of the plant to regulate its stomata greatly affects its ability to survive drought stress. Plants that close their stomata quickly when experiencing water shortages are usually much more likely to be damaged by drought than those that maintain partial stomatal opening during mild stress. Corn is an example of a crop that closes its stomata quickly, and sorghum is an example of a crop that can keep its stomata open during drought stress. Remember, when stomata close, photosynthesis stops.

Mechanism of Transpiration

Transpiration is the **diffusion** of water out of the plant. The interior of the leaf is water saturated, and as mentioned before, the mesophyll is composed of loosely packed cells with considerable air space between them. Therefore, the air within the leaf is water saturated, or it is at 100% relative humidity. Since the air outside the leaf is usually not water saturated, water will diffuse out of the leaf. Diffusion is a passive process. In other words, it requires no expenditure of energy. The water vapor is simply moving from an area of higher concentration to an area of lower concentration.

Factors Affecting Transpiration Rate

The rate of transpiration is affected by environmental and soil factors. Temperature, as mentioned before, can affect stomatal closure, but the biggest effect of temperature on the rate of transpiration is its effect on relative humidity. Relative humidity is the percent saturation of water vapor in the air. Warmer air holds more water, so the relative humidity (percent saturation) drops as the temperature of the air increases. As mentioned before, transpiration occurs because water diffuses out of the leaf in response to differences in concentration of water vapor. The rate of diffusion is directly related to the difference in humidity between the interior and exterior of the leaf. Lower relative humidity increases transpiration.

Wind can increase the rate of transpiration. As water vapor moves out of the leaf it disperses into the atmosphere. If there is no air movement on the leaf surface, this dispersal of water vapor is gradual which slows the rate of transpiration. However, if there is air movement, the removal of water vapor from the opening of the stomata is quite rapid allowing more water vapor to escape. Another environmental factor that affects transpiration is light. As mentioned before, stomata open in response to light and close during the night in most crop plants.

Soil factors can also affect the rate of transpiration. The most important of these is the amount of water in the soil. As discussed previously, stomata close when the leaves begin to lose more water from transpiration than is being supplied through the xylem. The amount of water that can be taken up by the roots is directly affected by the amount of water in the soil, which then affects the tension of the soil water. As discussed in Chapter 4, the tension of the soil water increases as the soil dries out. This increases the tension that the roots must exert to extract soil water, and this increased tension is reflected in the leaves. When the tension of the water in the leaves becomes so great that the cells cannot maintain full turgidity, the stomata will begin to close.

Soil fertility can also affect transpiration rate. Crops grown on fertile soils use more total water because there is more leaf area produced. However, the plants can use the water more efficiently by producing higher yields. **Water use efficiency** is discussed in Chapter 12. It is very important to match the level of fertilization to the amount of water that will be available to the crop. Too much fertilizer can stimulate excess vegetative growth that can cause excessive transpiration. Too little fertilizer results in unhealthy plants that cannot efficiently use the water available.

REVIEW QUESTIONS

1. How do monocot stems differ from dicot stems in their structure and growth? What are the advantages and disadvantages of each?
2. How do the intercalary meristems of grasses make them very adaptable to grazing and mowing?
3. What is the primary structural determination of plant height? What factors can affect it?
4. How are modified stems used by plants? By people?
5. What purpose does apical dominance serve in plants? How does apical dominance function?
6. Why or how are plants with erect stems more efficient at intercepting light? Why is that an advantage?
7. What effects do lodging have on crops?
8. What factors increase stem growth? What factors decrease it?
9. What factors affect the pattern of a crop canopy? How and why does each of these affect light interception within the canopy?
10. Why is it advantageous to close the plant canopy early in the growing season?
11. What factors cause stomata to open and close? How does each affect them?
12. Why is stomatal closure important to plants? How does it affect crop yield?
13. What is diffusion and how does it occur? What does it affect in the plant?
14. Is transpiration good or bad for plants? Why? What can a crop producer do to affect transpiration?

CHAPTER NINE

Photosynthesis and Respiration

INSTRUCTIONAL OBJECTIVES

Upon completion of this chapter, you should be able to:

1. Define photosynthesis and respiration. Discuss their function and importance to plants and people.
2. List the beginning and end products of photosynthesis and respiration. Discuss the logistics of each one.
3. Define light reaction, dark reaction, photophosphorylation, photolysis, electron transfer, carbon dioxide fixation, and sugar formation. Discuss their relationships with each other in photosynthesis.
4. Discuss differences between C_3 and C_4 plants regarding photosynthesis and respiration.
5. List seven factors that affect photosynthesis. Discuss the effect of each one on the rate of photosynthesis.
6. Define light saturation point, photosynthetically active radiation, and compensation point.
7. Define source-sink relationship. Give an example of a source and sink for photosynthate. Name the largest photosynthetic sink.
8. Define aerobic respiration, anaerobic respiration, and photorespiration. Discuss differences in efficiency between them.
9. List five factors that affect the rate of respiration. Discuss the effect of each one.

INTRODUCTION

Photosynthesis and respiration are two of the most important metabolic processes in the plant. Photosynthesis is the pathway that gives the plant the energy it needs for growth, development, and reproduction; and respiration is the pathway that results in that growth, development, and reproduction.

As you probably already know, these two systems are biochemical. However, you do not have to be an expert in biochemistry to understand how they work and comprehend their importance to plants and crop production. In this chapter we will study these important systems in the plant's metabolism without getting too deep into the chemistry involved.

LIGHT

Since light is the driving force behind photosynthesis, a brief discussion of how plants react to light will help in understanding photosynthesis and other plant processes. Visible light is only a small part of the electromagnetic spectrum as shown in Figure 9-1. Plants absorb mostly visible light for photosynthesis. That makes it easier to comprehend as we can relate to visible light since we sense it ourselves.

When light energy strikes any object three things can happen to it; it can be absorbed by the object, it can be reflected off the object, or it can be transmitted through the object. What happens is dependent on the properties of the object. Our eyes detect reflected light. The color of an object is deter-

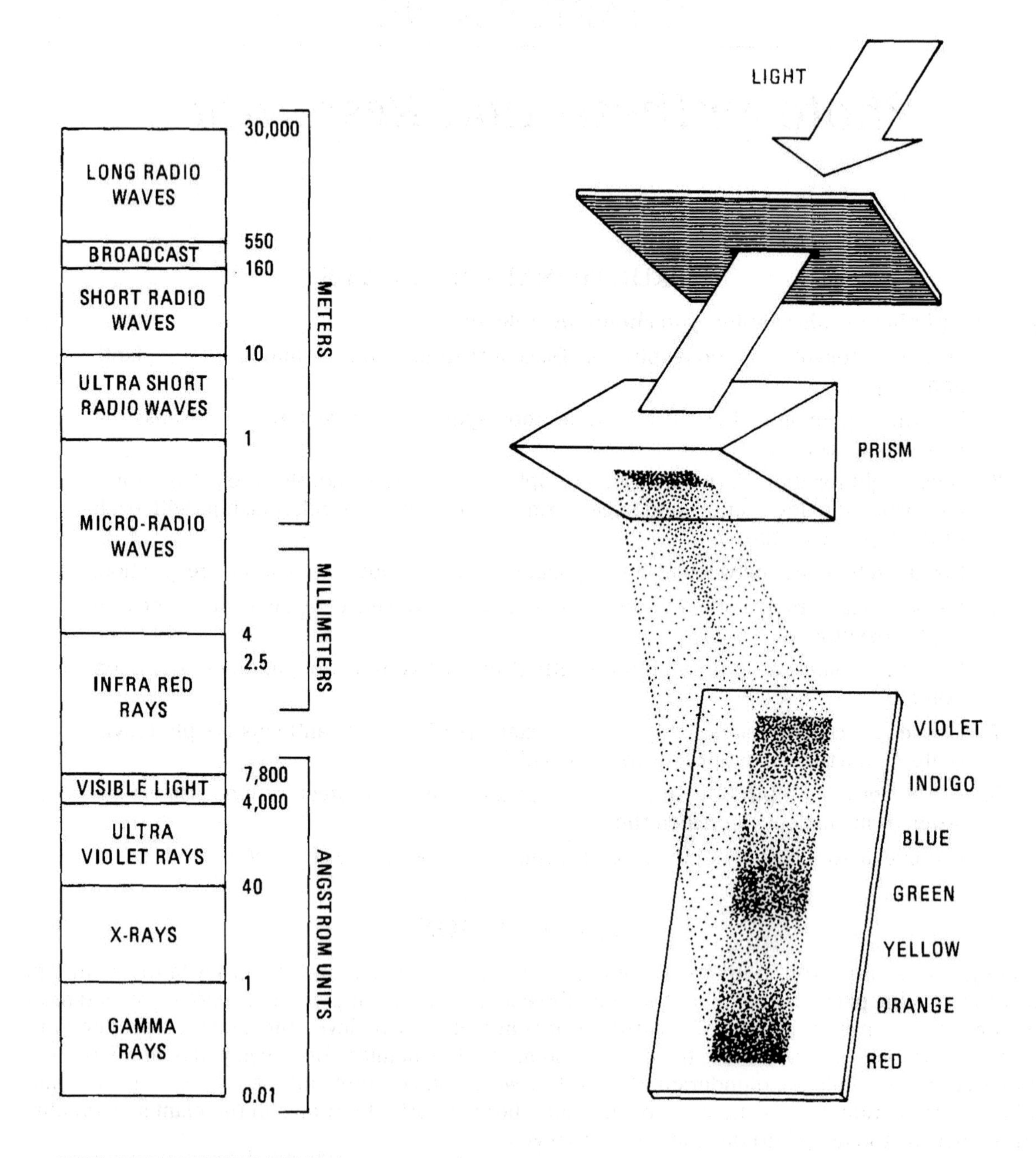

Figure 9-1 The electromagnetic spectrum.

mined by which colors or wavelengths are being reflected. A red object is reflecting mostly red light and is absorbing other wavelengths.

Since light striking leaves is either absorbed, reflected, or transmitted, we can use this information to determine leaf characteristics and to measure specific processes within the plant. For example, chlorophyll absorbs mostly blue and red wavelengths of light and reflects green. That is why leaves look green and the more chlorophyll in the leaf, the greener it looks to us. By measuring the amount of green color being reflected by the leaf, we can measure leaf chlorophyll. That in turn, allows us to determine if the plant has sufficient nutrients or if it is under stress from weather, pests, or diseases.

In recent years there has been increased interest in using remote sensing techniques to monitor crop conditions by measuring its reflectance, or specific wavelengths of light being reflected by the crop. Although it is possible to measure crop reflectance while standing in the field, most measuring is done with aircraft or even satellites orbiting the earth. These crafts are equipped with special cameras or sensors that measure specific wavelengths of light.

Much remote sensing involves wavelengths outside of the visible range especially infrared light. For example, plants that are experiencing drought stress will emit more energy in that range as the plant is getting less cooling from transpiration, and plants emitting less energy have less photosynthesis. Remote sensing is another tool we can use to monitor the crop and alert us to problems.

PHOTOSYNTHESIS

The word, photosynthesis, comes from the words *photos* (light) and *synthesis* (to put together). **Photosynthesis** is the conversion of light energy to chemical energy; and is one of the most significant life processes. Photosynthesis is the basis of all agriculture and specifically crop production. Life on earth as we now know it is almost totally the result of photosynthesis. Photosynthesis is the beginning of the food chain for practically all of the organisms on earth. Photosynthesis also supplies practically all of the oxygen that is needed for life.

Reactions in Photosynthesis

The overall reaction of photosynthesis, as shown in Equation 9-1, tells us that carbon dioxide and water are inputs, which, when reacted in the presence of light and chlorophyll, combine to produce a simple sugar, oxygen, and water.

$$6CO_2 + 12\,H_2O + LIGHT \rightarrow C_6H_{12}O_6 + 6O2 + 6H_2O$$

Equation 9-1. Overall reaction for photosynthesis.

However, this equation only shows the beginning and end products of photosynthesis and gives no hint of what happens in between. That is like saying that if you combine metal, plastic, and rubber you get an automobile. The process of photosynthesis is a biochemical pathway that involves several distinct, yet closely interrelated steps. We will now look at some of the more important steps in photosynthesis so that you have a better understanding of how it works.

The Light Reactions

The **light reactions** of photosynthesis are so named because they directly use radiant energy and convert it to metabolically useful energy. These reactions all occur simultaneously in the presence of light and chlorophyll. The first of these is **electron transfer**. As you probably know, atoms are composed of a nucleus surrounded by several electrons that are orbiting around it. When radiant energy strikes the magnesium atom in the center of a chlorophyll molecule, this energy is absorbed by an electron of the magnesium atom and it is displaced from its orbit. Thus, the radiant energy has been captured and can be used by the plant as you will see shortly.

The next light reaction is **photophosphorylation**. This reaction uses light to create a high-energy bond in the formation of ATP. In this reaction, **ADP**, or adenosine diphosphate, is connected to an inorganic phosphate radicle by the infusion of energy from light and the activated electron from the previous reaction. This results in the formation of a high-energy bond in **ATP**, or adenosine triphosphate. The last light reaction is **photolysis**. Photolysis results in the splitting of water and the capture of additional radiant energy. In this reaction, two water molecules are split in the presence of light and **NADP**, or nicotinamide adenine dinucleotide phosphate, resulting in free oxygen and captured energy in **$NADPH_2$**. Photolysis is also called the **Hill Reaction.**

ATP and $NADPH_2$ are common energy carriers used by both plants and animals. ATP is a common and versatile compound that is used to store metabolic energy until it is needed for other reactions. The energy stored in the high-energy bond is released to drive other reactions when the bond is broken and the compound is changed back to ADP. NADP accepts hydrogen and energy that are transferred to other compounds in many metabolic reactions. However, in photosynthesis, the light reactions are designed to capture radiant energy in ATP and $NADPH_2$ to be used in further reactions. The light reactions are dependent on the amount of light that is being intercepted and are independent of temperature.

The Dark Reactions

The **dark reactions** of photosynthesis are also called the **Calvin Cycle**. These reactions are called dark reactions because they do not directly use light energy. Instead, they use the energy captured in the previously discussed light reactions. However, do not be misled into thinking that the light reactions occur in the light and the dark reactions occur during the night. Both sets of reactions occur simultaneously, and the dark reactions will stop within minutes after the light reactions stop. The dark reactions are very temperature dependent; that is, they progress faster as temperature increases.

The first dark reaction is **carbon dioxide fixation**. This reaction involves the incorporation of a carbon dioxide molecule into a carbon chain. It uses energy from ATP and energy and hydrogens from $NADPH_2$. A new water molecule is also formed at this time. The next dark reaction is **sugar formation**. In this reaction, the carbon chain formed in the previous reaction is given additional energy from ATP and $NADPH_2$ and is rearranged into a simple sugar molecule called **glucose** ($C_6H_{12}O_6$). Figure 9-2 shows an overall schematic of photosynthesis.

The actual compounds involved in the dark reactions differ between types of plants. **C_4 plants** use a slightly different pathway that allows them to fix more carbon dioxide than **C_3 plants**. There are distinct differences between these types of plants; not only in their biochemistry, but also in their productivity. These differences will be discussed later in this chapter.

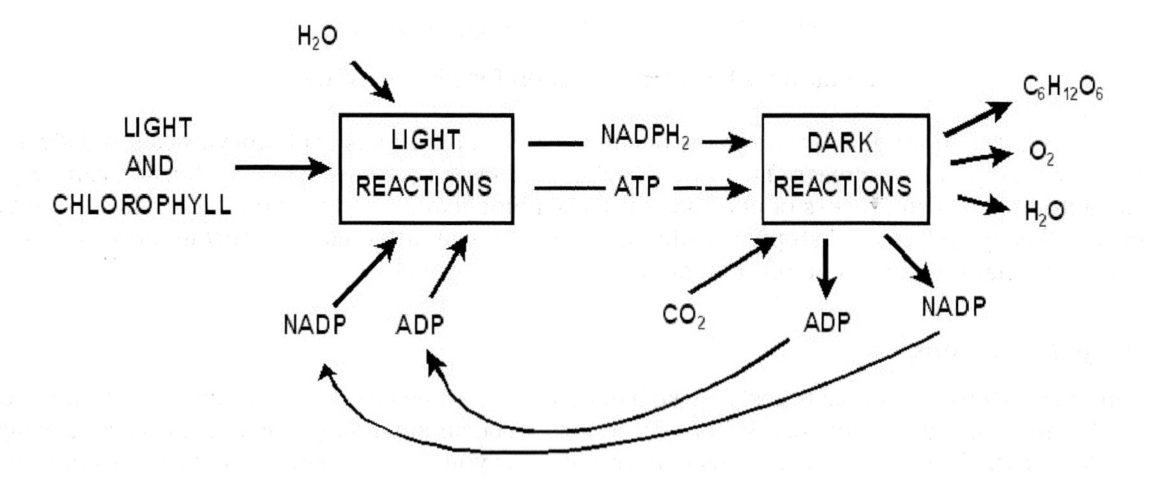

Figure 9-2 Overall schematic of photosynthesis.

Logistics of Photosynthesis

Now that you have a better understanding of how photosynthesis works, let's look at how the plant supplies the compounds used in photosynthesis and uses the products. Figure 9-3 illustrates the logistics of photosynthesis.

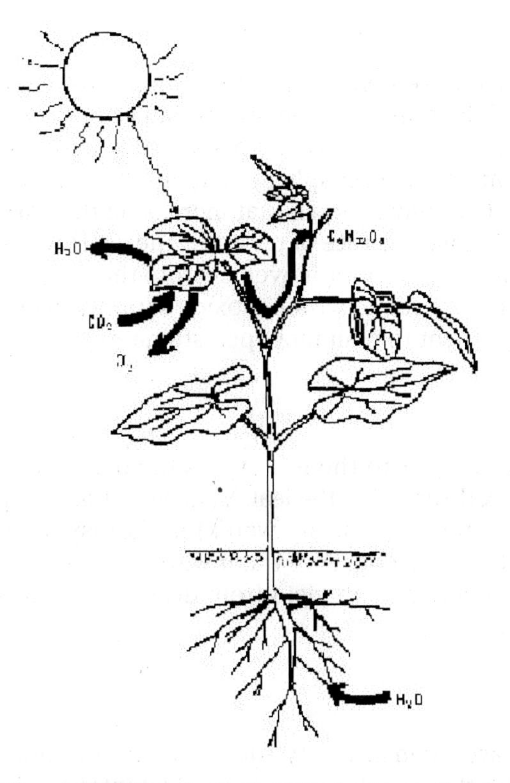

Figure 9-3 The logistics of photosynthesis.

Carbon dioxide from the atmosphere enters the leaves through the stomata. It is then dissolved in the water in the leaf and used in photosynthesis as the **bicarbonate ion** (HCO_3^-). The movement of carbon dioxide into the leaf is a simple **diffusion** process. No metabolic energy is expended. Since carbon dioxide is being used in the leaf there will be less in the leaf air spaces than is in the air outside the stomata. If allowed, gases and liquids will move (diffuse) from areas of higher concentration to areas of lower concentration in an effort to reach equilibrium. Water from the soil is absorbed by the roots and moves to the leaves through the xylem tissue. Water uptake by the roots is an active process; that is, it requires metabolic energy.

The light energy from the sun is absorbed by the chlorophyll in the leaf and converted to metabolically useful energy. The oxygen produced in the leaves diffuses out of the leaf through the stomata, or it can be used by the plant in respiration or other metabolic reactions. Most of the time during photosynthesis, the plant is producing more oxygen than it is using, and oxygen is released to the atmosphere. In like manner, the water produced in photosynthesis is used in other reactions, or it diffuses out of the leaf through the stomata.

Several things can happen to the sugars (glucose) formed in photosynthesis. Most of the sugar is converted to **sucrose**, a 12-carbon sugar, and immediately translocated to other parts of the plant through the phloem tissue. Sucrose is common table sugar and is the building blocks upon which the plant grows and reproduces. Some sugar, however, can be stored in the leaf mesophyll and may be transformed into starch, which are long chains of glucose. Starch is also formed in other parts of the plant such as the roots and grain.

Factors Affecting Photosynthesis

Photosynthesis, as you now know, is a very complex system of reactions that depend on each other and other metabolic processes in the plant. Because of this, there are many factors that can affect the rate of photosynthesis. Some of these factors can be easily controlled by the producer, while others are controlled only by nature. Many of the cultural practices used in crop production are aimed at optimizing photosynthesis through the factors that affect it.

Carbon Dioxide

As mentioned before, carbon dioxide enters the leaf through the stomata. When the stomata close, the supply of carbon dioxide is deficient and photosynthesis slows or ceases. Stomatal opening and closing are largely controlled by the water status of the plant as discussed in Chapter 8.

The carbon dioxide content of the atmosphere is about 400 parts per million (0.04%). Studies in greenhouses and growth chambers have shown that increasing the carbon dioxide content of the air entering the leaf can greatly increase the rate of photosynthesis. However, attempts to do this in field crops have not been successful because of air movement into the leaf canopy. Therefore, the only thing that can be done with field crops regarding carbon dioxide supply is using cultural practices that result in good supplies of water to the plant to maintain open stomata.

Water

As just mentioned, the supply of water to the leaf affects the degree that the stomata are open, regulating the amount of carbon dioxide entering the leaf. Since less than one percent of the water absorbed by plants is used as a reactant in photosynthesis, water for photosynthesis itself is never lacking. With a water deficit, the stomata will close to conserve water within the plant long before water would be limiting photosynthesis. The reduction in crop yield from drought stress is thus due to limiting carbon dioxide after stomatal closure.

Light

Light for photosynthesis is measured in intensity and duration. Duration is the length of time that sunlight strikes the leaf and its longest period is in late June. Light intensity is the amount of radiant energy that strikes the leaf. About 0.6 cal/cm(/min of usable energy for photosynthesis reaches the earth's surface on a sunny summer day at solar noon.

As discussed previously, plants do not absorb all the solar energy that strikes the leaf. Chlorophyll absorbs mostly blue (about 450 nm) and red (about 660 nm) wavelengths of light as shown in Figure 9-4. Radiant energy that is absorbed by chlorophyll and used to drive photosynthesis is called **photosynthetically active radiation (PAR)**. Notice that chlorophyll absorbs very little green light. That is why leaves look green.

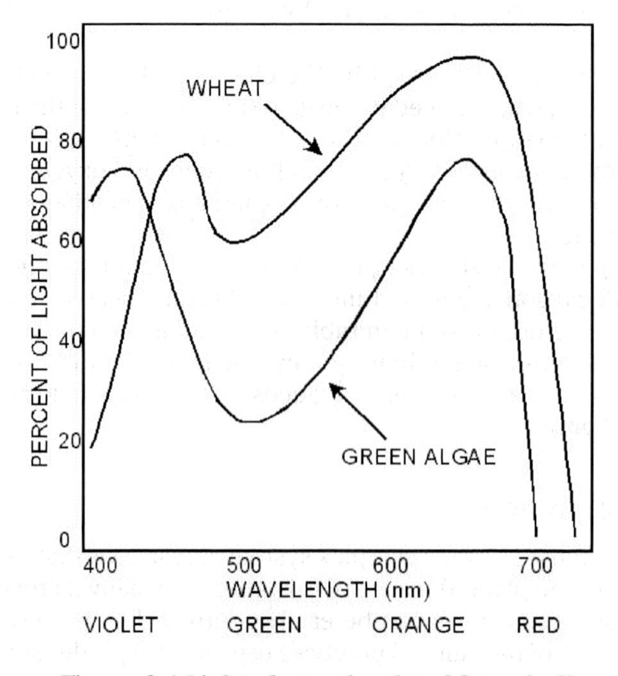

Figure 9-4 Light absorption by chlorophyll.

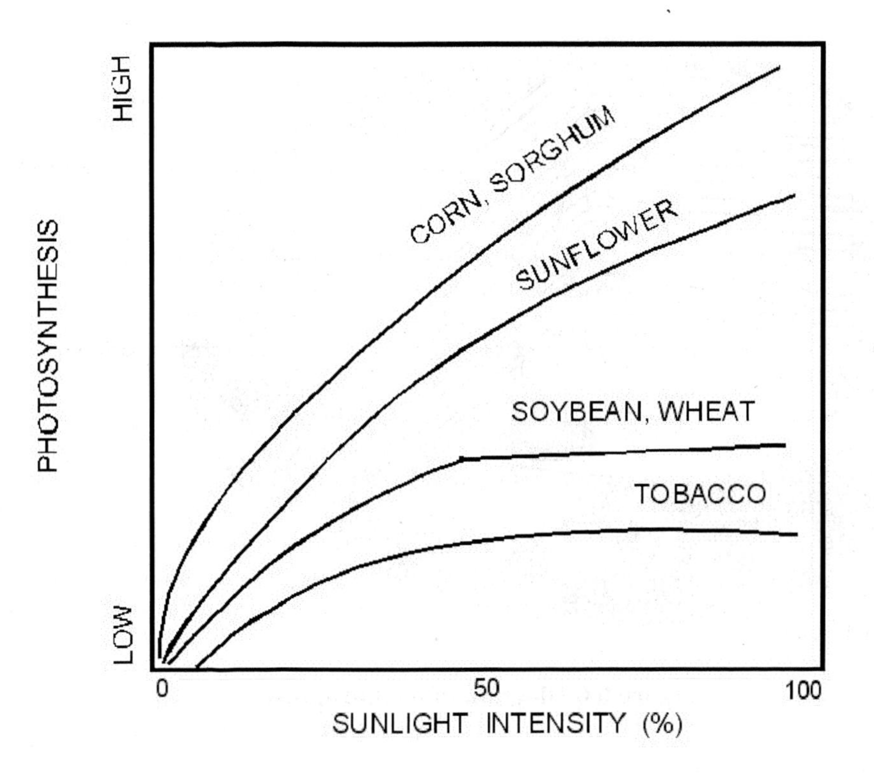

Figure 9-5 Photosynthetic rates of various crops.

The light intensity when photosynthesis reaches its maximum rate is called the **light saturation point**. The degree that plants can use solar radiation depends largely on whether or not they light saturate at full sunlight.

Earlier in this chapter we mentioned that there are C_3 and C_4 plants. Many C_3 plants such as soybeans, cotton, potatoes, alfalfa, and cool-season grasses will not increase their rate of photosynthesis when the light intensity is greater than 1/3 to 1/3 full sunlight. In other words, these plants light saturate at 1/3 to 1/2 full sunlight. Most C_4 plants do not light saturate even at full sunlight. These include corn, sorghum, sugarcane and most warm season grasses. Some C_3 plants, such as peanut and sunflower also do not light saturate. Plants that light saturate will produce much lower yields than plants that do not saturate because of their lower rates of photosynthesis. Figure 9-5 shows the rate of photosynthesis of several different crops under different light intensities. Notice the light saturation point of some of them.

A crop producer has no control over the intensity or duration of sunlight that reaches the crop. However, cultural practices used in crop production can greatly affect the amount of incoming light energy intercepted by the crop and used in photosynthesis. Higher plant populations increase leaf area for light interception. Narrowing the row spacing can increase the penetration of light into the leaf canopy by improving the spatial arrangement of plants and spreading out the leaves. Proper soil fertility produces healthy, vigorous plants with lots of chlorophyll. Pest control prevents loss of leaf area from insects and diseases; and competition from weeds for light, water, and nutrients. Good water management keeps the stomata open. Factors that affect leaf area, leaf duration, and leaf distribution were discussed in Chapter 8.

Leaf Chlorophyll

Chlorophyll is a green pigment that is found in a specialized cell structure called a **chloroplast** that is illustrated in Figure 9-6. Chloroplasts are designed to efficiently capture radiant energy and convert it to chemical energy. The entire process of photosynthesis occurs in the chloroplast and a typical leaf

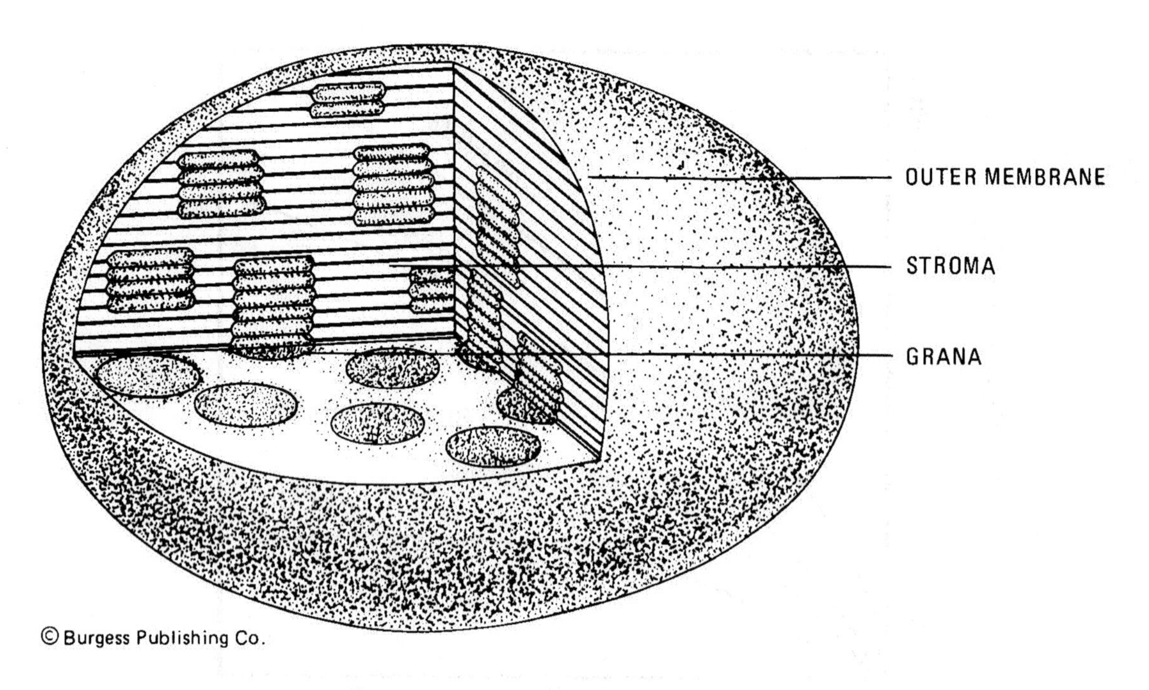

Figure 9-6 Diagram of a chloroplast.

mesophyll cell contains from 20 to 100 chloroplasts. Since light absorption occurs in the chlorophyll of the leaf, the amount of chlorophyll can greatly determine the rate of photosynthesis.

Chlorosis is a lack of chlorophyll or a deterioration of chlorophyll. Many factors can cause chlorosis. Deficiencies of many nutrients can cause chlorosis, including nitrogen, potassium, magnesium, iron, zinc, copper, and manganese. Nitrogen and magnesium are constituents of chlorophyll. Potassium starts the enzyme system that produces chlorophyll. Other nutrients are essential for formation and functioning of chlorophyll. Many plant diseases can also cause chlorosis. Again, good cultural practices can insure that adequate chlorophyll will be produced to maximize photosynthesis.

Temperature

Temperature affects many metabolic reactions in the plant and photosynthesis is no exception. The term, Q_{10}, is commonly used as an expression of reaction velocity or the rate of change in reaction activity that results form a change in temperature. Van't Hoff's Law states that the rate of a chemical reaction approximately doubles for every 10°C (18°F) rise in temperature, or, the Q_{10} is about 2.4. However, the Q_{10} for catalyzed biochemical reactions in the plant such as photosynthesis is only 1.2 to 1.3 because there are many factors that can limit the reaction rate in a living system. This does tell us, however, that increasing the temperature will usually increase the rate of photosynthesis. It is the dark reactions that are affected by temperature. The light reactions are primarily affected by light intensity.

Temperature affects the viscosity of water, cytoplasm, and other fluids in the plant. Higher temperatures can increase the streaming of the **cytoplasm**, or the fluid contents of the cell, which speeds the movement of compounds from one metabolic reaction to another. Higher temperatures also increase the permeability of the cell membrane to the transfer of water and other substances thus increasing the availability of compounds necessary for metabolic reactions. Extremely high temperatures can interrupt the metabolic processes within the plant and stop photosynthesis. Extremely low temperatures can also disrupt photosynthesis and, if freezing occurs, cells can rupture resulting in tissue death. There is little that can be done about temperatures in the field. However, producers can reduce the chances of temperature extremes affecting a crop by seeding at the proper time and selecting varieties that mature before frost.

Carbohydrate Translocation

As mentioned before, the accumulation of the sugars produced by photosynthesis in the mesophyll causes a feedback inhibition of photosynthesis. Conversely, an increase in the demand for photosynthate (sugar) can cause an increase in the rate of photosynthesis. This relationship of the rate of photosynthesis to the demand for the products of photosynthesis is called a **source-sink relationship**. The source is the supplier of sugar, namely the leaves. The sink is the demand for sugar and can be the roots or growing point. The developing seeds, however, are by far the largest sink for photosynthesis. When the plant is filling its grain or seeds, the rate of photosynthesis will be higher than it was during the vegetative stage and the plant is also more sensitive to stress that can limit photosynthesis during this time. Conversely, the loss of a major sink such as the grain will greatly reduce the rate of photosynthesis.

Leaf Age

The rate of photosynthesis will decline as the leaf ages. This is usually of no importance since the younger leaves are nearer the top of the canopy and intercept most of the light. However, this factor can become important if loss of the upper leaves should occur from damage by insects, weather, etc. Even if exposed to full sunlight, these lower leaves will not be able to totally compensate for the loss of leaf area in the plant and yield will decrease.

RESPIRATION

Plant cells require energy for cell division, elongation, and maintenance, as well as for nutrient and water uptake. This energy comes from the low temperature oxidation of sugars called **respiration**. Respiration differs from types of high temperature oxidation such as fire because the oxidation occurs at low temperatures due to the action of enzymes that catalyze the reactions.

Respiration occurs in the plant at the same time as photosynthesis is occurring and is often referred to as the opposite of photosynthesis since the beginning and end products of both pathways are essentially

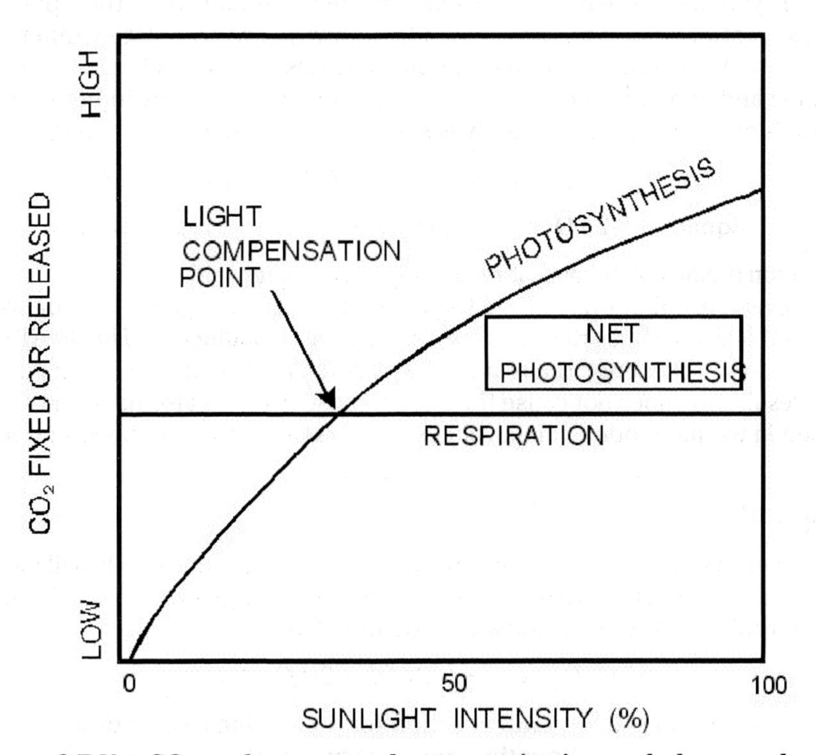

Figure 9-7 Net CO_2 exchange rate due to respiration and photosynthesis.

opposite each other. However, the pathways involved in respiration are entirely different and separate from those involved in photosynthesis.

Respiration occurs at all times in all living cells in the plant. The energy from respiration may be used for many purposes including those already mentioned and for the synthesis of more complex compounds, such as proteins, fats, and oils. Since respiration releases carbon dioxide and photosynthesis uses it, it is possible to measure the net input or output of carbon dioxide and decide whether the plant is experiencing net growth. This is illustrated in Figure 9-7.

During the night, respiration will continue and the leaves will be using sugars and releasing carbon dioxide. After sunrise, the light intensity will increase causing a subsequent increase in photosynthesis and carbon dioxide fixation. When carbon dioxide fixed in photosynthesis equals the carbon dioxide released in respiration, the plant has reached the **compensation point**. Further increases in light intensity will result in more carbon dioxide being fixed than is being released. The light intensity must be above the compensation point before the plant can increase its dry matter and grow. The actual light intensity at the compensation point varies with the type of crop, and other factors that influence the rates of photosynthesis and respiration. For most crop plants, the compensation point is easily reached except on extremely cloudy days.

Kinds of Respiration

Respiration can occur via different pathways in the plant cell depending on the availability of oxygen and the metabolism of the plant. As you will see, these different forms of respiration are not all equally efficient in converting sugars to metabolically useful energy.

Aerobic Respiration

Aerobic respiration occurs when adequate amounts of oxygen are present. It is the most efficient type of respiration and therefore the most common in the plant. The overall reaction for aerobic respiration is shown in Equation 9-2.

In this reaction, a glucose sugar ($C_6H_{12}O_6$) is broken down in the presence of adequate oxygen to carbon dioxide and water, and the energy in the glucose is captured in 40 ATP compounds. In comparing this to Equation 9-1, you can see why respiration is sometimes said to be the opposite of photosynthesis. However, as mentioned before, the metabolic pathways are completely different and separate. Again, as discussed with the overall reaction for photosynthesis, this overall reaction for respiration only shows the beginning and end products and gives no hint of what happens in between. The process of aerobic respiration is quite complex and involves several steps that are referred to as the **Krebs Cycle**.

$$C_6H_{12}O_6 + 6O_2 + 40ADP + 40P_i \rightarrow 6CO_2 + 6H_2O + 40ATP$$

Equation 9-2 . Overall reaction for aerobic respiration.

Aerobic respiration is efficient because it results in the complete oxidation of glucose into carbon dioxide and water. However, only about half of the total energy released in glucose oxidation is captured by ATP. There are a total of 686 kilocalories of energy in each mole of glucose. The 40 ATP formed capture only about 340 kilocalories of energy; the rest is mostly converted to heat. In plants this heat is quickly dissipated so that respiration does not cause the plant to heat. Animals also have aerobic respiration and this heat is retained in warm-blooded animals such as yourself to maintain body temperature.

Anaerobic Respiration

When adequate oxygen is not available for aerobic respiration, the plant's cells will switch to a less efficient form of respiration in an effort to get the energy it needs for growth and maintenance. The overall reaction for anaerobic respiration is shown in Equation 9-3.

$$C_6H_{12}O_6 + O_2 \rightarrow 2C_2H_5OH + 2H_2O + 2ATP$$

Equation 9-3 . Overall reaction for anaerobic respiration.
(Equation does not balance)

Notice that in this reaction, the glucose was not completely oxidized to carbon dioxide and water. Instead, only two carbon dioxide and water molecules each were formed and the remaining carbon was converted into **ethyl alcohol** (C_2H_5OH). The net result is that only two ATP molecules are formed instead of forty. The plant's cells can use anaerobic respiration for only a short time before they experience a shortage of metabolic energy from lack of adequate ATP and begin to suffer from the buildup of toxic compounds such as the ethyl alcohol. This reaction is valuable in producing alcohol for fuel and brewing, but is not desirable when it occurs in the plant. There are many reasons why a plant would not receive adequate oxygen and these are discussed later in this chapter.

Photorespiration

As mentioned before, photosynthesis and respiration occur concurrently. In the light, photosynthesis is greater than respiration and the plant increases in weight. In the dark, respiration continues but photosynthesis stops. The plant may increase in size during the night due to cell division and elongation but there is not an increase in weight, and it may even decrease in weight.

It has been discovered that some plants have a respiration rate in the light that is higher than in the dark. This is due to a different type of respiration that occurs only in the chloroplasts and other structures in the green cell. Normal respiration occurs in all living cells. This different type of respiration is called **photorespiration** because it uses light energy to drive the reaction in much the same way that photosynthesis uses light energy.

Photorespiration is considered to be detrimental to the plant because it results in the oxidation of the sugars just produced in photosynthesis before the plant can use them for cell maintenance or translocate them to other parts for growth. Photorespiration increases with increasing light intensity, temperature, and oxygen; and with decreasing carbon dioxide. A cell that is actively photosynthesizing will be low in carbon dioxide and high in oxygen. The plant gets no usable energy from photorespiration and it is a mystery why such a system ever developed in plants.

Plants that have photorespiration easily light saturate because photorespiration will increase at the same rate as photosynthesis at higher light intensities. This results in a much lower rate of net photosynthesis and subsequent dry matter production. **Net photosynthesis** is the rate of photosynthesis minus the rate of respiration from all forms and gives an accurate account of the rate of growth. Net photosynthesis is also called the **net assimilation rate.**

There is very little, if any, photorespiration in C_4 plants while many C_3 plants have high rates of photorespiration. This accounts in large part for the increased dry matter production of C_4 plants as discussed earlier. The reason photorespiration is found only in C_3 plants is not fully known. C_3 plants with little or no photorespiration, such as sunflower and peanut, have almost as high a rate of photosynthesis as C_4 plants. When an inhibitor of photorespiration is applied to crops that have high rates of photorespiration, the rate of photosynthesis jumps dramatically and the leaves will usually not light saturate. At the present, it is not possible to apply photorespiration inhibitors to field crops, and the only solution is to develop crop varieties that have lower rates of photorespiration. However, development of varieties with little or no photorespiration is a very slow process that may or may not succeed in the near future.

Factors Affecting Respiration

As with photosynthesis and other metabolic processes in the plant, there are several factors that can affect the rate of respiration in the plant. Some of these have already been mentioned briefly such as temperature and oxygen. We will now look at these and other factors in more detail.

Kind of Tissue

Some kinds of tissues in the plant have high rates of respiration while others have little, if any, respiration. Some cells, once they are formed, may die or at least will be almost dormant in the plant. These include tissues that function in providing mechanical strength and support for the stem, and many cells in the vascular bundles in the older parts of the stems and roots. Cells and tissues that have high rates of respiration include meristematic cells of the growing points of the plant and the developing seeds.

The rate of respiration in any cell is dependent on the activities and functions of the cell, the availability of sugars for oxidation, the degree of hydration (water content) of the cytoplasm (cell liquid), and the activity of the enzymes that catalyze the reactions of respiration.

Temperature

Within the limits of minimum to optimum temperatures for growth, an increase in the temperature will increase the respiration rate. This is due primarily to an increase in enzyme activity, and an increase in cell wall permeability to the movement of sugars and other compounds as temperature increases. However, increasing the temperature above the optimum for growth causes a decrease in the rate of respiration because of a decline in enzyme activity. Extremely high temperatures can disrupt the plant's metabolism even to the point of death.

If all other factors are not limiting, temperature has the greatest effect on the rate of crop growth. Practical applications of this fact have been developed by the use of growing degree days that is discussed in detail in Chapter 12.

Oxygen

As already discussed, the availability of oxygen is essential for efficient respiration. Lack of adequate oxygen can cause the cell to shift to less efficient pathways that greatly decrease the usable metabolic energy coming from respiration which can then lead to the accumulation of toxic substances.

Adequate oxygen is rarely a problem in the shoot of a crop plant but can be a problem in the roots. As discussed in previous chapters on seed germination and roots, oxygen can easily be lacking in the soil air. Soil crusting, compaction, and water saturation can cause anaerobic conditions in the soil that can ultimately lead to death of root tissue.

Light

When photorespiration is present, light has a direct effect on the total respiration rate in the leaf. Light, however, does not directly affect other forms of respiration in the plant. Sunlight can have an indirect effect through its effect on air temperature, since much of the total energy load from solar radiation is converted to heat.

Concentrations of Other Compounds

The concentration or supply of compounds other than oxygen can also affect the rate of respiration. An inadequate supply of glucose sugar lowers the respiration rate because there is no reacting compound to oxidize. This can occur when there is a disruption in the translocation of photosynthate within the plant, or when photosynthesis has been disrupted. Disruption of photosynthesis due to drought or other stress, however, does not affect the rate of respiration until the readily usable supplies of carbohydrates have been exhausted. This causes the plant to lose considerable weight that can be very detrimental. The effect of drought stress on photosynthesis and respiration is shown in Figure 9-8.

There is also a feedback inhibition of respiration when the products of respiration begin to accumulate. The most notable of these is carbon dioxide and ATP. High concentrations of these compounds cause respiration to slow and even stop. An accumulation of ATP signals the cell that its metabolism has slowed or has been somehow interrupted. The factors that affect other pathways in the cell's metabolism are basically the same as those for respiration so there usually is no imbalance between ATP being produced in respiration and being used in other pathways. An example of an imbalance would be a deficit of nitrogen compounds for amino acid and protein synthesis in the cell that uses substantial quantities of ATP. A buildup of carbon dioxide can also slow the rate of respiration. This is usually not a problem by itself. In most cases, carbon dioxide will accumulate only if there is inadequate gaseous exchange near the tissue involved, which also causes a deficiency of oxygen.

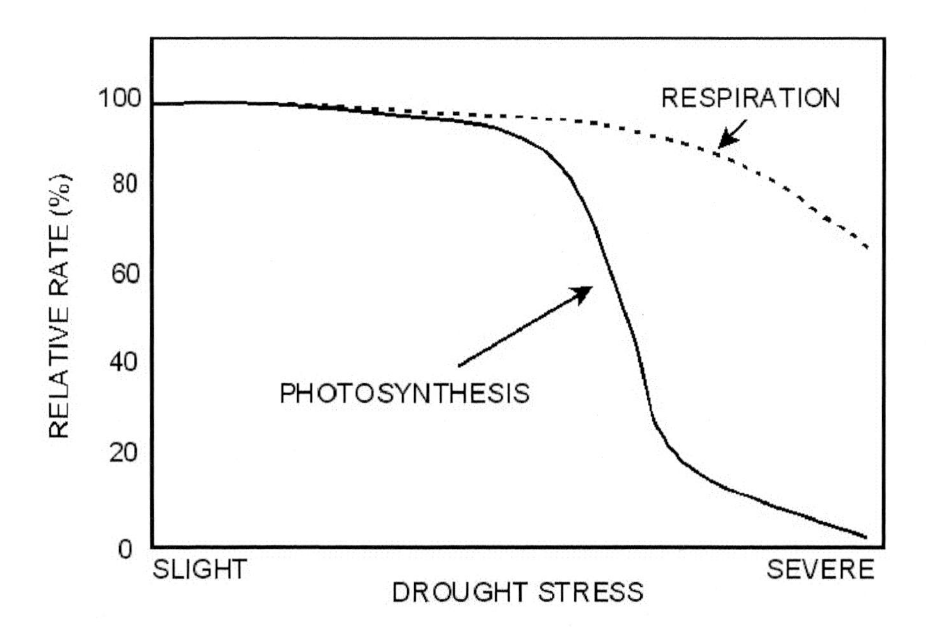

Figure 9-8 Effect of drought on photosynthesis and respiration.

Injury

Injury to any tissue in the plant usually results in an increase in the rate of respiration in the affected tissue as the plant attempts to repair the damage or at least close the wound. Examples of physical or mechanical damage are hail, wind, pest feeding, disease infestation, and root pruning from cultivation.

Chemical damage can also affect respiration rate. Damage from herbicides or insecticides can raise or lower the rate of respiration depending on their effect on the plant's metabolism. These chemicals can either affect respiration directly through their effect on enzyme activity, or they can indirectly affect respiration by affecting other metabolic processes in the plant. The toxicity of many herbicides is based on disrupting respiration.

REVIEW QUESTIONS

1. Why do the overall reactions for photosynthesis and respiration fail to give a complete picture of these processes?
2. When do the light reactions and the dark reactions occur? What is their relationship to each other?
3. What is a feedback inhibition? How does it relate to a source-sink relationship?
4. How and why does stomatal opening affect photosynthesis?
5. How and why does chlorosis affect photosynthesis?
6. What effects, if any, can crop scientists and producers have on the factors that affect the rate of photosynthesis? What effects can they have on the factors that affect the rate of respiration?
7. How does photorespiration affect net photosynthesis? What are the effects of the plant?
8. What factors determine the relative rates of aerobic, anaerobic, and photorespiration in a plant?

CHAPTER TEN

Flowering and Reproduction

INSTRUCTIONAL OBJECTIVES

Upon completion of this chapter, you should be able to:

1. Define flowering. Discuss its importance to plants.
2. Define determinate and indeterminate type of growth.
3. Define essential flower parts and accessory flower parts.
4. Define sepal, calyx, petal, corolla, and perianth. Discuss the function of each one.
5. Define stamen, anther, filament, pistil, stigma, style, ovule, and ovary. Discuss the function of each one.
6. Define peduncle, pedicel, and receptacle. Discuss the function of each.
7. Define bract, spikelet, glume, floret, awn, lemma, and palea. Discuss the function of each.
8. List the structures in a typical dicot flower and a typical grass flower.
9. Define complete flower, incomplete flower, imperfect flower, perfect flower, pistillate flower, staminate flower, monoecious plant, dioecious plant, and synoecious plant. Give an example of each.
10. Define inflorescence, and identify six types of inflorescences. Give an example of each type.
11. Define floral initiation. List at least four factors that affect it.
12. Define photoperiodism. Define long-day plant, short-day plant, and day-neutral plant. Give an example of each.
13. Define vernalization and devernalization. Give an example of each.
14. Define reproduction and propagation.
15. Define asexual reproduction. Discuss its advantages and disadvantages.
16. Define apomixis, cutting, layering, division, and grafting. Give an example of each.
17. Define sexual reproduction. Discuss its advantages and disadvantages.
18. Define chromosome, gene, mitosis, meiosis, megaspore mother cell and microspore mother cell. Discuss their functions in sexual reproduction.
19. Define pollination, fertilization, triple fusion, and double fertilization.
20. List and discuss the steps involved in pollination and fertilization.
21. Define self pollination and cross pollination. List at least three examples of each type.

INTRODUCTION

The cycle of crop growth, development, and differentiation begins with seed germination and progresses through vegetative juvenility and maturity and finally flowering. A plant is **vegetatively mature** when it is potentially capable of reproduction. All development before flowering occurs during the plant's **juvenile period**.

Flowering represents a wide spectrum of physiological and morphological events. The first step is the changing of the vegetative stem primordium, or growing point, into **floral primordium**. In other words, the **growing point** stops producing new stem and leaf tissue, and begins producing flower tissue. This change is called **floral initiation**. In **determinate** plants, this change is very abrupt and the plant ceases to produce vegetative growth at floral initiation. Determinate plants flower only once. Examples of crops with determinate flowering are all grass crops including small grains, corn, and sorghum. In other plants, the plant continues to grow vegetatively after floral initiation, and these plants are called **indeterminate**. Indeterminate plants flower over a long period of time. Examples of crops with indeterminate flowering are some varieties of soybeans and cotton.

FLOWER STRUCTURE

Flowers come in many shapes, sizes, and colors, as you already know. Flower characteristics are probably the most common and easiest method of identification of plant species. Understanding flower structure is important in crop science; not only for identification, but also because of its effect on pollination, seed harvest, and other agronomic concerns.

Flower structures can be divided into two main categories. The **reproductive organs** are the **pistil** and **stamens** that are directly involved in sexual reproduction of the species. These are called the **essential flower parts**. The **accessory flower parts** are the **petals** and **sepals** that can be involved in protection of the reproductive parts, in attracting insects and other pollination agents, and for show and color. Some flowers, such as those found in grasses, have no accessory flower parts. These are discussed later in this chapter.

The Dicot Flower

Figure 10-1 shows the parts of a typical dicot flower. **Sepals** are small, green, leaf-like structures that are found below the outermost whorl of petals. They serve to enclose the flower when it is in a bud. All of the sepals are collectively called the **calyx**. The **petals** are located above or inside the sepals. They are usually conspicuous and highly colored and evolved primarily to attract bees and other insect pollinators through their color. Ornamental flowers, of course, have been selected and bred for their large colorful petals. All the petals are collectively called the corolla, and may be composed of one to many whorls of petals. The collective term for the calyx and corolla is the **perianth**. As mentioned before, the calyx and corolla are the accessory flower parts.

Stamens are the male reproductive structures of the flower. They usually number three or more; and each is composed of an **anther** that produces the pollen and a **filament** or stalk that supports the anther. Pollen production is discussed in more detail later in this chapter. There may be special glands at the base of the stamens, called **nectaries,** that produce a sticky, sugary substance that may also be fragrant to attract insect pollinators. All of the stamens are collectively called the **androecium**.

The **pistil** is the female reproductive structure of the flower. Although most flowers have only one pistil, there may be more than one per flower. The pistil is composed of the **stigma** that receives the pollen, a **style** that connects the stigma to the ovary, and the **ovary** that contains the egg within one or more **ovules**. If there is more than one pistil, they are collectively called the **gynoecium**.

Other structures shown in Figure 10-1 are the **receptacle** and **pedicel**. These are not considered to be parts of the flower but are closely associated with it by providing support and connective tissue to the rest of the plant. The receptacle is located at the base of the flower just below the calyx. In some flowers, the ovary is recessed into the receptacle such as the apple. A structure not shown, the **peduncle**, is a stalk that supports the flower and connects it to the stem. When there is more than one flower on a peduncle, the branches that support the individual flowers are called pedicels.

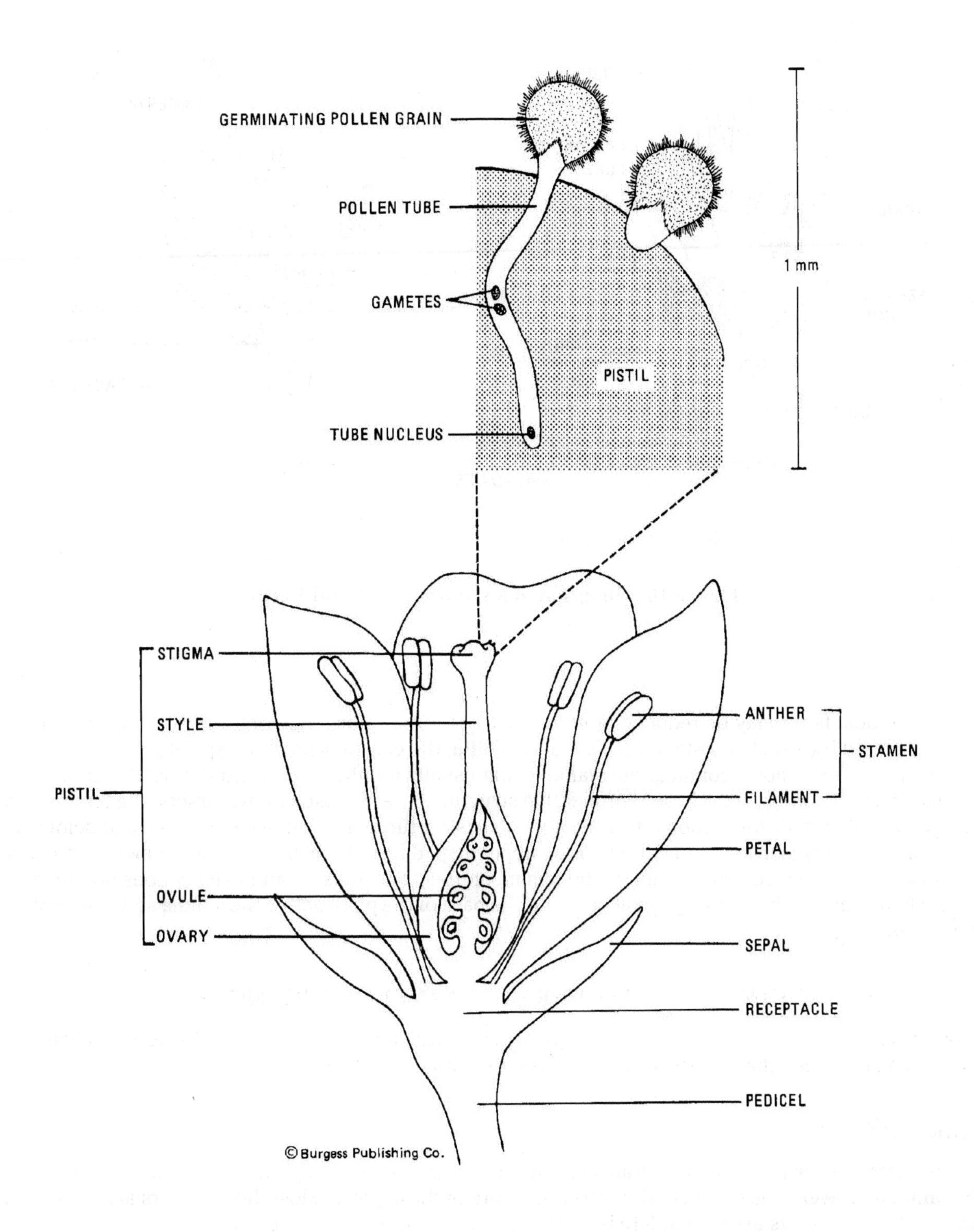

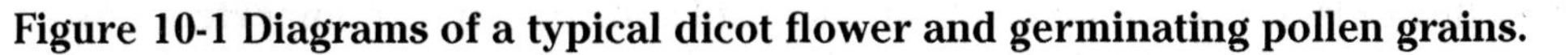

Figure 10-1 Diagrams of a typical dicot flower and germinating pollen grains.

The Grass Flower

All grasses have modified flowers in which the calyx and the corolla have been replaced with special modified leaves called **bracts**. Figure 10-2 shows a typical grass flower. An individual grass flower is called a **spikelet**. At the base of each spikelet are two **glumes** or leaflike bracts. In some species, the glumes completely enclose the spikelet. Above the glumes, there will be one or more **florets** that contain the stamens and pistil together with other bracts. In some grasses with several florets per spikelet,

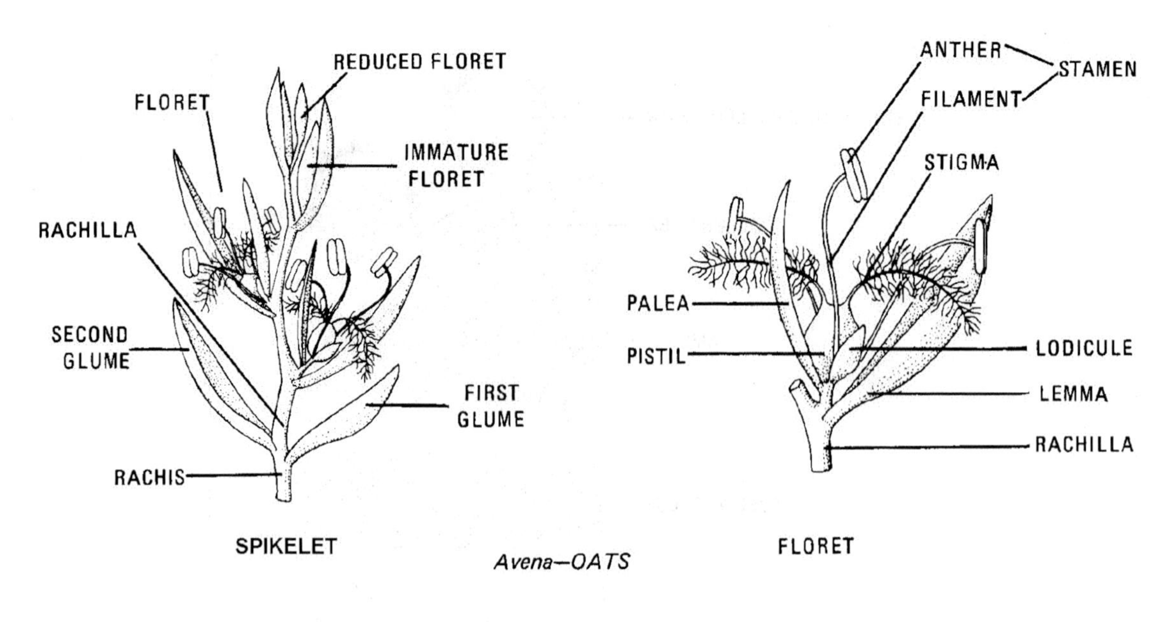

Figure 10-2 Diagram of a grass spikelet and floret.

the uppermost floret may be reduced or sterile. When there is more than one floret in a spikelet, they are connected by a stalk-like structure called a **rachilla**, the central axis of the spikelet.

Each individual floret contains the stamens that usually number three, and the pistil that has two stigmas but only one ovary. These reproductive structures are enclosed in two bracts, the **lemma** and the **palea** that provide protection for the reproductive structures. The lemma is attached just below the palea. It is usually larger and may contain an appendage on its tip or back called an **awn** that is not shown. The palea is located just above the lemma on the opposite side and usually does not have an awn. There may also be two very small bracts at the base of the pistil above the lemma and palea called **lodicules**.

KINDS OF FLOWERS AND INFLORESCENCES

Flowers can be classified according to the presence or absence of the different flower parts just discussed. Again, these characteristics are important identification tools.

Kinds of Flowers

A **complete flower** is one that contains all four flower parts: sepals, petals, pistil, and stamens. An **incomplete flower** is one that is missing one or more of these parts. Most dicot flowers are complete, while all grass flowers are incomplete because they have no sepals and petals.

A **perfect flower** is one that contains both the pistil and stamens. An **imperfect flower** is one that is missing either stamens or pistil. An imperfect flower that contains only the pistil is called a **pistillate** or **female flower**. A flower that contains only stamens is called a **staminate** or **male flower**. Plants with imperfect flowers can also be classified according to what kinds of flowers they contain. Plants with both pistillate and staminate flowers on the same plant are called **monoecious**. An example of a monoecious plant is corn. The ears contain pistillate flowers and the tassel contains staminate flowers. When a plant has only staminate or pistillate flowers it is called **dioecious**, and the plants are either male or female. An example of a dioecious plant is buffalograss. Plants that contain perfect flowers are called **synoecious**.

Kinds of Inflorescences

An **inflorescence** is a group or cluster of flowers on the same peduncle. Most crop plants have their flowers in inflorescences. Like flower type, the type of inflorescence is a good identification tool. Figure 10-3 shows some common kinds of inflorescences.

An inflorescence in which all of the flowers are attached directly to a central axis is called a **spike**. The central axis of a spike is called a **rachis**. When the flowers are attached directly to a rachis or other main supporting stalk they are said to be **sessile**. Examples of crops with spike inflorescences are wheat, barley, and rye. The ear of corn is a modified spike in which the cob is a modified fleshy rachis. A **raceme** is similar to a spike except that the individual flowers are connected to the rachis by a **pedicel**. These flowers are said to be **pedicellate**. Examples of crops with raceme inflorescences are alfalfa, sweetclover, and soybean. A **panicle** is an inflorescence that contains many branches that connect the flowers. The

INFLORESCENCE TYPES

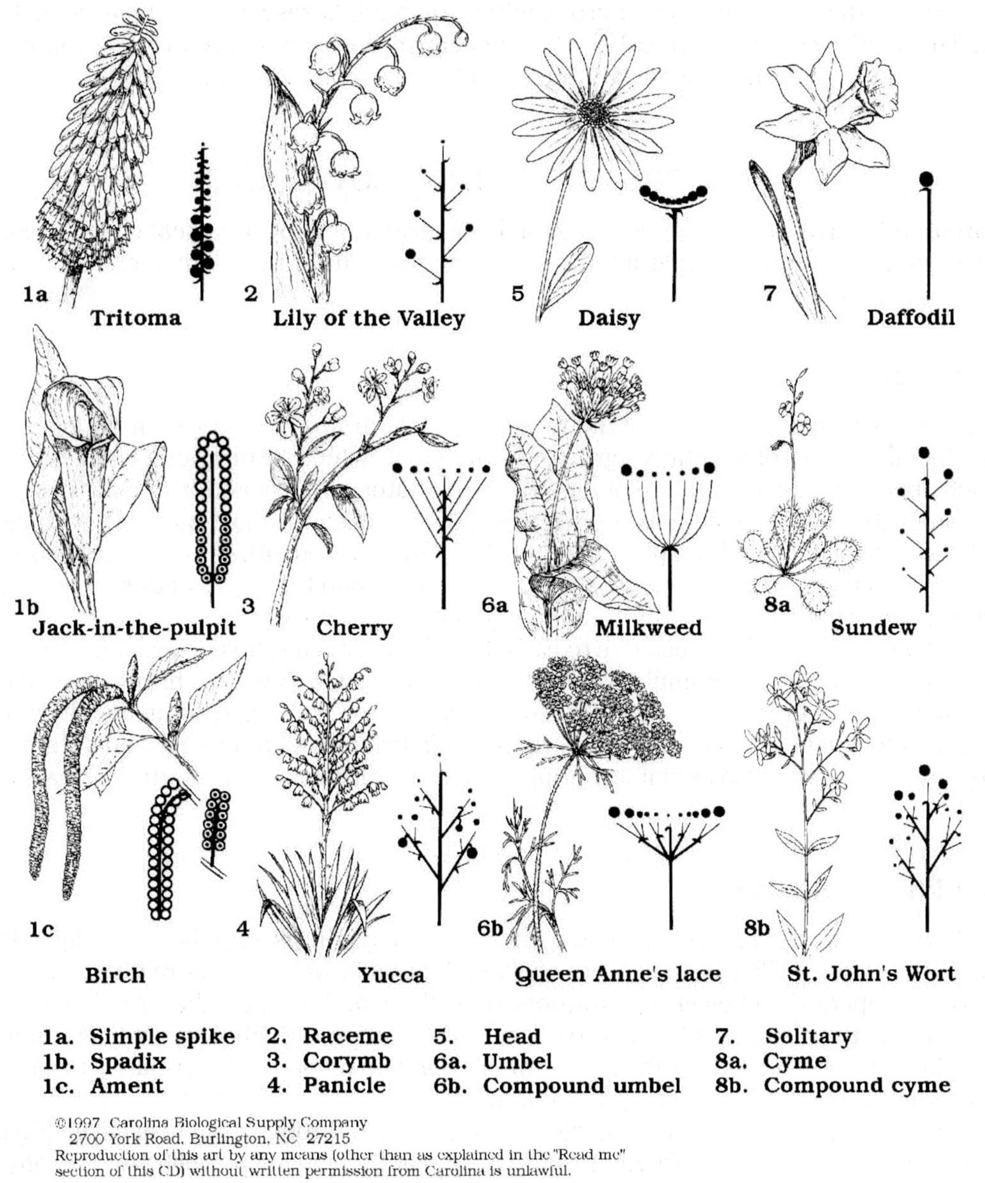

Figure 10-3 Types of inflorescences.

flowers themselves can be sessile or pedicellate. Examples are sorghum, switchgrass, and millet. Most crop plants have spike, raceme, or panicle inflorescences.

Other types of inflorescences found on crops are also shown in Figure 10-3. A **corymb** contains a main axis with pedicels of varying lengths to create a flat-topped inflorescence. It differs from a panicle in that it is not extensively branched, and it differs from a raceme in that the pedicels of a raceme are all about the same length. An example of a crop with a corymb inflorescence is potato. Another type of flat-topped inflorescence is an **umbel**. An umbel differs from a corymb in that it has no central rachis. All of the pedicels originate from the same point at the base of the inflorescence. An example of a crop with an umbel inflorescence is onion. A **head inflorescence** has many sessile flowers that are tightly clustered on a flattened receptacle. Most heads have two types of flowers. **Ray flowers** contain petals and may only be found around the outside of the inflorescence. **Disk flowers** are located in the center of the inflorescence and usually have no petals. In most heads, the ray flowers are usually sterile and the disk flowers are fertile. An example of a crop with a head inflorescence is sunflower. A **capitulum** inflorescence is similar to a head except that the receptacle is rounded instead of flat. The flowers can be sessile or pedicellate. Examples of crops with capitulums are red clover and white clover.

It is important not to confuse a head type of inflorescence with the term "head" that is commonly used to designate the inflorescences of crops with terminal inflorescences such as wheat, oats, and sorghum. These inflorescences are called heads because they are located on the top of the plant. Although this use of head is not botanically correct, it is a very common term used by professionals and others.

FACTORS AFFECTING FLOWERING

Floral initiation involves a wide spectrum of physiological and morphological changes. Factors that affect flower initiation are those that influence the environment of the plant such as plant nutrition, light, and temperature.

Plant Nutrition

Flowering in many plants is affected by the ratio of carbohydrates to nitrogen in the plant. This ratio primarily affects the rate of vegetative growth in the plant, which in turn affects flowering. In these plants, such as tomato, a combination of high rates of photosynthesis and high soil nitrogen can prevent flowering with subsequent excess vegetative growth. When nitrogen supply is reduced, vegetative growth is also reduced and the plants flower abundantly. When either photosynthesis or nitrogen is restricted the plants are stunted and will not flower. It is beyond the scope of this book to explore the causes of this factor but it is important to understand its implications.

Even in plants that are not as sensitive to carbohydrate:nitrogen ratios, which includes most crops, any environmental change that stimulates vegetative growth, such as excess nitrogen, tends to reduce or delay flowering. Conversely, anything that reduces vegetative growth, without affecting the health of the plant, tends to stimulate flowering. Indeterminate plants are affected more than determinate. Deficiencies of any nutrient can also inhibit flowering, so proper soil fertility is important in maximizing crop yields.

Light and Photoperiodism

In the previous section we discussed how high rates of photosynthesis, together with high nitrogen, can inhibit or delay flowering. This is an indirect effect of light on flowering. A more direct effect of light is the effect of photoperiod on flowering. **Photoperiod** is the length of the light period in a 24-hour day. **Photoperiodism** is the response of plants to changes in photoperiod. Photoperiodism is a remarkable system by which the plant can sense the season of the year and respond appropriately to insure the success of flowering and subsequent seed production.

Photoperiodism involves hormonal control of floral initiation through a pigment called **phytochrome**. Phytochrome occurs in two forms. Phytochrome red (P_r) absorbs red light (660 nm) and phytochrome far-red (P_{fr}) absorbs far-red light (730 nm). Phytochrome reacts to light and darkness by changing its form between the two forms.

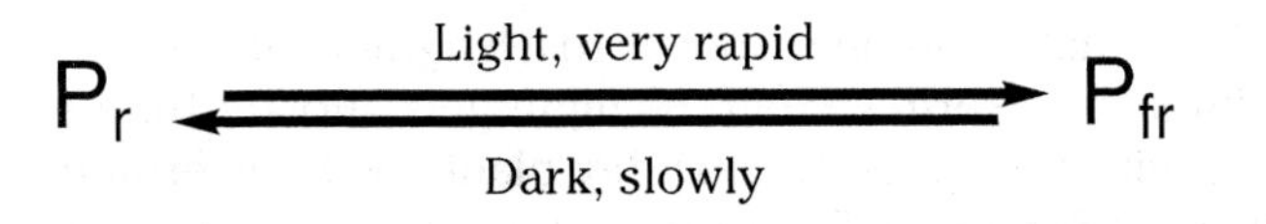

During the day, the phytochrome is in the P_{fr} form. During the night, it slowly changes back to the P_r form. With longer dark periods (shorter day length), more P_{fr} will be changed back to P_r. With shorter nights (longer days), less P_{fr} will be changed to P_r. Plants sensitive to changes in photoperiod can measure the length of the dark period by the ratio of P_{fr} to P_r at sunrise. When the night reaches a genetically determined critical length, the plant initiates flowering or other processes. The actual **critical night length** depends on the particular species and even variety. Using this system, plants can time critical events, such as flowering, to avoid unfavorable weather conditions, such as freezing temperatures.

Some weed seeds are photoperiod sensitive and will not germinate before the days are long enough to assure proper temperatures for growth. In perennial plants, photoperiodism is involved in many physiological processes besides flowering. Hardening which involves decreased growth and increased carbohydrate accumulation prepares plants for winter dormancy. Bud activation in the spring is triggered by longer photoperiods and temperature. Leaf senescence in the fall is triggered by shorter photoperiods.

Some plants are not photoperiod sensitive until a certain vegetative stage. Other plants, including many weeds, are photoperiod sensitive at emergence. Temperature can affect photoperiod sensitivity in some plants as discussed later in this chapter.

Plants can be classified as short-day, long-day, or day-neutral, depending on their response to changes in the photoperiod. Figure 10-4 shows a diagrammatic summary of photoperiod response of short-day and long-day plants where 12 hours is the critical night length.

Short-day plants are those that flower in response to a shortening photoperiod. Actually the plant is responding to a lengthening night period, and it initiates flowering when the night gets longer than a critical length. Short-day plants will delay flowering under continuous light or if the night is broken by a flash of light. There have been cases where lights from automobiles have interfered with flowering but this is usually not a problem. Examples of short-day crops are soybeans, sorghum, corn, tobacco, and other warm-season plants.

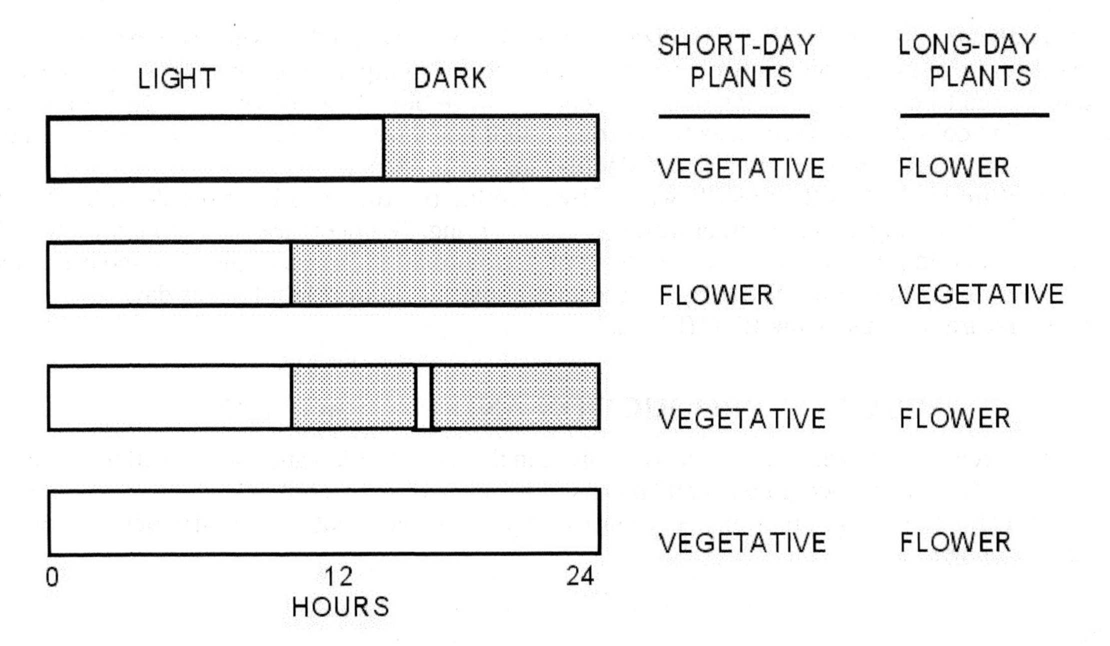

Figure 10-4 Response of plants to changes in photoperiod.

Long-day plants are those that flower in response to a lengthening photoperiod, or a shortening night period. Long-day plants initiate flowering when the night gets shorter than a critical length. Again, the critical night length varies with the species. Long-day plants will flower in continuous light, and stray light sources have no effect on flowering. Examples of long-day crops are small grains and cool-season plants.

There are some plants, called **day-neutral plants,** which do not respond to changes in photoperiod. These plants will flower after a certain amount of vegetative growth regardless of the photoperiod. Examples of day-neutral crops are alfalfa, cotton, tomato, and some varieties of field beans. There are some hybrids of corn, particularly early maturing ones, which behave like day-neutral plants but corn is still generally classified as a short-day plant.

Soybean is an example of a plant that is very sensitive to changes in photoperiod during the growing season. Flowering is initiated when the daylength becomes shorter than a certain number of hours. The exact daylength that triggers flowering depends on the variety. Since change in photoperiod during the growing season increases as you travel north, a particular soybean variety will be adapted within long belts east and west but which will be only about 170 to 240 km (100–150 miles) wide north and south.

Varieties are classified into maturity groups depending on their photoperiodic response. These groups range from 00 for northern Minnesota and North Dakota, to VIII for the southernmost part of the U.S. Most varieties that are classified as 00 to IV are indeterminate; those V to VIII are determinate. It is important to select varieties that are adapted to your area.

Since flowering is triggered by daylength, a later planted variety will grow vegetatively for a shorter time before flowering begins. Later seeded soybeans will usually be shorter than earlier seeded soybeans and have fewer nodes and internodes.

Temperature

Temperature can have either a direct or an indirect effect on flowering. Temperature directly affects flowering through vernalization and devernalization. **Vernalization** is a requirement for cold temperatures before a plant can initiate flowering. Most winter annuals and biennials must be vernalized before they can flower. Usually these plants must be exposed to at least six to eight weeks of temperatures near or below freezing to initiate flowering. Again, as with photoperiodism, hormonal control is involved. When a vernalized plant is grafted to a nonvernalized plant, the nonvernalized part will also flower. Examples of crops that require vernalization are winter wheat and barley, rye, and sugarbeet. When a winter annual crop is planted in the spring, it will only grow vegetatively, and will not flower under normal conditions.

Devernalization is a reversal of vernalization from flowering to nonflowering by exposure to warm temperatures. Devernalization rarely occurs naturally because temperatures warm up gradually in the spring. It is used to prevent flowering of onion sets that have been kept in cold storage before sale in the spring. The cold storage vernalizes the onions, and if not devernalized they would not produce a bulb, just a seed stalk that would not be satisfactory for the gardener.

Temperature can indirectly affect flowering by affecting the time needed from floral initiation to actual flowering. Usually, cooler temperatures delay flowering. Temperature can also indirectly affect flowering by affecting the photoperiod response of certain plants. For example, June-bearing strawberries act as short-day plants if the temperature goes above 19° C (67° F), but act as day-neutral plants if the temperature remains below 19° C (67° F).

PLANT REPRODUCTION AND PROPAGATION

Reproduction refers to the sequence of steps involved in the perpetuation and multiplication of the plant species. Reproduction can occur by natural means or it can be directly controlled by people. When the perpetuation and multiplication of plants occur through the direct control of reproductive processes it is called **propagation**.

Asexual Reproduction

Asexual reproduction involves reproduction and propagation from vegetative parts of plants and is possible because the parts of many plants have the capacity for regeneration. New plants can start from a single cell because each cell contains all the genetic information needed for a complete organism. One of life's great mysteries is exactly how certain cells know how and when to differentiate into specialized tissues when all cells contain the same genetic information.

Asexual reproduction is controlled by growers to serve a variety of purposes and alleviate a variety of problems. Although asexual reproduction is not widely used in field crops, it is important to understand the various methods used because they are commonly used in plant reproduction and propagation.

Asexual reproduction can be used to maintain genetic purity. Many plants are extremely **heterozygous** (genetically variable) when reproduced sexually. Also some plants would lose some of their unique characteristics when reproduced sexually. For instance, most varieties of apples are propagated asexually to maintain their characteristics. When apple seeds are planted, the resulting trees will be quite variable and quite unlike the parents because of the mixing of two sets of genes from different parents. This principle is explained more fully later in this chapter.

In some plants, it is more economical to propagate asexually. Some plants produce no seed, few seeds, or seeds of low quality and germination. In some plants, seedlings grow slower than plants propagated asexually. Also, in some plants, seedlings have a longer juvenile period than those begun asexually. In other words, they take longer to flower and produce fruit.

Some plants have more resistance to plant pests if grown on related root systems. For example, if an apple variety that has desirable fruit quality but is susceptible to certain diseases is grafted to a resistant rootstock, the resistance will be transmitted to the desirable plant parts. Asexual reproduction can also be used to maintain disease-free plants. In some plants, such as sugarcane and potatoes, diseases can be transmitted from generation to generation through the parts used in asexual propagation. By carefully maintaining disease-free plants, the incidence of diseases is greatly reduced.

Asexual reproduction is also used to increase ease of harvesting. Many dwarf fruit trees are propagated asexually. These plants are much easier to harvest while still maintaining high yields.

Methods of Asexual Reproduction

Apomixis is the production of seed without sexual fertilization. Plants with **apomictic seed** have the same characteristics as the mother plant. Apomixis is a natural phenomenon that is used by growers to maintain genetic purity. Examples of plants that produce apomictic seed are bluegrass, buffalograss, most citrus, and dandelion.

Apomixis has the potential to retain genetic traits in improved hybrids and varieties. For example, there are apomictic wild types of pearl millet. If they could be crossed with improved pearl millet hybrids to develop apomictic hybrids, the plants would retain their improved traits for several generations. It may also be possible to develop apomixis in other crops through genetic engineering. Apomictic hybrids or varieties would be especially advantageous in less developed countries where growers do not have the economic resources to purchase seed every year. This technique could revolutionize food production around the world.

Most methods of asexual reproduction involve the use of plant parts. These are illustrated in Figure 10-5. **Cuttings** involve the production of new individuals from plant parts that have been severed from the parent plant. Various plant parts can be used for cuttings but most plants are specific for the part or parts that work best. Stem cuttings can be used either with or without leaves, and either herbaceous or woody, depending on the species involved. Sugarcane is an example of a crop that is propagated by stem cuttings. In other plants, roots or leaves can be used to produce new plants. Cuttings are the most common method of propagating house plants and ornamentals.

When the method of asexual reproduction involves the production of new roots from stem or leaf tissue and the wounding of the plant, certain chemicals are used to promote rooting and healing. The most common are synthetic growth hormones such as indoleacetic acid, indolebutyric acid, and naphthaleneacetic acid. These are called **rooting powders** and can be purchased under a variety of names.

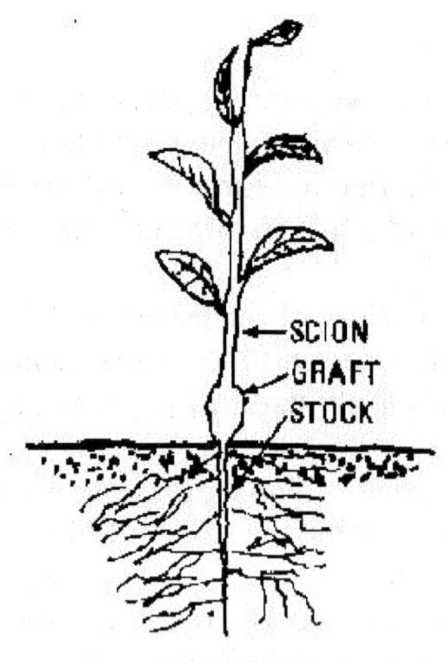

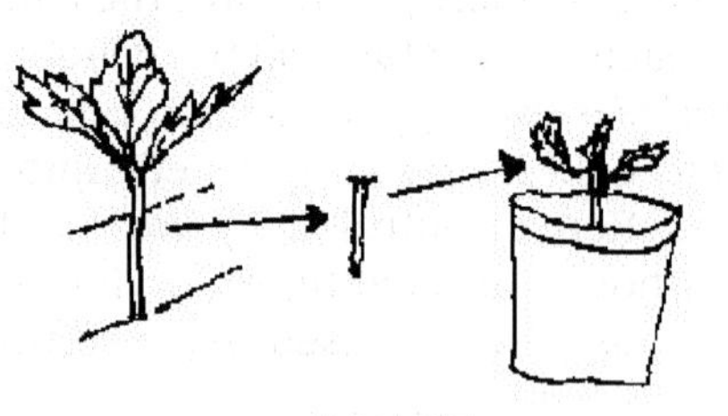

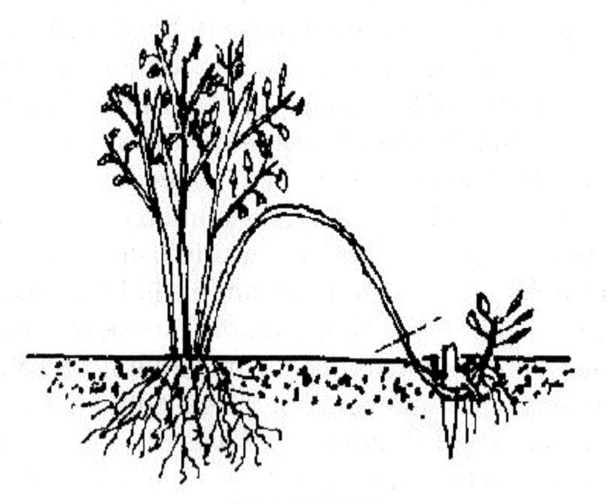

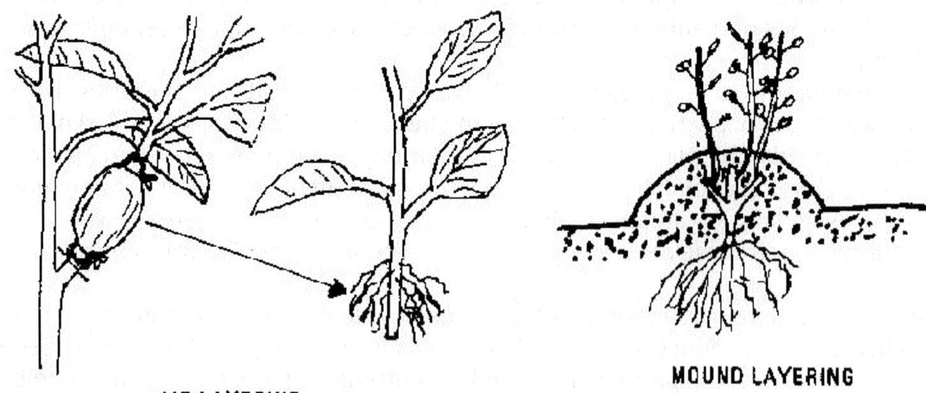

Figure 10-5 Methods of asexual reproduction.

Layering is the development of roots and a new plant while still attached to the parent. As Figure 10-5 shows, there are several different types of layering. **Tip layering** involves covering the stem tip with soil; and, when new roots and leaves emerge, the new plant is severed from the parent. Tip layering is used on plants with flexible stems such as raspberry and blackberry. **Compound** or **serpentine layering** involves alternately covering and exposing a long flexible stem. This method allows the production of several new plants from each stem. It is used on wisteria and clematis. **Mound layering** is used with some plants with stems not flexible enough for the previous layering methods. The stems are cut and soil is mounded at the base of the plant. Sometimes, several two to three inch layers of soil are added in succession as the stems produce new roots and shoots, thus producing several plants from each stem. **Air layering** is the production of a new plant above the soil. In this method, a cut is made in the stem and peat moss or another rooting medium is packed around the wound and held in place with polyethylene or other plastic wrap that allows oxygen and carbon dioxide to pass through but not water. When roots are formed the stem is severed and the new plant is repotted. Air layering is used with some house plants such as the India rubber tree.

Another method of asexual propagation is **division**, which uses specialized vegetative structures. Division differs from cuttings in that specialized plant parts are used and these parts are usually divided into several pieces. Examples of division include tubers (potato), stolons (bermudagrass), rhizomes (bluegrass), bulbs (tulips), corms (gladiolus), and crowns (chrysanthemum). Pineapple is propagated by division using special stems called ratoons.

Grafting is the joining of plant parts by tissue regeneration. It differs from the methods discussed previously because it involves the union of two plant parts instead of the separating of parts. Grafting is used primarily with woody dicot tissue because the presence of an active cambium is necessary for tissue regeneration. The upper plant part used in a graft is called a **scion**. It can be a main stem, twig, or bud. The lower plant part used is called a **stock**. It provides the base for the scion.

Grafting is used extensively in fruit trees. The most common uses of grafting are to maintain genetic purity, incorporate resistance to diseases and other problems, produce dwarf trees, produce several varieties on the same tree, and to repair damage. The scion and stock must be compatible for a graft to be successful. Although they do not have to be the same species, in most cases they must be closely related.

Sexual Reproduction

Sexual reproduction involves the fusion of male and female **gametes** resulting in a fertilized egg that develops into a new plant. The nucleus of a cell contains the hereditary blueprint of the plant. This blueprint is contained in the chromosomes that are linear structures composed mostly of DNA. **Chromosomes,** shown in Figure 10-6, occur in pairs in most cells, with the number of pairs varying between species. **Genes** are specific sites on the chromosomes that control particular genetically heritable traits.

Sexual reproduction involves a special type of cell division called **meiosis**. In meiosis, the chromosome pairs split apart and one of each pair goes to each daughter cell as shown in Figure 10-7. There are four daughter cells produced from each original mother cell.

When the nucleus contains complete pairs of chromosomes, it is said to be **diploid**, or **2N**, with N being the number of chromosomes. After meiosis, the daughter cells are **haploid**, or **1N**. In other words, they have only half as much genetic material in their nuclei as the mother cell had. Meiosis only occurs in sexual reproduction.

Another type of cell division illustrated in Figure 10-8 is **mitosis**. In mitosis, the chromosomes duplicate themselves, and a complete set of pairs goes to each daughter cell. During mitosis the chromosome number does not change. The two daughter cells are diploid just like the mother cell. Mitosis is cell division that results in growth. Meiosis occurs only in the flower during sexual reproduction. All other cell division that occurs in the plant for its entire life is mitosis.

Figure 10-9 diagrams the steps involved in sexual reproduction in both the ovary and the anthers of the flower. The **megaspore mother cell** is located in the ovary of the pistil. It goes through meiotic division resulting in four haploid daughter cells called **megaspores**. It is important to keep in mind that this is the only time meiosis occurs in the ovary. All subsequent cell divisions are mitosis because the chromosomes are duplicating themselves and the chromosome number is not changing. Haploid cells cannot go through meiosis as there are no chromosome pairs to divide.

Only one of the megaspores develops further, the other cells die. This active cell goes through four mitotic divisions to produce the **egg**, two **polar nuclei,** and five other nuclei, all of which are located within the ovule of the ovary. The functions of these nuclei will be discussed later.

Now let's look at what happens in the anther as shown in Figure 10-9. The **microspore mother cell** in the anther also goes through a meiotic division to produce four haploid daughter cells called **microspores**. Just like in the ovary, meiosis occurs only once. All subsequent cell divisions are mitosis.

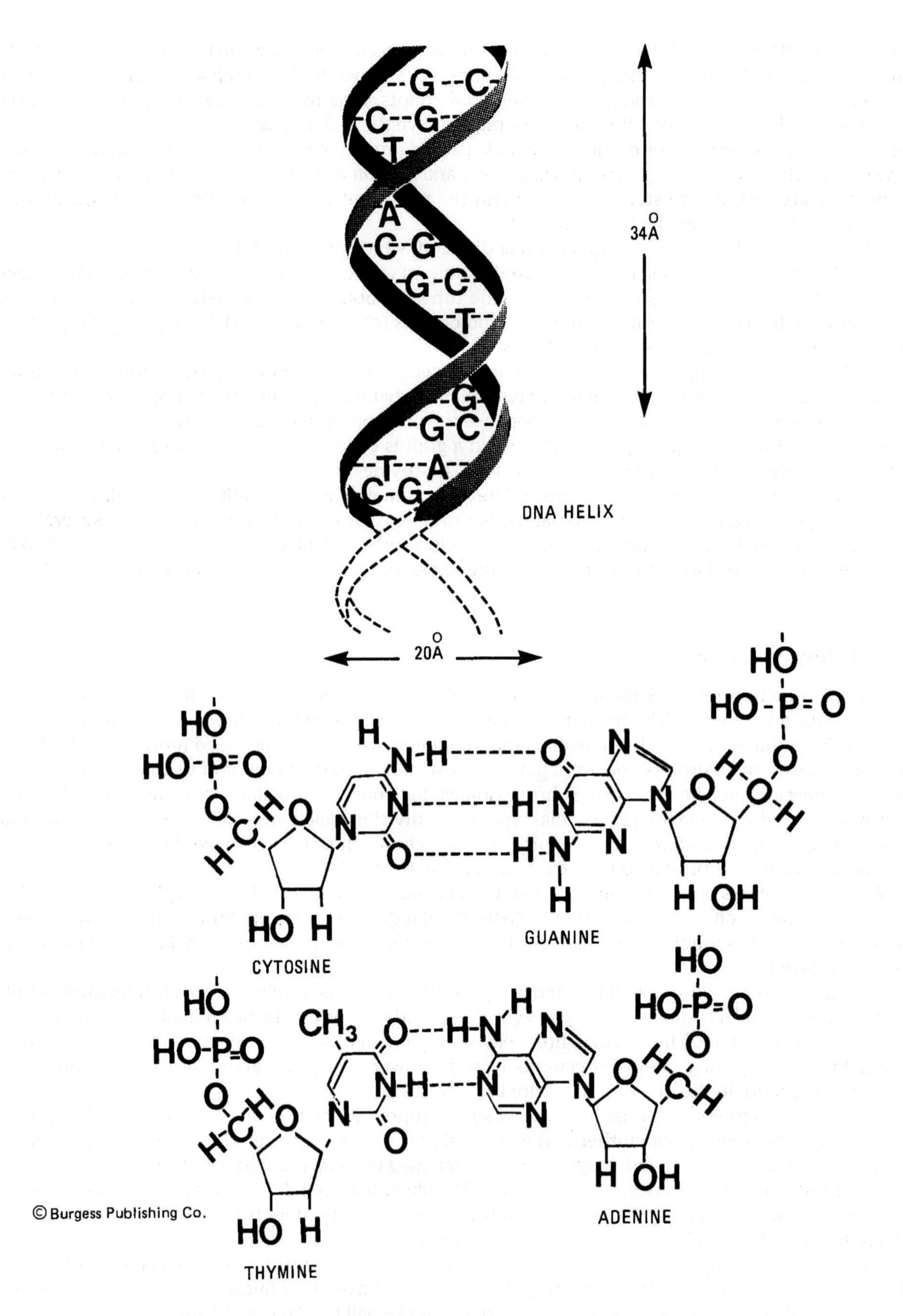

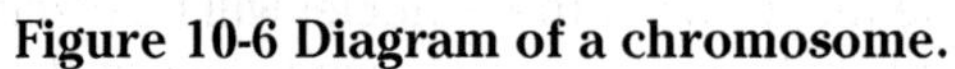

Figure 10-6 Diagram of a chromosome.

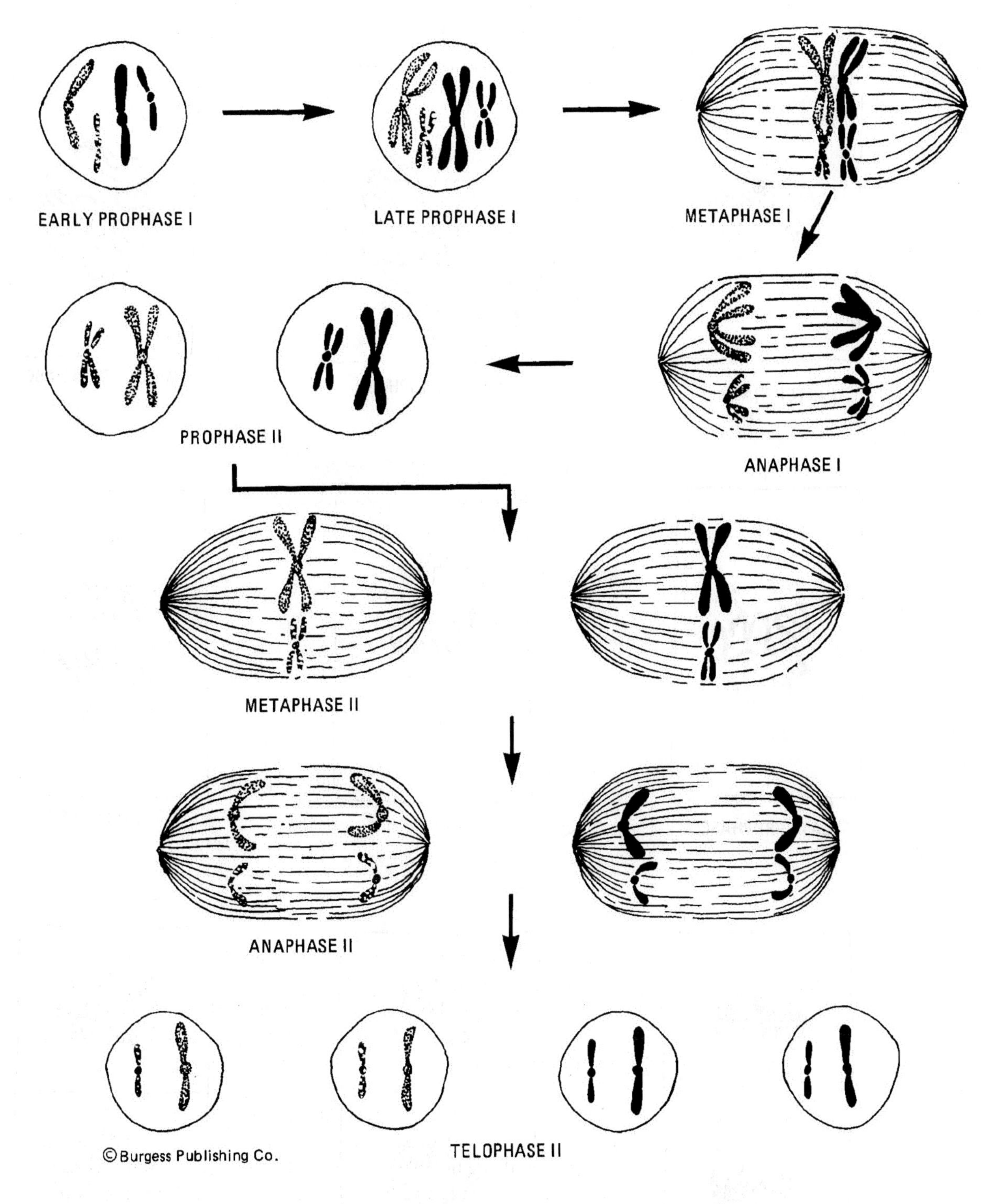

Figure 10-7 Cell division by meiosis.

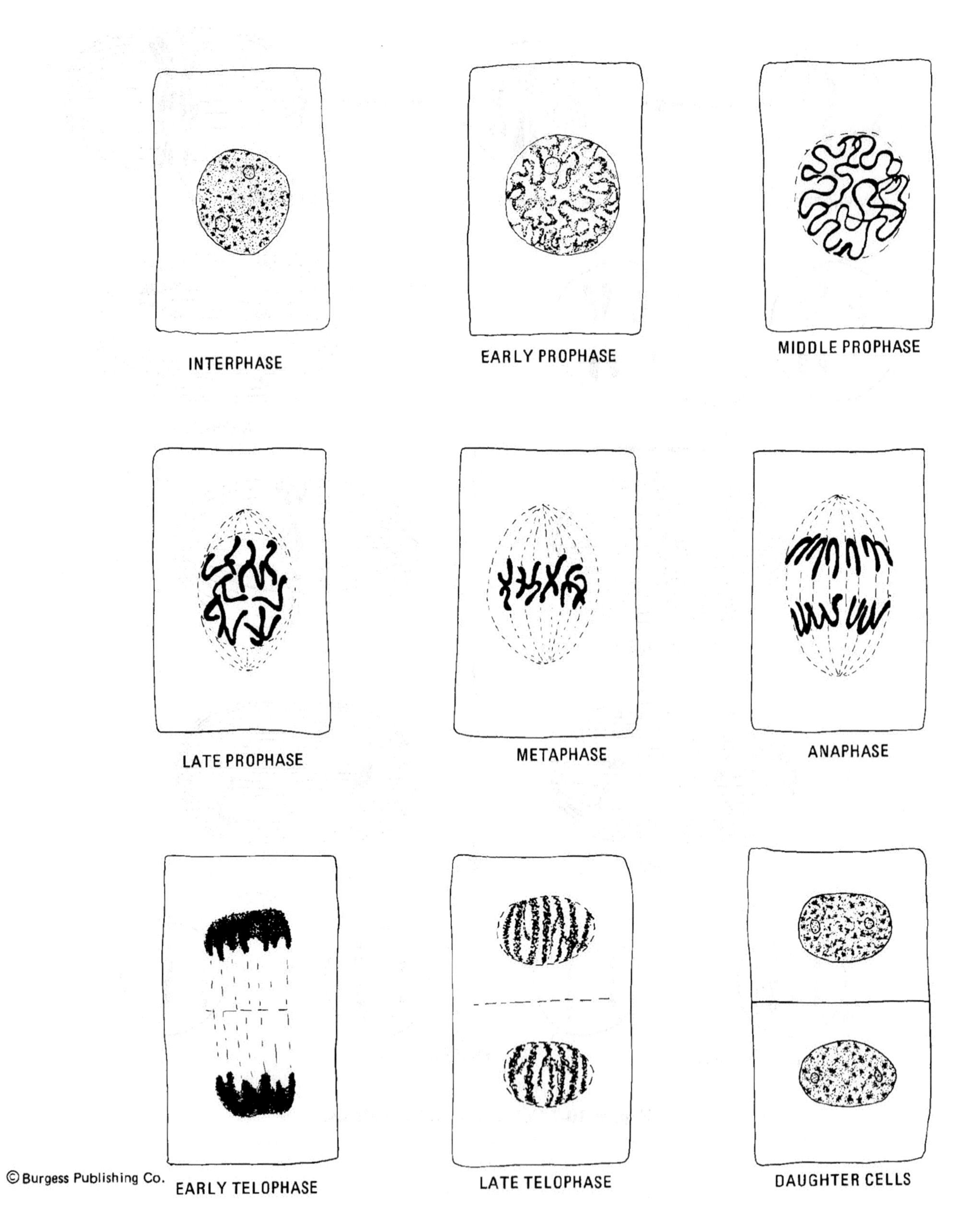

Figure 10-8 Cell division by mitosis.

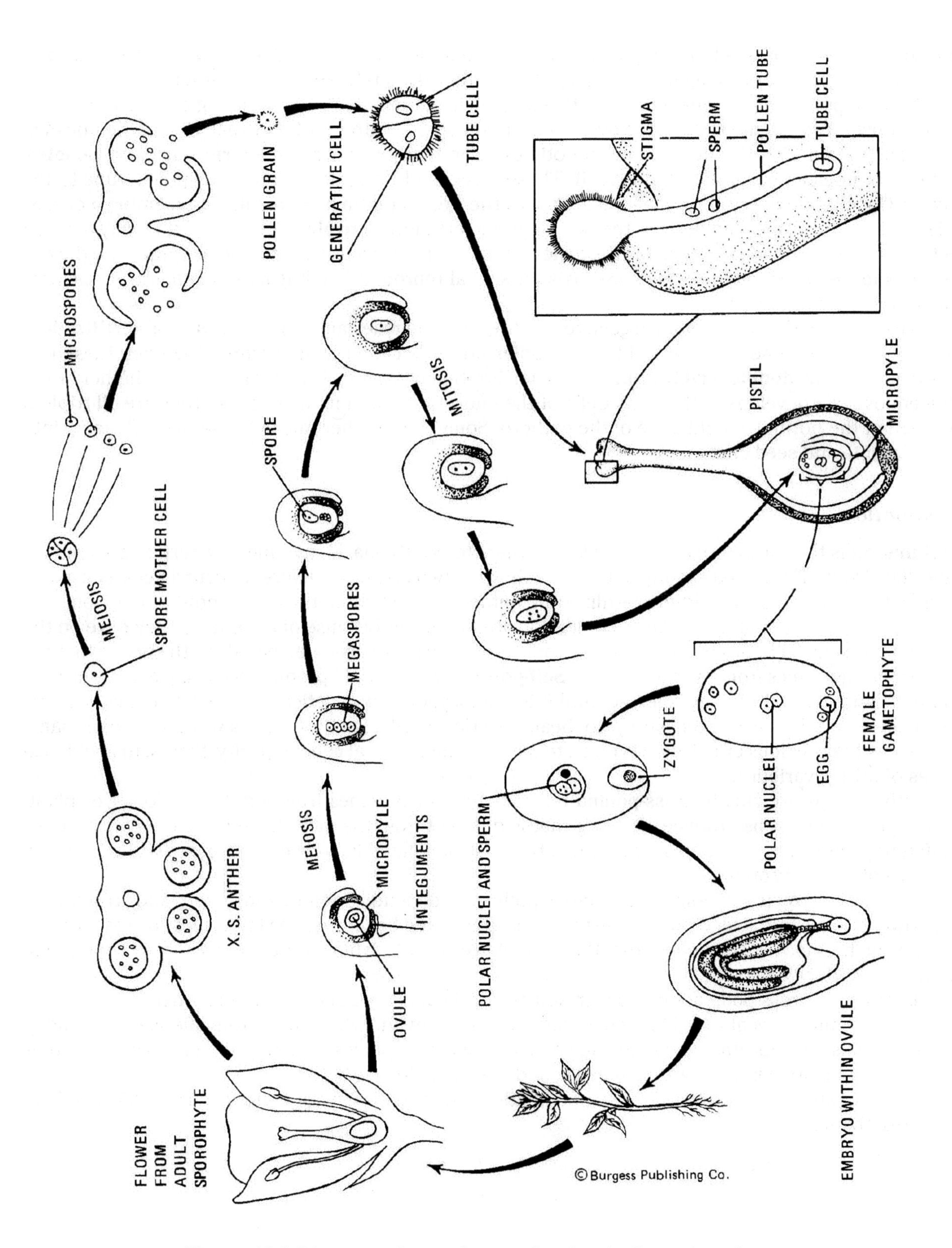

Figure 10-9 Diagram of sexual reproduction in flowering plants.

Unlike the megaspores, all four microspores continue to develop. Each microspore divides again and goes through structural changes to form a **pollen grain** that is released by the anther.

When a pollen grain lands on a receptive stigma, it "germinates" by producing a **pollen tube** that grows down through the style to the ovary. This is shown in Figures 10-1 and 10-9. One of the nuclei in the pollen grain goes through another mitotic division that produces two **sperm nuclei** or **gametes**. The other nucleus becomes the **tube cell**. These three nuclei migrate through the pollen tube to the ovary. Once in the ovary, one gamete unites with the egg that ultimately results in an **embryo** or new plant. The other gamete unites with the two polar nuclei to form a **triploid**, or **3N**, cell that may develop into the **endosperm** of the seed. It is the fusion of the sperm nucleus from the pollen grain and the egg nucleus in the ovary that provides the basis for sexual reproduction. It is a recombination of genetic material from two parents.

The union of the sperm or male gamete and the egg or female gamete is called **sexual fertilization**. The union of the other gamete and the two polar nuclei is called **triple fusion**. Together, these two fusions are called **double fertilization**. Double fertilization occurs in all flowering plants, whether or not the endosperm develops in the seed. Cells of the endosperm are unique because they are all triploid instead of the normal diploid state of the embryo. Some of the other nuclei in the ovule develop into the placenta and seed coat.

Pollination

Pollination is the transfer of pollen from the anthers to the stigma. Many times the terms "pollination" and "fertilization" are used synonymously, but they are two distinctly different occurrences as you can see. Pollination can occur without fertilization, but fertilization cannot occur without pollination.

In crops, the type and method of pollination are important because of the effects they have on the hereditary characteristics of the next crop. Some crops are mostly self-pollinated, or, the pollen that fertilizes the egg comes from the same flower. **Self-pollinated crops** will produce offspring exactly like the parent if the parent has been selected until it is homozygous, or genetically pure. Crops that are self-pollinated include wheat, oats, barley, soybean, and rice. Seed from these crops will produce the same variety as the previous crop if care is taken to maintain mechanical genetic purity, that is, to avoid mixtures of others varieties.

Other crops are mostly cross-pollinated, and the pollen comes from a different flower or plant. **Cross-pollinated crops** produce offspring that is different from the parents because the parents have different genetic material. Crops that are mostly cross-pollinated include corn, rye, sugarbeet, clovers, alfalfa, cotton, and sorghum.

Whether a crop is self-pollinated or cross-pollinated depends on the arrangement and structure of its flowers. Corn, by its very nature, must be cross-pollinated because it has imperfect flowers. In some plants, such as clover, the pollen from the same flower is not viable on the stigma and will not germinate and produce a pollen tube.

In order for cross-pollination to occur, something must carry pollen from the anther to the stigma. The most common agent of pollination is wind, to which anyone who suffers from allergies will readily agree. Insects are important in transferring pollen with many species including alfalfa and clover. Other agents of pollination include water, animals, and even people.

Genetic purity and the effects of self pollination and cross pollination are discussed more fully in the next chapter.

REVIEW QUESTIONS

1. What roles do the sepals and petals serve in flowers?
2. What is the difference between a spikelet and a floret?
3. What combinations of complete, incomplete, perfect, and imperfect flowers are possible?
4. What types of inflorescences are most commonly found on field crops? How can you tell them apart?
5. How does a short period of light at night prevent a short-day plant from flowering? Why doesn't it bother a long-day plant? What effect would it have on a day-neutral plant?
6. How does reproduction differ from propagation?
7. What type of asexual reproduction is not directly controlled by human means? Does asexual reproduction ever occur naturally? What are some examples?
8. What is the difference between mitosis and meiosis? When does each occur in plants?
9. What is the difference between pollination and fertilization? Which occurs first? Can you have one without the other?
10. What plants have double fertilization? What if the seed has no endosperm?
11. What roles do the two sperm nuclei and the tube nucleus play in fertilization?
12. What roles do the egg nucleus, the two polar nuclei, and the other five nuclei play in fertilization?
13. What are some examples of agents of pollination? Which are common in field crops?

CHAPTER ELEVEN

Crop Improvement

INSTRUCTIONAL OBJECTIVES

Upon completion of this chapter, you should be able to:

1. Define variety, cultivar, and hybrid.
2. Discuss the advantages of crop improvement over other methods of increasing yields.
3. Define resistance, including the two types of resistance.
4. List five additional factors that the plant breeder can use to increase crop yield or quality. List at least two examples of each.
5. Define crop introduction, mass selection, and pedigree selection. Diagram methods for the two types of selection.
6. Define hybridization, inbred line, sister line, and homozygous line.
7. Define emasculation, genetic male sterility, and detasseling. Give an example of each.
8. Diagram the production of a single cross, three way cross, modified cross, and double cross hybrid.
9. List the advantages and disadvantages of the different types of hybrids.
10. Define a backcross, including recurrent and nonrecurrent parents. Diagram how a backcross is accomplished.
11. Define genetic engineering. Describe three methods used to incorporate genes from different organisms.
12. Define recombinant DNA and gene mapping.
13. List three ways that crop plants have been genetically altered which reduce their ability to survive under natural conditions.

INTRODUCTION

Crop improvement began when our ancestors began to domesticate wild plants for their own use. How long this process of domestication took is not known, but, since the dawn of civilization, people have been constantly striving to improve the plants that produce their food. This process of crop improvement almost certainly first involved the selection of superior plants; but later, and to the present time, crop improvement also involves plant breeding. **Plant breeding** is the art and science of changing and improving the heredity of plants to suit our needs. Before scientific knowledge, plant breeders used skill and judgment to select superior types of plants. With the development of genetics, plant breeding has become more of a science and less of an art.

Plant breeders have been successful in improving yield and other characteristics such as resistance to pests and environmental stress, crop quality, and harvestability. These successes have been largely due to a team approach that incorporates the sciences of genetics and cytogenetics, plant pathology, plant physiology, entomology, agronomy, botany, statistics, and computer science.

A crop **variety** is a group of plants with the same genetic composition. People have always strived to improve crop plants in regard to desirable traits. A **cultivar** is a variety that will remain genetically pure for several generations. Examples are normally self-pollinated crops, such as wheat and soybeans. A **hybrid** is a variety whose offspring will not be the same as the parents because of the genetic diversity of the parents. Examples are normally cross-pollinated crops such as corn and sorghum.

The development of a more productive hybrid or variety is a semipermanent advance in increased crop productivity. As important as improved tillage, fertilization, and pest control are in increasing productivity, these are annual procedures. An improved variety represents a more stable step toward increased yield. However, since conditions and standards are constantly changing, the efforts of plant breeders to maintain or increase crop yields through crop improvement are a continual process. New diseases and pests occur, or change from existing populations, which require breeding for resistance. Food and feed requirements may change and thus require altering the chemical composition of the grain, forage, fiber, or fruit produced. Finally, people's innate desires for self-improvement push economic and physical standards constantly higher which may change dietary habits and standards.

Crop improvement involves both sexual and asexual reproduction and propagation that were discussed in the previous chapter. Plant breeding, however, involves only sexual reproduction, although asexual propagation may be used once the improved variety has been developed. A good understanding of sexual and asexual reproduction is necessary to fully grasp the ideas discussed in this chapter.

FACTORS INVOLVED IN IMPROVING CROPS

An increase in yield and quality is the ultimate goal of crop improvement. However, attaining this goal may involve breeding and selecting for many factors that can contribute to the yield and quality of the crop. Although some of these factors are more readily noticeable than others, all are equally important. For instance, a crop or variety that is very resistant to a certain disease will be of no value if its grain is of poor quality.

Resistance

One factor in which considerable progress has been made in recent years is crop resistance. **Resistance** is the ability of the plant to withstand adverse environmental conditions and still produce an economic product. The last part of that definition is very important. A variety that is resistant, but which cannot produce something of economic value is virtually worthless.

Resistance comes in two forms, avoidance and tolerance. **Avoidance** is the ability of the plant to avoid the unfavorable situation. For instance, a variety that is resistant to a particular insect pest may contain a substance that makes it unpalatable or toxic to the pest, thus avoiding its feeding activity and subsequent damage. The second type of resistance is **tolerance** that enables the plant to tolerate or withstand an unfavorable situation. For instance, if the variety in the previous example had the ability to withstand the feeding of the insect pest with minimum damage, it would exhibit tolerance.

Resistance can involve many factors. A plant can be resistant to disease, pests, drought, heat, cold, alkali soil, and so on. The plant breeder tries to develop plants that fit certain conditions and can cope with specific problems. It is important that the crop producer be aware of problems that are likely to cause yield losses in specific fields and select those varieties that are resistant to those problems. However, planting varieties that are resistant to problems that are not likely to occur is not advisable as there is an inverse relationship between yield potential and resistance. In other words, breeding resistance to pests and problems can lower yield potential under ideal conditions. However, when the problem exists, the increased yields of resistant varieties make them vastly superior.

Adaptation

Increased yield can also be accomplished by developing varieties that are adaptable to a variety of growing conditions. The environment in which a crop plant grows varies widely within a field and between fields. Soil characteristics usually vary from one location to another in a field, as can the microclimate in which the plant is living. Other factors for which the plant breeder may be selecting could include adaptability to the length of the growing season, ability to withstand heavy grazing, the ability to produce a large, vigorous root system, or winter hardiness, just to mention a few.

Plant breeders try to develop varieties that are adapted to the differing conditions that occur in a certain area or region. However, there is also an inverse relationship between adaptation and yield potential under ideal conditions. Breeders are faced with the problem of attaining as wide a range of adaptation as possible while still maintaining high yield potential under those conditions. That is why different varieties do the best as you travel from one region of the country to another.

Nutritional and Market Quality

The quality or value of a crop for livestock feed or human food can also affect the yield, or at least the value of the yield. In livestock feed, the breeders select for such traits as high palatability, leafiness, nutritional value, digestibility, etc. For human consumption, they may be looking for such traits as high protein, high oil content, milling and baking qualities, high sugar or starch content, or high or low fiber content.

Of course, it is hard to separate nutritional quality from market quality as any variety without high quality will have little market value. Other factors that can affect market quality include the ease of handling of the crop, and the ability of the variety to maintain its nutritional quality during prolonged storage. With grains and other commodities, the variety must also meet the quality standards for marketing that are used throughout the world, such as weight per volume, color, shape, and moisture.

Seed Quality

The importance of the ability of a seed to germinate and produce a vigorous seedling was discussed in Chapter 6. Plant breeders develop varieties that have large and uniform seeds with high rates of germination. They also select plants that exhibit genetic purity so that the next generation of plants will maintain the high standards of quality, resistance, and adaptation that the present generation has.

Harvest Quality

The ease with which a crop can be harvested with a minimum of loss of the product in the field can also greatly affect the yield of that crop. Plant breeders select for such characteristics as uniform head height, uniform maturity, rapid loss of grain moisture after maturity, and ease of threshing. Other factors could include the ability of the plant to avoid dropping its grain before harvest (shattering), resistance to lodging, and maintenance of high crop quality from maturity to harvest.

To accomplish these goals, plant breeders have greatly changed the growth habit of some crops to ease their harvesting. For example, most hybrids of grain sorghum grown today are much shorter than their ancestors, which makes them easier to harvest mechanically; and their panicle inflorescences are much more open, which enables the grain to dry down faster after maturity. Plant breeders have changed the flowering of some vegetable crops, such as tomato, from indeterminate to determinate that permits harvest of the entire plant at once, thus greatly lowering labor costs.

Productive Capacity

With some perennial plants, such as forages, the ability of the plant to resume growth after harvest can also greatly affect its yield. Plant breeders develop varieties that can recover rapidly after grazing or cutting, as well as maintain a longer life span. To accomplish these goals, breeders select plants that have the ability to maintain large reserves of stored food that they can use to initiate new growth. They may also develop plants that have large, vigorous root systems, stolons, crowns, and/or rhizomes.

METHODS OF CROP IMPROVEMENT

There are four basic techniques used in crop improvement. They are **introduction, selection, hybridization**, and **genetic engineering**. These methods are used individually and collectively to accomplish the goals of crop improvement.

Crop Introduction

Many of the economically important crops produced in the United States have been introduced from other countries or continents. Corn and cotton came from Mexico, soybeans from the Orient, wheat from Europe, sorghum from Africa, and alfalfa from Asia. Korean lespedeza was introduced from Korea and sudangrass from Sudan in Africa.

An important crop introduction to the winter wheat area of the Great Plains occurred in 1873 when a group of German-Russian Mennonites brought Turkey Red wheat to Central Kansas. This introduction was the basis for the vast hard red winter wheat industry of the central Great Plains. Most of the varieties currently grown today are derived from selections and crosses from the original Turkey Red wheat. Similarly, Marquis spring wheat was introduced from Canada in 1913 and has been used in breeding programs for the development of current varieties. Bond oats from Austria were used in 1929 to develop disease resistant varieties of spring oats for use in the U.S. Many forage grasses have been introduced from Europe, Asia, and Africa for use in pasture and range seeding.

Less spectacular, but equally as important and productive, is the current exchange of crop varieties, strains, and selections between plant breeders at Land Grant universities, and the interchange of breeding material between public and private researchers. A new variety or hybrid may do poorly under one set of environmental conditions, but will be well adapted in another state or region.

Scientists in different parts of the world also exchange plant breeding material. In recent years there has been an exchange of soybean varieties and strains between scientists in the United States and the Peoples Republic of China. Since soybeans originated in that part of the world, plant breeders in the U.S. should be able to greatly increase the resistance of local varieties to insects and diseases since the ancestors of the soybean have evolved considerable resistance by natural selection over the years.

Crop Selection

The selection method of crop improvement involves a system of increasing the desirable individuals in a plant population, and decreasing or eliminating the undesirable plant types. There are two basic kinds of selection, mass selection and pedigree selection as diagramed in Figure 11-1.

Mass Selection

Mass selection involves the selection of desirable plants from a large population of plants. With mass selection, individual plants are selected from a crop field (population) based on appearance, head size, straw strength, early maturity, disease resistance, or whatever traits the plant breeder is looking for. The selected individuals are harvested, threshed, and all the seed is combined. Selection may also be accomplished by eliminating or roguing out the undesirable plant types and harvesting the remaining plants. The mass selection procedure is usually continued for several generations until a reasonable level of uniformity and yield increase are achieved. If the selected variety yields more, has greater disease and insect resistance, is less susceptible to lodging, or matures more satisfactorily than the current variety, it replaces the older variety.

Mass selection is a simple technique for crop improvement, but it takes longer to make stable advances since some of the poorer individuals are more likely to be retained in the population with the more desirable types. Two examples of varieties developed through mass selection are Grim alfalfa and Kentucky 31 fescue. Wendelin Grim brought alfalfa seeds from Germany to Minnesota in 1858. Those plants that survived the extremely cold Minnesota winters formed the foundation for Grim alfalfa by the process of natural selection. Wild fescue that remained green all winter was observed growing on a hillside in Kentucky. Some of the best plants were selected in 1931 and were the original source of Kentucky 31 fescue, a well known and adapted variety.

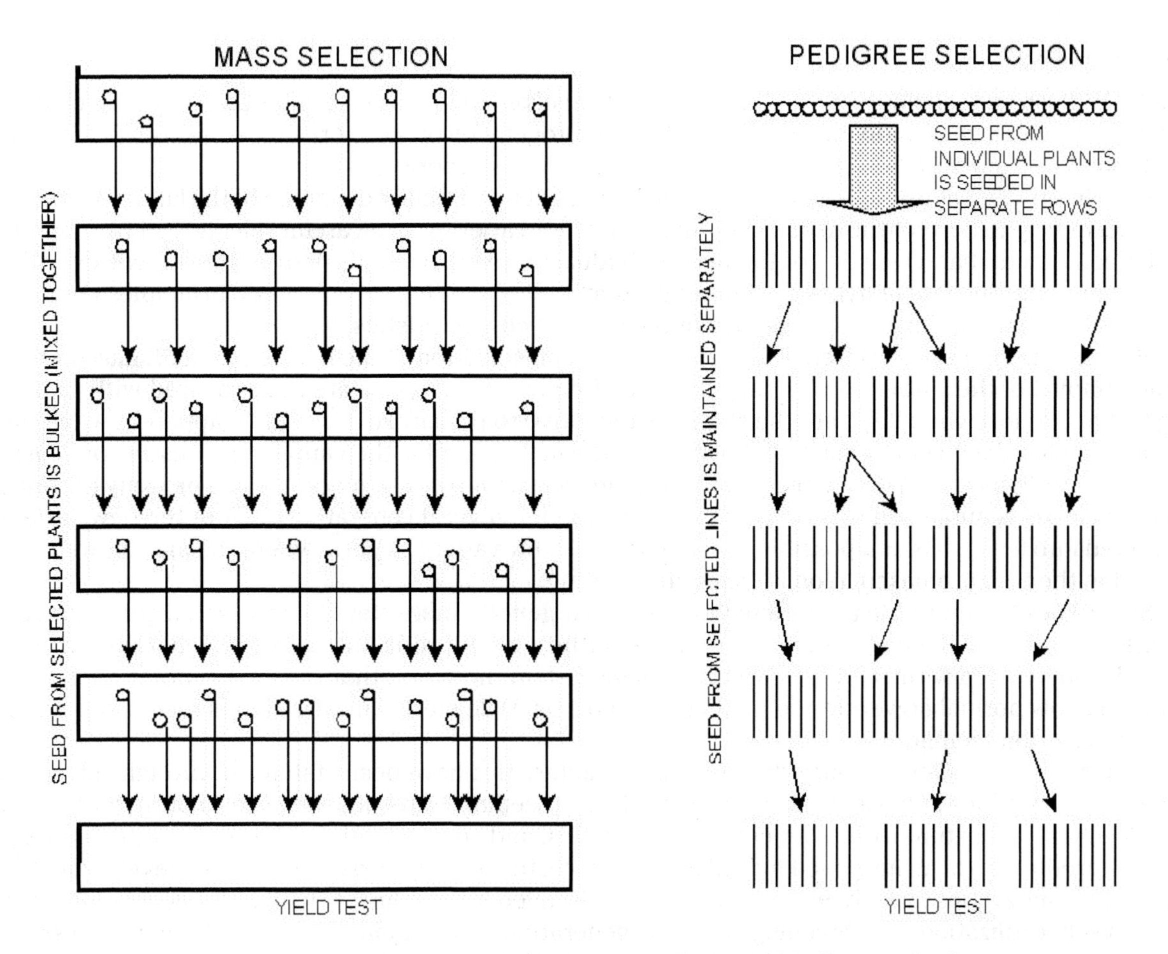

Figure 11-1 Diagrams of mass and pedigree selection methods.

Pedigree Selection

Pedigree selection involves selecting individual plants and growing out the progeny of each selected plant in a separate row, called a **head row**. As shown in Figure 11-1, plants from each selection are carefully observed and only outstanding individual plants or head rows are retained for planting the following year. The breeder can select individual plants from the head rows and plant them into head rows the next year, or select an entire head row and advance the plants into a **rod row**. An entire rod row can be retained, thus continuing to increase the number of plants. However, at any time, the breeder can select individual plants for planting into head rows and repeat the process. This procedure is continued for several generations with continued testing for desirable characteristics and retention of only the best plant populations. Eventually, only the best individuals are left for final yield testing and varietal release.

Pedigree selection is a time and labor intensive program that requires rigid selection criteria to keep the number of genetic lines from becoming too large. Rapid progress can be made because the poor individuals are eliminated early in the program, and only the best individuals are retained for further observation and selection. Pedigree selection is used extensively with self-pollinated crops, such as wheat and soybeans, since the different selections can be planted next to each other with little cross-pollination occurring.

The selection method of crop improvement has been, and continues to be one of the most effective and efficient techniques of the plant breeder. Selection does not, however, create new individuals with new characteristics, but only identifies those plants with desirable traits. Therefore, the crop improvement possibilities of a variety are limited by the best individuals in the population from which the selection is made.

Hybridization

The hybridization technique of crop improvement involves the combination of the desirable characteristics of two or more individuals. A **hybrid** is the progeny or offspring resulting from the cross-fertilization of two unrelated crop strains, varieties, species, or genera.

Crop improvement through hybridization is dependent upon a clear definition of the kind of improvement necessary to correct the deficiencies in the present varieties, and careful selection of the parents so the combined characteristics of the new individuals represent the desired improvement over the original parents. Success in hybridization requires an understanding of the genetic principles involved and information regarding the inheritance of the desired characteristics.

Hybridization in self-fertilized crops requires the emasculation of the female plants. **Emasculation** is the removal of the anthers before they have shed pollen. The flowers are then covered with a small bag to exclude foreign pollen until the stigma is receptive to pollen. This is usually apparent when the stigma is covered with sticky exudate. Pollen from the male parent is then applied by dusting or brushing it on the stigma. The flower is recovered with the bag and fertilization occurs. The progeny (hybrid) from this cross is subjected to rigid pedigree selection for several generations before the new variety is entered into field trials and possible final release as a new variety. In corn, where the male flowers are located in the tassel, emasculation is called **detasseling.**

Sometimes it is possible to use a technique called **genetic male sterility** in which a gene is incorporated into the female plants that prevents the anthers from producing viable pollen. Genetic male sterility can be used with any type of plant. It is important, however, that the next generation is male fertile to allow normal flowering and pollination to occur. Male sterile lines are developed using crossing and selection techniques.

With self-fertilized crops, once the varietal characteristics have been stabilized through selection, farmers may produce and use their own seed for future seeding. Care must be used to protect the seed from mechanical mixture with other crops and varieties, and contamination with seed borne diseases.

The production of an improved hybrid from cross-fertilized crops first requires the development of desirable homozygous lines. A **homozygous line**, also called **inbred line**, is one that is genetically pure through self-fertilization, or inbreeding, for many generations. This homozygous condition is necessary to assure the expression of the desirable characteristics of the parent lines in the resulting progeny. Self-fertilized crops are naturally homozygous (inbred) and can be crossed directly to produce hybrids. Inbred lines of cross-fertilized crops are produced through seven to nine generations of self-fertilization by protecting the stigma from foreign pollen as described previously. This "selfing" tends to reduce the vigor of the inbred. As in any plant breeding program, careful selection is exercised during the inbreeding process to maintain only the most desirable individuals. Not all inbreds produced are immediately useful in a breeding program.

The second step in the production of a hybrid with cross-fertilized crops is the crossing of two carefully selected inbred lines to produce the desired characteristics in the new hybrid progeny. This cross between two inbreds gives the resulting plants **heterosis**, or hybrid vigor. Heterosis is the increase in growth or vigor of the hybrid compared to the average of the parents. The amount or degree of hybrid vigor in the progeny is decided by the combining ability of the two inbred lines.

Types of Hybrids

The most common example of hybrid production is the commercial hybrid seed corn industry in the United States. Four or six rows of the female inbred line are planted in alternate strips with one to two rows of the male inbred line in well isolated fields. If the female plants are male fertile (capable of pollen production), they will be emasculated by removing their tassels before pollen shedding. This insures fertilization from only the male plants. Hybrid seed is produced on the female rows and the male inbred line is perpetuated on the male rows.

If two inbreds lines are crossed, the resulting progeny is called a **single cross hybrid,** or F_1 hybrid, as shown in Table 11-1. If two F_1 hybrids are crossed, the result is a **double cross hybrid,** or F_2 hybrid. A **three-way cross hybrid** results from the cross between an F_1 hybrid and an inbred line. These three types of hybrids are commonly used in commercial seed corn production.

Table 11-1 Types of crosses used in hybrid seed production.

			Number of Inbreds used	Hybrid Name	
Inbred	X	Inbred	2	Single Cross	(SX)
Inbred	X	SX Hyrid	3	Three-Way Cross	(3X)
Inbred	X	Sister line	3	Modified Cross	(MX)
SX Hybrid	X	SX Hybrid	4	Double Cross	(DX)

A **modified cross hybrid** results from the crossing of three inbreds, two of which are closely related to each other. These closely related inbred lines are called **sister lines** and are the result of separate selections from the same parent that were then developed as unique inbred lines. Modified crosses are used to enhance the characteristics of the sister lines' parent. Usually, modified cross hybrids are developed to enhance a specific trait such as disease or pest resistance.

There are distinct differences between the types of hybrids. Single cross hybrids usually have the highest yield potential since they are a first generation (F_1) hybrid. Also, since there are only two inbreds in their parentage, the plants are very uniform in growth habit, pollination, and maturity. Single cross hybrids also have disadvantages. They have a short pollination period because they are so uniform, which makes them more susceptible to damage if weather or pests interfere with pollination. Single cross hybrids have less adaptability because their genetic base is only two inbreds that limit the possible genetic traits that can be used.

Double cross hybrids have a wider range of adaptability because they have four inbred lines in their genetic background. This also makes the plants less uniform which lengthens the pollination period but also results in less uniformity of maturity and other characteristics. Double cross hybrids usually have a lower yield potential than single cross hybrids because they are second generation (F_2) hybrids. Three-way cross hybrids are intermediate between single cross hybrids and double cross hybrids in yield potential, uniformity, and adaptability.

Each type of hybrid has its advantages and disadvantages. By developing a variety of hybrids, plant breeders are able to provide crops that are suitable for the wide range of soil and environmental conditions encountered in agriculture.

Backcross

A variation of hybridization is backcrossing. The **backcross** plant breeding technique is used when an otherwise desirable variety is lacking in only one or two simply inherited characteristics, for example, resistance to a new disease type or race. The desirable variety, called the **recurrent parent,** is crossed to the donor parent, called the **nonrecurrent parent,** which has the characteristic lacking in the recurrent parent. The resulting hybrid is then crossed back to the recurrent parent for several generations, resulting in the designated backcross. Selection for the desired characteristic, such as disease resistance, is made after each successive backcross. The backcrossing procedure usually requires five to seven generations resulting in a new variety with at least 95% or more of the genes of the original variety plus the desirable characteristics of the nonrecurrent parent. Figure 11-2 provides a diagram of the backcross technique. In this example, Variety A is the recurrent parent.

The backcross technique has been most useful with self-fertilized crops, and with plant characteristics that are simply inherited and controlled by a single gene. The incorporation of resistance to new races of phytopthora root rot in current soybean varieties is an example of the success of the backcross method of crop improvement.

Variety A X Variety B = Hybrid AB
100 % Variety A genes

Hybrid AB X Variety A = Backcross 1
50% Variety A genes

B1 X Variety A = Backcross 2
75% Variety A genes

B2 X Variety A = Backcross 3
88% Variety A genes

B3 X Variety A = Backcross 4
94% Variety A genes

B4 X Variety A = Backcross 5
97% Variety A genes

B5 X Variety A = Improved variety released
99% Variety A genes

Figure 11-2 Backcross method of crop improvement.

Genetic Engineering

A plant breeding technique that has recently become important is genetic engineering. In classical plant breeding, the entire genetic makeup of the plant is involved and the breeder is limited to only the genetic combinations possible within the existing populations of plants available. **Genetic engineering** is the direct manipulation of genetic material within the cell that can result in the development of totally new and different combinations of genetic traits never before possible. Genetic engineering allows the incorporation of genes from vastly different species such as bacteria, weeds, or even animals into plants.

Genetic engineering has the potential to greatly revolutionize crop production. Incorporation of natural defenses has already been done and will undoubtedly continue. An example is *Bt* corn and cotton in which genes from *Bacillus thuringiensis* were incorporated to make the plants toxic to certain larval pests. Genetic engineering will also allow the development of plants to fit specific markets. Examples are increased production of certain amino acids, proteins, or enzymes to increase nutritive value for specific types of livestock or poultry; correct genetic deficiencies in certain humans; and production of pharmaceuticals for incorporation into the diet or for extraction and subsequent sale. In the future many growers will no longer simply produce a crop; instead they will produce products for specific uses.

Important developments that have made genetic engineering possible are **gene mapping** techniques that allow the breeder to identify the location of specific genes on the chromosomes. With this information, the breeder can isolate and transfer specific genetic traits between organisms. Gene mapping requires intense research. For example, corn has over 100,000 genes.

Genetic engineering uses **recombinant DNA,** or genetic material produced in a laboratory by combining parts of chromosomes from different cells. Presently, three methods for introducing recombinant DNA into plant cells are used.

The Plasmid Method uses a bacterium, *Agrobacterium tumefaciens,* which has the natural ability to cause tumors by transferring a piece of its genetic material into the chromosomes of another organism. This bacterium uses a plasmid, to transfer its genes. A **plasmid** is a genetic element that replicates itself in the cell independently of the chromosomes. The plant breeder will remove a plasmid from *Agrobacterium,* cut it open using a special enzyme, and splice a gene with the desired trait into it. The altered plasmid is then put back into the bacterium. When the bacterium is mixed with plant cells it will duplicate its plasmid and transfer the new gene into a chromosome of the plant. The result is a genetically altered plant that carries the trait.

The Vector Method is similar to the Plasmid Method except that it uses a viral vector to insert the genetic material into the chromosome.

The Biolistic Method uses a gene gun. In this method, microscopic tungsten beads (bullets) are coated with segments of DNA that contain the desired genes. These bullets are then placed in a device that shoots them into the cell where the new genes will combine with the chromosomes already present.

In all methods, the altered cells are grown in the laboratory until they develop into plants and can be transferred to a growth chamber or greenhouse. Once they flower, they can be used in hybrid crosses like other plants followed by selection for the desirable traits of both crossed plants

Differences between Crops and Wild Plants

People have changed the genetic characteristics of crop plants over thousands of years by selecting superior plants; and more recently, by using advanced technology such as hybridization and genetic engineering. Many crop plants in common use today have been altered to such an extent that they may not be able to successfully survive without the care given them by crop producers.

One important way that crop plants have been altered is the reduction, and sometimes complete inhibition of seed dispersal at maturity. Natural plant species have evolved many ways of dispersing seeds within a region by animals, wind, water, etc. Crop plants have been designed to hold their seed until harvest to reduce losses to the producer. Probably no crop has become more removed from its ancestors than corn. The corn ear is the most inefficient way to produce seed for dispersal. In fact, little seed is dispersed from a mature corn plant by natural means.

Crop plants have also been bred for maximum productivity under controlled conditions; and many times will not survive under natural conditions. For example, many crop plants have less ability to compete with other types of plants because the producer removes all other competition through weed control. Most crop plants have lost the trait of seed dormancy which protects the species from extinction if a disaster occurs, such as drought or flood. Individual cultivars of crops may be very resistant to a particular pest or disease, but wild plants will have more complete resistance because they are more genetically diverse. Table 11-2 compares differences between crop plants and wild plants.

Table 11-2 Comparisons between populations of annual crop plants and wild plants.

Crop Populations	Wild Populations
Seed brought into system by manager	Seed produced mostly by local population
Seed viability high	Seed viability varies
Seed dispersal is uniform	Seed dispersal nonuniform
No seed dormancy or carryover	Seed dormancy and carryover present
Soil environment mostly uniform	Soil environment very nonuniform
Population very uniform:	Population more diverse:
Population even-aged	More variability in age
Uniform genetic characteristics	Less uniform genetic characteristics
Intraspecific competition reduced:	Intraspecific competition intense:
Plants evenly spaced	Plant density highly variable
Interspecific competition reduced	Interspecific competition may be intense
Population very uniform genetically	Population more genetically diverse
Reproductive allocation very high	Reproductive allocation lower

REVIEW QUESTIONS

1. What techniques, if any, did our ancestors use in improving their crops?
2. What characteristics might a plant breeder use in improving crops or varieties? Are any more important than others? Why, or why not?
3. What method of crop improvement is used more than any other? Should a plant breeder use just one method? Why, or why not?
4. What is the ultimate limitation of selection as a method of crop improvement? What are the limits of hybridization?
5. Why are inbred lines needed for hybridization?
6. What techniques would you use to develop a single cross hybrid of corn? Would it be more difficult to develop a single cross hybrid of wheat? Why?
7. What are the advantages and disadvantages of each different type of hybrid? Is any one type clearly better?
8. How can two sister lines come from the same parent and still be different?
9. How does the backcross method maintain genetic purity and still get the desired characteristic of the other parent?
10. What is the most significant difference between genetic engineering and other methods of crop improvement?

Part III

Environmental Factors Affecting Crops

CHAPTER TWELVE

Climate, Weather, and Crops

INSTRUCTIONAL OBJECTIVES

Upon completion of this chapter, you should be able to:

1. Define weather, climate, meteorology, and climatology.
2. Discuss the different layers of the atmosphere and their effects on weather and climate.
3. List the gaseous components of the atmosphere. Discuss the importance of each to plants.
4. List the major components of weather and climate.
5. Discuss the effects of solar radiation, convection, conduction, latitude, and altitude on air temperature.
6. Define precipitation, precipitation distribution, precipitation intensity, and precipitation effectiveness. Discuss their effects on cropping practices in a region.
7. Define arid, semiarid, subhumid, humid, and wet regions. Discuss the cropping systems possible in each one.
8. Discuss the environmental factors that affect the water requirement of crops.
9. Define water requirement and water use efficiency.
10. Define transpiration ratio. Discuss factors that affect it.
11. Diagram the major air pressure belts on the earth. Discuss their effects on wind movement. Discuss the effects of prevailing wind direction on the climate in a given area.
12. Define growing degree days. Discuss the effect of temperature on crop growth. Calculate growing degree days given the daily maximum and minimum temperatures.
13. Define cardinal temperatures. Discuss their effect on crop growth.
14. Discuss the effects of low temperatures on crop plants including differences between crops in their susceptibility to freezing temperatures.
15. Discuss the effects of high temperatures and drought stress on crop growth.
16. Discuss the effects of excess water on crop growth.
17. Discuss the beneficial and detrimental effects of wind on crop growth.

INTRODUCTION

The old adage "Everyone talks about the weather, but no one ever does anything about it.", is essentially true. People's efforts at weather modification, particularly with respect to crop production, have generally been less than successful. However, we have been able to adapt and modify crop production practices to adjust to the adversities of weather and climate. For instance, precipitation has been supplemented with irrigation, and precipitation effectiveness has been enhanced by mulch tillage and fallow. Wind movements have been altered by windbreaks and surface mulches. The adverse effects of temperature extremes have been avoided by earlier or later planting dates. Although we cannot change the weather, we can take advantage of its limits to maximize crop production.

The Atmosphere

The **atmosphere** is the layer of air surrounding the surface of the earth and is divided into four layers. The **troposphere** is the layer nearest the earth's surface, and most of the weather of interest to the agriculturist originates in this layer. It is about 8 km (5 miles) thick at the poles, and about 19 km (11 miles) thick at the equator. The troposphere is characterized by a rapid decrease in temperature with height, and strong horizontal and vertical air movement.

The air layer just above the troposphere is called the **stratosphere** and extends upward to approximately 32 km (20 miles). The temperature is relatively constant and clouds and turbulence are rare. The third layer that extends from 32-80 km (20-50 miles) above the earth is called the **ozonosphere**. It is important because it screens out a high percentage of the ultraviolet light and other potentially lethal radiation. The final layer of the atmosphere is called the **ionosphere** because it is a layer of ionized gases. This is the highest layer above the earth and extends an estimated 80-805 km (50-500 miles) upward.

Weather is the point condition, both in time and location, of the atmosphere with respect to temperature, light intensity and duration, wind direction and velocity, relative humidity, cloudiness, air pressure, and any other characteristics capable of measurement. Weather conditions may change on an hourly or daily basis but generally have a weekly and/or monthly pattern.

Meteorology is the science of atmospheric conditions especially as they relate to weather and changes in weather. Although people have probably always been conscious of, and made observations of, weather changes, meteorology is a relatively new science. Accurate measurements of atmospheric conditions were not possible until the invention of such meteorological instruments as the thermometer, barometer, anemometer, and psychrometer for measuring temperature, air pressure, wind speed, and relative humidity, respectively. Even after accurate weather records were begun, the prediction of weather required the accumulation of weather data over a number of years.

The composite or generalization of the weather conditions over years is the **climate**. Weather may fluctuate from day to day, week to week, month to month, and even season to season; but from these variations a weather pattern or climate evolves. The climate is the average of decades of weather and the science and study of the climate is called **climatology**.

Although the same atmospheric components, namely temperature, moisture, air pressure, and wind movement, occur in both weather and climate, their effect on crop production is slightly different. Climate tends to delineate crop adaptation and production areas while weather influences the daily and seasonal growth and development of crops. The length of the growing season, mean monthly temperature, seasonal amount and distribution of precipitation, and other climatic factors decide the limits of the Cotton Belt of the South, the Corn Belt of the Midwest, the Winter Wheat Region of the Great Plains, the Hay and Dairy Region of the Northeast, and other types of farming areas. The daily, weekly, or monthly weather influence rate of germination and crop growth, amount of evapotranspiration, and crop damage due to drought, frost, hail, or high winds.

The air of the troposphere is composed of gases, water vapor, and particulate matter, with each component having an effect on the weather. The gases found in the atmosphere are mostly nitrogen (78%), oxygen (21%), argon (0.9%), and carbon dioxide (0.04%). The remainder of the gaseous atmosphere is occupied by minute amounts of neon, helium, krypton, xenon, radon, and ozone.

The water vapor content of the air is extremely variable, with a maximum of four percent occurring in warm tropical air and only minute amounts are found in desert air. The water vapor content of the air is influenced by air temperature and density.

Particles in the air commonly include soil particles, dust, pollen, smoke, volcanic ash, and salts from ocean spray. Air particles form the nuclei for the condensation of water vapor and the occurrence of precipitation. Rain making efforts are based on the "seeding" of particulate matter in clouds so that condensation will occur and droplet size will be large enough to cause rain to fall.

COMPONENTS OF WEATHER

The components of weather are as many as the observer can measure and record, but in this chapter only the major weather elements of temperature, moisture, air pressure, and wind movement will be discussed.

Temperature

Temperature is the most apparent weather component because it has the most profound effects on living organisms. It is the primary factor determining crop adaptation in belts north and south of the equator. Of course, the source of the heat, or the temperature effect, is solar radiation (insolation) from the sun. Solar radiation is short wave radiation that is largely responsible for heating the surface of the earth. Heat is also transferred to the earth's surface by convection, or movement of large air masses, and conduction, or heat transfer through some medium. Less than half of the solar radiation is absorbed by the earth. The remainder is reflected back into the atmosphere as shown in Figure 12-1.

In contrast, water absorbs about 70 to 75% of the radiation that strikes its surface. Water has a very high specific heat, which is the amount of heat energy needed to raise its temperature. Although water absorbs more solar radiation, it also requires more heat to raise its temperature, which can have a moderating effect on the air masses above it as will be discussed later.

The amount of solar radiation is decided primarily by latitude and seasonal change, but the final effect on air temperature can be modified by altitude and the location of large bodies of water. Solar radiation varies little from month to month at the equator because the angle of the sun does not deviate

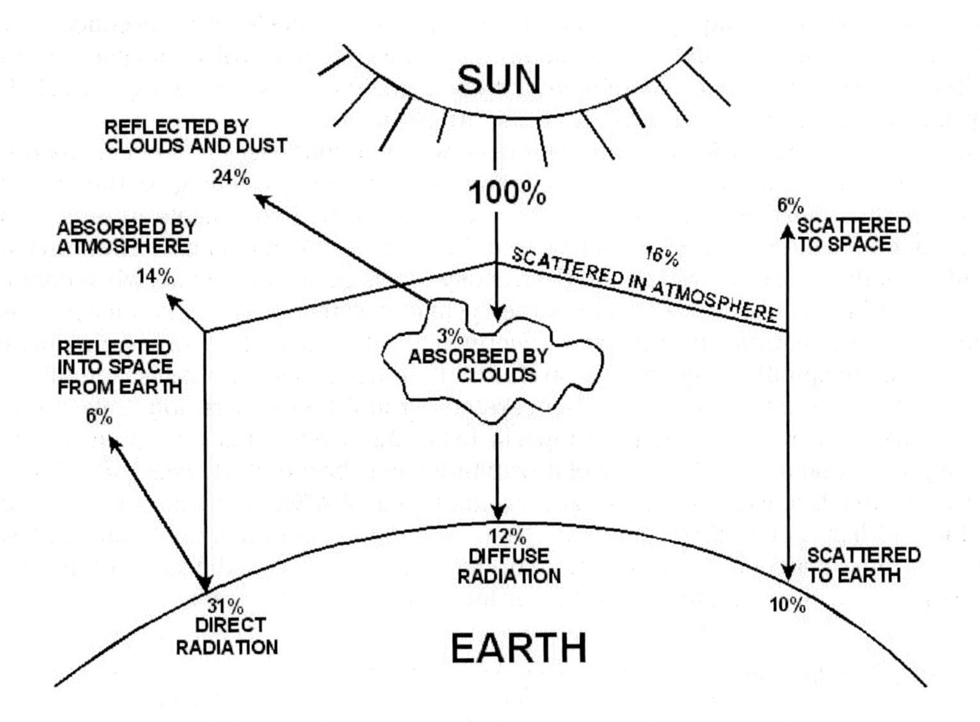

Figure 12-1 Solar radiation of the earth.

greatly from the vertical. At increasing latitudes, however, the amount of radiation received decreases because the angle of incidence to the surface is greater. However, total radiation during the growing season may be similar at different latitudes since days are longer at higher latitudes as shown in Table 12-1. The earth is tilted 23.5° on its axis and requires 365 days to rotate around the sun. The tilt changes the earth's exposure with respect to the sun and consequently the light exposure and the seasonal changes during the year.

Since temperature varies with the altitude, increases in elevation tend to have the same effect as increases in latitude. As altitude increases, the mean annual temperature decreases; a decrease of 1.6° C for every 300 meters increase in altitude (3° F/1000 ft). Increases in the altitude shorten the length of the growing season with later spring frosts and earlier fall frosts.

Table 12-1 Comparison of approximate sunlight hours at various latitudes and seasons of the year.

	Hours of Sunlight Each Date			
	Mar. 21	Jun. 21	Sep. 21	Dec. 21
North Pole	12	24	12	0
45° N. Latitude	12	16	12	9
Equator	12	12	12	12
45° S Latitude	12	9	12	16
South Pole	12	0	12	24

Moisture

Moisture as water vapor is always present in the atmosphere at some level. If the concentration is high enough and the atmospheric conditions are favorable, the water vapor will condense and precipitation occurs. Although we usually think of precipitation as rain, it also occurs as dew, fog, snow, hail, and frost.

The primary source of the water vapor in the atmosphere is evaporation from oceans. Large air masses carry the water vapor inland from the ocean where it condenses and occurs as precipitation. Some evaporation occurs from continental lakes and rivers but the percentage of the totals from these sources is quite small. Smaller yet is the evaporation or transpiration from plants and soil. Although transpiration accounts for more than 99% of the water used by plants and has a major effect on plant growth, the percentage of the total evaporation into the atmosphere is exceedingly small when compared to the total water vapor in the air. Figure 12-2 shows the cycling of water and water vapor on the earth.

The amount, distribution, intensity, and effectiveness of precipitation have a dominant effect on crop production. Precipitation influences crop choice, seeding and fertilizer rates, and tillage practices; essentially the entire crop and soil management system. Cropping system regions have been delineated based on average annual precipitation as shown in Table 2-2. Even though precipitation amounts vary greatly from year to year and the influence of precipitation may be modified by latitude and altitude, this classification system is a useful guide. Of the total land area, 55% lies in the arid and semiarid precipitation regions, with 20% occurring in subhumid and 11% and 14% found in the humid and wet regions, respectively. Approximately 70% of the earth's surface requires some modification of the water regime, either by irrigation or drainage, to be suitable for intensive crop production.

Precipitation Distribution

Precipitation distribution refers to the season or month of the year when precipitation occurs. It is often as much of a factor in deciding cropping systems as the amount of precipitation. Precipitation that occurs during the period of peak crop water use is most beneficial. For example, July and August are

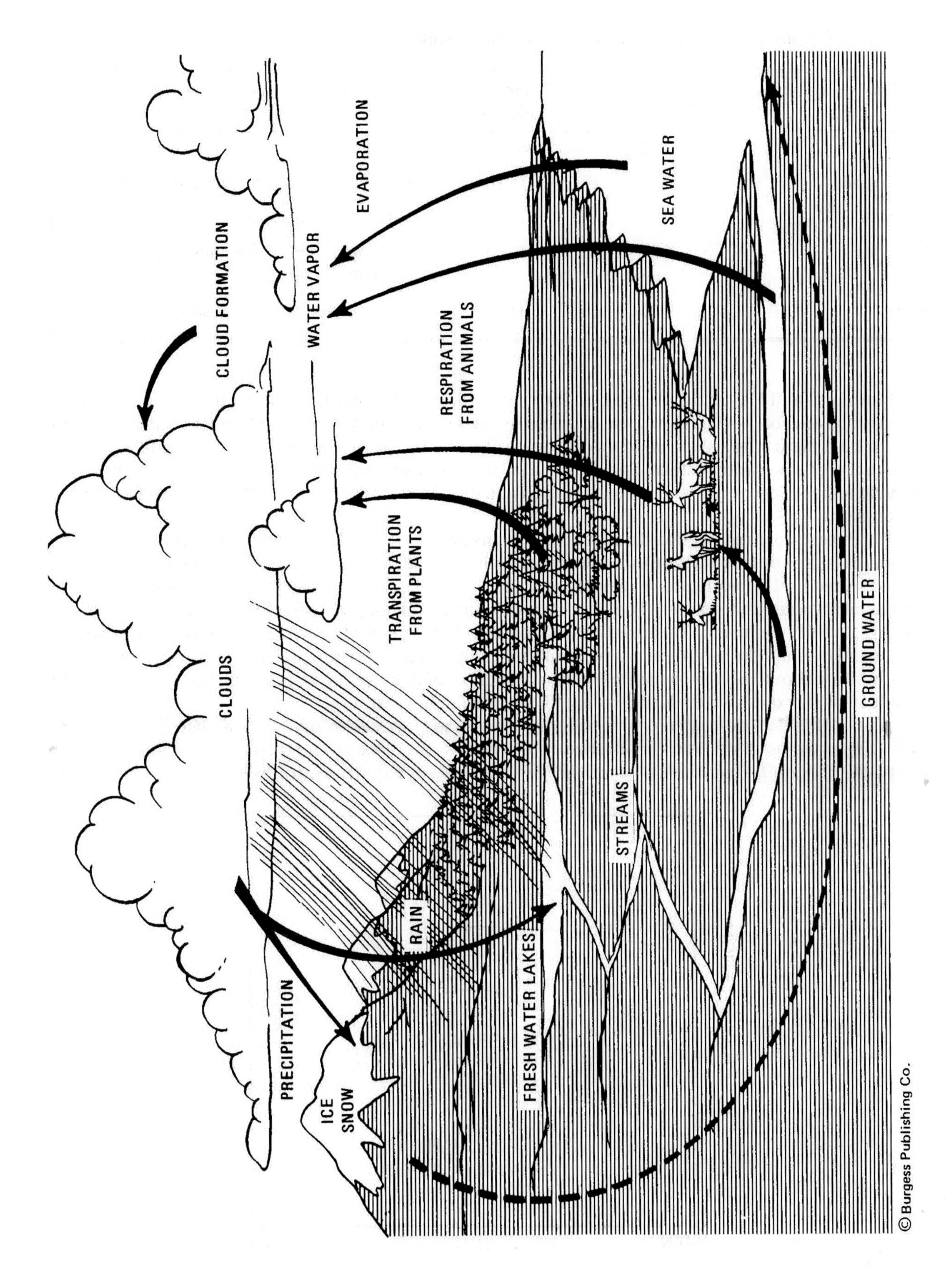

Figure 12-2 The cycle of water on earth.

Table 12-2 Cropping systems based on average annual precipitation.

Climatic Region	Annual Precipitation (cm)	(in)	Crop Management System
Arid	<25	<10	No crop production without irrigation
Semiarid	25–50	10–20	Crop Production with fallow or irrigaiton
Subhumid	50–100	20–40	Variety of systems possible
Humid	100–150	40–60	Crops and systems determined by precipitation distribution
Wet	>150	>60	Drainage frequently necessary

best for corn, and May and June are best for winter wheat. Precipitation that occurs in the off-season when crops are not growing is less useful unless it can be stored in the soil for future use. Soil water storage in the nongrowing season is dependent upon the water storage capacity of the soil and a soil surface that will ease water intake. Snow that falls on frozen soil and melts before the soil thaws may provide temporary cover to reduce evaporation and crop desiccation but has little direct value for crop production.

Three types of precipitation distribution patterns are shown in Figure 12-3. The pattern for Lincoln, Nebraska is the continental type that features dry winters and moist summers. This type of distribution is best suited for the production of cereal grains, row crops such as corn and grain sorghum, and hardy perennial legumes and grasses. The pattern for Pullman, Washington is just the opposite with moist winters and dry summers. This precipitation pattern is best suited to winter annuals such as winter wheat and rye, and drought resistant perennials. At Columbus, Ohio, the precipitation occurs more uniformly throughout the year which permits production of a wider range of crops but presents some problems with excess water during spring tillage and fall harvest.

The effectiveness of a given amount of precipitation is dependent on seasonal distribution, intensity or rate of precipitation, the infiltration rate of water into the soil; and the amount of surface evaporation and plant transpiration, commonly referred to as **evapotranspiration** (ET). Any cultural or

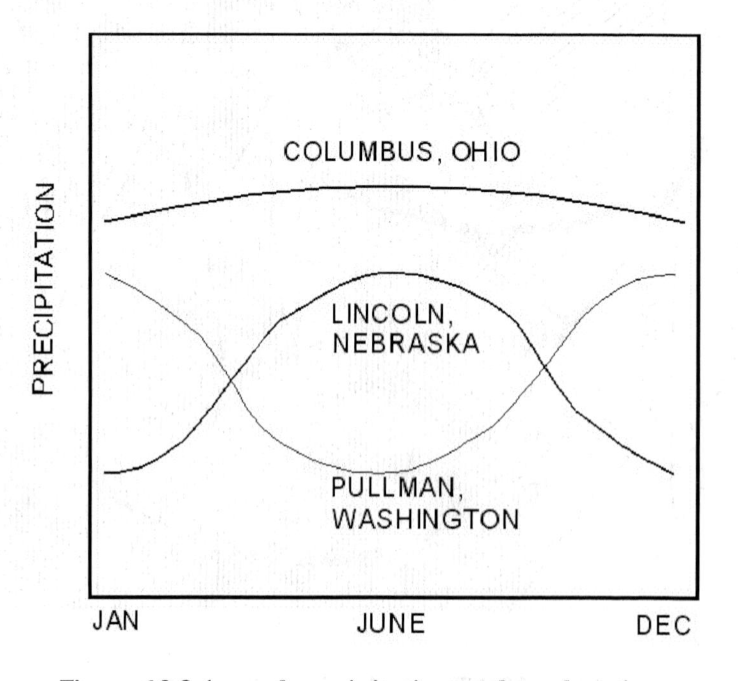

Figure 12-3 Annual precipitation at three locations.

mechanical practice that reduces the amount of runoff greatly increases precipitation effectiveness. Mulch tillage, reduced tillage, contour farming, and terracing all increase soil water intake and reduce runoff. A crop residue mulch also decreases the amount of evaporation. Good water conservation management increases precipitation effectiveness.

Precipitation Intensity

The precipitation intensity, or the rate at which precipitation falls, also influences the effectiveness of precipitation. When the intensity exceeds the soil intake rate and the rainfall continues for any period of time, runoff and erosion are likely to occur. Rainfall that occurs at the rate of 6 mm/hr (0.25 in/hr) is considered to be low intensity, while 25 mm/hr (1 in/hr) is rated high, and 125 mm/hr (5 in/hr) would be a very high intensity rainfall. Very high intensity rain storms are common but generally very short in duration. For example, assume two storms of 125 mm/hr, one for a duration of 10 minutes and the other for a duration of 30 minutes. The first storm would drop 22 mm (0.83 in) of rain and the second would drop 66 mm (2.5 in). On a bare soil, the first storm would pack and puddle the soil but would cause very little runoff or erosion. However, the second storm would cause considerable runoff and soil erosion. On a soil surface with adequate crop residue, neither storm would be likely to reach erosion capacity.

The frequency of high or very high intensity rainfall decides cropping practices and conservation measures. During periods when high intensity rains are likely to occur, the soil surface should be covered with a crop canopy or crop residue. Mechanical conservation structures such as terraces, waterways, and dams are designed to withstand the maximum high intensity rainfall that is likely to occur once in 25, 50, or 100 years.

Water Requirement

Precipitation effectiveness affects the total amount of water that will be available to a crop. However, there are differences between crops in the efficiency with which they use stored soil water. Since the amount of water that evaporates from the soil is a small proportion of the total plant water used, transpiration is frequently used as a measure of water use. The ratio of the weight of water used to the weight of dry matter produced is called the **transpiration ratio** and is a measure of the relative efficiency of crop water use. Table 12-3 shows the transpiration ratio of several crops. Transpiration ratio is not a measure of the drought tolerance of a crop or an indication of the total seasonal water use, but merely shows relative efficiency of water use. Total water use will be affected by the total amount of dry matter produced by the crop. For example, corn has a lower transpiration ratio than wheat but will use more total water because it produces more dry matter.

Temperature and relative humidity influence transpiration, and consequently, the transpiration ratio. Corn grown under an average relative humidity of 43% had a transpiration ratio of 340 but when the average relative humidity was 70%, the transpiration ratio was reduced to 191. Similarly, the transpiration ratio for alfalfa grown in Arizona is commonly around one thousand, but only one-half that amount when grown in Minnesota where average temperatures are lower.

Another measure of crop water use is **water use efficiency**. Water use efficiency is derived by dividing the crop yield by total water use. There are differences between transpiration ratio and water use efficiency. Water use efficiency uses only crop yield, not total dry matter production. Also, water use efficiency is the inverse of transpiration ratio since total water use is the divider in the ratio. High soil fertility usually increases the total water use by a crop but increases water use efficiency because there is a greater increase in yield than the increase in water use. Crop diseases, particularly leaf diseases, can decrease water use efficiency by 50–100%.

Air Pressure

Air pressure is the weight of the air on a given surface and is measured in atmospheres or bars. One atmosphere is equal to 760 mm of mercury and a bar is slightly less, 750 mm of mercury. There are two basic principles involved in air pressure. First, air at higher altitudes is less dense, and therefore air pressure is less. Secondly, warm air is lighter than cool air that reduces air pressure.

Table 12-3 Transpiration ration (kg water/kg dry matter) of several crops.

Crop	Transpiration Ratio	Relative Ratio
Proso millet	267	1.00
Common millet	287	1.07
Sorghum	304	1.14
Corn	350	1.31
Barley	518	1.94
Wheat	558	2.09
Oats	585	2.18
Rye	633	2.37
Soybean	744	2.79
Alfalfa	858	3.21
Bromegrass	1016	3.80

There are four general air pressure belts around the earth as shown in Figure 12-4. There is a constant low pressure belt at the equator where warm air rises, and constant high pressure areas at the poles where cold air settles. There are also subtropical belts of high pressure at 30–35° north and south of the equator, and polar belts of low pressure 60–65° latitudes. The differences in air pressure between these belts cause surface winds to blow north and south from areas of high pressure to areas of low pressure. The air masses do not move directly north and south, however, because the rotation of the earth creates the Coriolis Force. This forces the winds to the right of the direction they are moving in the Northern Hemisphere, and to the left in the Southern Hemisphere. The result is the primary air circulation patterns and wind movements of the earth shown in Figure 12-4.

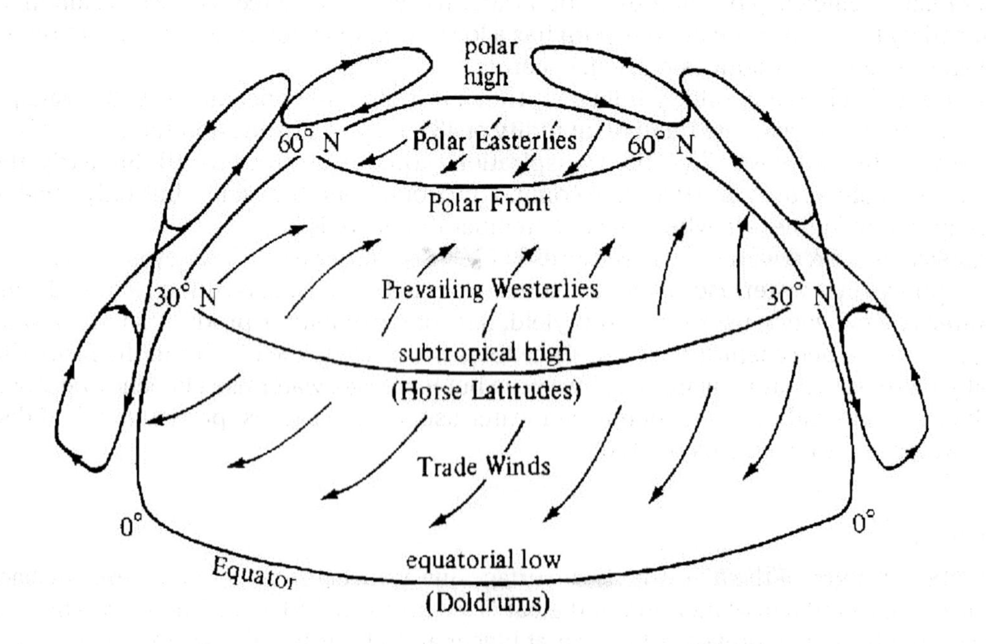

Figure 12-4 Global circulation of the atmosphere.

Wind

The winds just described in the previous section are responsible for movement of warm air masses from lower to higher latitudes and the movement of moisture laden air masses to land areas. The United States lies in the general path of the Prevailing Westerlies, so winds blow primarily from west to east. In the Central U.S., winds move primarily from the north and northwest from October through April, and primarily from the south and southwest from May through September.

EFFECTS OF WEATHER COMPONENTS ON CROP GROWTH

Temperature

The subject of **cardinal temperatures** on seed germination was introduced in Chapter 6. Briefly, there is a minimum temperature below which plant growth ceases, a maximum temperature above which growth also ceases, and an optimum temperature where growth is most rapid. There is a range of cardinal temperatures for each crop, variety, and growth stage in the life of the crop. The optimum temperature that produces the most rapid vegetative growth may not always produce optimum yields. Too rapid vegetative growth may prevent or delay flowering or produce plant stems that are structurally weak and susceptible to wind damage or disease infection.

Crop plants are very sensitive to temperature effects; and temperature and light are primary factors in crop growth and development. One of the more apparent temperature effects is the frost-free period, or the length of the growing season. The annual **growing season** is the number of consecutive days without a killing frost measured from the last killing frost in the spring to the first killing frost in the fall. The frost-free period is useful in a general way but does not consider the facts that different crops are killed at different temperatures and have different optimum temperatures for growth.

The correlation of the temperature influence on crop growth and development has been studied and quantified as the concept of **growing degree days (GDD),** or **heat units**. The basic premises of growing degree days are: first, that plant growth is largely a response to temperature and not time, such as weeks or months; second, that there is a base temperature below which crop growth essentially stops; and third, that the rate of crop growth increases as temperature increases.

$$GDD = \frac{Min.\ Temperature + Max.\ Temperature}{2} - Base\ Temperature$$

Equation 12-1. Formula for calculating growing degree days.

Growing degrees days are calculated by taking the average temperature for the day and subtracting the base temperature for the crop as shown in Equation 12-1. The GDD for each day is added to the accumulated total from the previous day so that GDD accumulates during the growing season. Figure 12-5 shows GDD accumulations in the U.S.

The temperature limits used by the National Weather Service are 86° F for the maximum temperature and 50° F for the minimum and base temperatures. In the calculation of GDD, if the maximum temperature is above 86° F, 86 is used, and if the minimum temperature is below 50° F, 50 is used. This tends to disregard the possible negative effects of temperature extremes on crop growth. Another limitation is the use of the same base temperature for all crops. A base of 40° F should be used for cool season crops such as wheat, oats, barley, peas, and potatoes. The 50° F base temperature is very satisfactory for warm season crops such as corn and soybean, but there is evidence that a base temperature of 60° F is more appropriate for sorghum and sudangrass.

In spite of these deficiencies, GDD is a valuable tool for farmers, crop consultants, researchers, and all persons concerned with crop production. Growing degree days are particularly useful in predicting crop maturity when the planting date and accumulated GDD are known. Many hybrid seed corn companies publish the GDD requirement of their hybrids to enable crop producers to accurately match hybrid maturity to their planting date and the average accumulated GDD in an area.

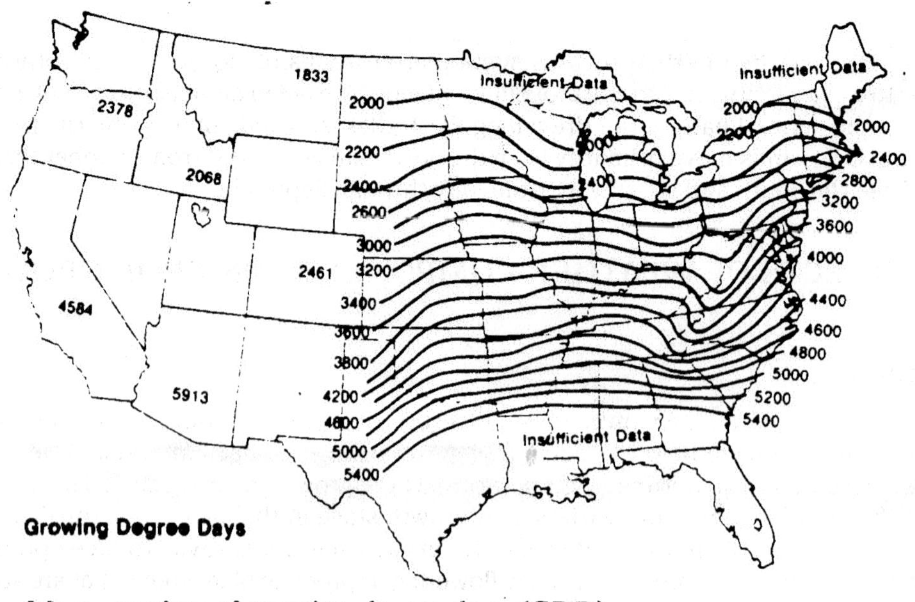

Mean number of growing degree days (GDD)

Figure 12-5 Growing Degree Day accumulations during the growing season.

Low Temperature Effects on Crops

Inherent in the GDD concept is the fact that cooler temperatures cause slower crop growth. Cool spring weather produces poor emergence, reduces seedling vigor and early growth and results in seedlings that are poor weed competitors. Cool summer and fall temperatures also cause slower growth and may delay maturity so that the crop is damaged by frost.

Freezing temperatures in the fall can cause damage if they occur before the plant has reached **physiological maturity.** Physiological maturity occurs when the plant has completed development of the grain or seed. If the crop has not fully matured when killed by a freeze, the grain or seed will not receive its full measure of dry matter. This can result in a poorly developed embryo and reduced food storage for the seed. Seed that has been produced on plants killed prematurely for any reason should never be used for seeding the next year's crop.

Even if the grain or seed will not be used for seeding it will still have reduced quality. The seed or grain will be shriveled and have a lower test weight. Total nutritive value will also be reduced, although percent protein may be higher. Potential for threshing losses may be higher if the grain or seed is significantly lighter. The degree of damage from an early freeze is directly proportional to the time that the damage occurs. The earlier the damage occurs, the greater the loss. If corn kernels have begun denting, damage will be small. The same is true if soybean leaves have begun to lose leaf color or sorghum has reached hard dough.

Severe freezing of an immature plant can also increase plant lodging as the formation of ice crystals in the wet stem damages the stem cell walls. Freezing temperature after physiological maturity can be beneficial as it kills the plant and hastens water loss from the seeds, thus speeding up harvest.

A late spring freeze may damage or kill emerging seedlings. Freezing kills plant tissue, particularly the leaves. However, the location of the **growing point** on a young crop plant greatly affects the amount of permanent damage that might occur. The growing point is the tissue where cell division occurs. In grass crops such as corn and sorghum, the growing point is still below the soil surface until the plant has produced five or six leaves with their collars visible. These seedlings are less likely to be killed by frost than dicot plants with apical growing points such as cotton, soybeans, field beans, or seedling alfalfa. In these plants, the growing point is above the soil and exposed to the freezing temperatures. Frost at or near the bloom stage of wheat, oats, or barley frequently kills the pollen or damages the pistil that reduces fertilization and seed production.

Winter injury may result in death or may just kill the crown buds and roots of winter wheat, winter barley, and perennial legumes and grasses. Winter hardiness (cold tolerance) has been correlated to certain plant characteristics such as the soil depth of the crown and certain chemical and physical properties of the plant cell. However, none of these characteristics is common to all cold tolerant plants and it appears that the true nature of winter hardiness lies in the enzyme and/or hormone systems of the plant.

High Temperature Effects on Crops

It is difficult to separate the effects of high temperature from the accompanying factors of high light intensity and rapid transpiration. The effects of drought stress are closely related to the effects of high temperature. The sensitivity of crops to high temperatures and water stress varies with the growth stage of the crop. For instance, corn is very sensitive to high temperatures during tasseling which causes poor pollen viability and shedding. Above normal temperatures during the early vegetative growth of corn and soybean can be beneficial to the crop due to the increased rate of growth that aids in the competition with weeds. Although most crop plants can survive temperatures up to 43° C (110° F), or more, yield loss will occur before it gets that hot as shown in Figure 12-6. High temperatures affect the plant's metabolism including photosynthesis. Figure 12-6 also shows the interaction of water supply with temperature.

Corn yields will decline when the temperature exceeds 35° C (95° F) during late vegetative growth through grain filling, especially during flowering, as the pollen grains may be damaged resulting in poor pollination and subsequent seed set. Soybeans are also sensitive to high temperatures at flowering. However, if the soybean is the indeterminate type, it may simply delay flowering until cooler weather and yield loss may be minimized. Sorghum is more tolerant of high temperatures than corn or soybean but high temperatures at flowering can cause damage. Wheat prefers cooler temperatures and can be damaged if the temperature exceeds 32° C (90° F). The high temperatures that usually occur when wheat is filling its grain are one of the major limiting factors to wheat yields in the Great Plains.

Humidity

The primary consequence of humidity on crop plants is its effect on the rate of transpiration or water loss from the leaves. While the stomata are open, lower humidity increases water loss. Lower humidity is usually associated with warmer temperatures. If transpiration from the leaves exceeds the plant's

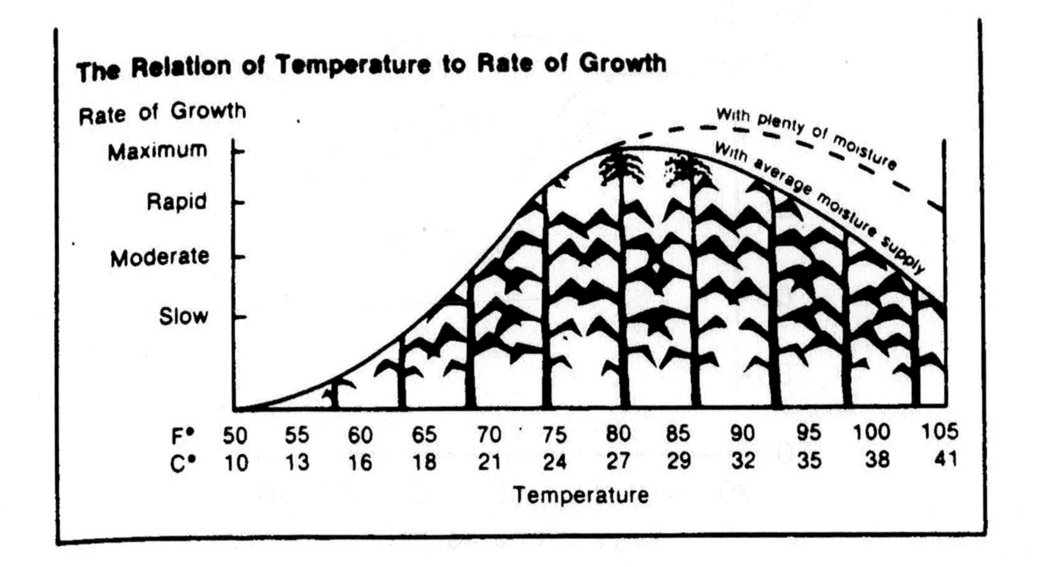

Figure 12-6 Effect of temperature on growth of corn.

ability to take up soil water and transport it to the leaves, the stomata may close and photosynthesis may greatly slow or stop resulting in loss of yield.

Water

Adequate water supply is the single most limiting factor to crop yield in most years in most parts of the world. Plants use tremendous quantities of water during the growing season and a lack of water, even for a short time can affect yield. For example, every inch of stored soil water above that needed for vegetative growth produces about seven bushels per acre of winter wheat.

As discussed in Chapter 6, roots need a moist, well-aerated soil for proper growth. Roots cannot grow into dry soil; there is no water for cell expansion and elongation. Therefore, the depth of the root system is decided by the depth to which the soil has been wetted by rainfall or other precipitation. Good soil management assures that high infiltration rates are maintained at the soil surface. This is accomplished by providing proper cover of crop residue and maintaining good soil structure.

Drought

The effect of increasing drought stress on crop growth is essentially that of decreasing the production of fresh and dry weight due to a decrease in the rate of photosynthesis. When leaves lose turgidity there is less leaf area exposed to intercept light, and when the stomata close the inward movement of carbon dioxide is stopped and photosynthesis ceases. Also, as mentioned above, the root system may be restricted if only a part of the rooting zone has been wetted. The result is shorter plants due to less internode elongation, reduced leaf area, and less lateral root growth.

Although any drought stress is harmful to yield, the growth stage at which the plant is subjected to drought greatly affects the amount of yield loss. Drought stress that occurs during early vegetative growth, unless prolonged, will have little effect on yield. As growth progresses, damage from drought will increase.

The most critical time for damage from drought is during flowering and pollination. Seven days of drought stress during this time can result in up to 50% yield loss as shown in Figure 12-7. Crops of the grass family such as corn, sorghum, and wheat are much more likely to sustain damage because they are **determinate** and flower only during a short period of time as discussed in Chapter 10. If drought stress interferes with proper flowering and pollination in these crops, there is no chance for recovery even if adequate rainfall occurs later.

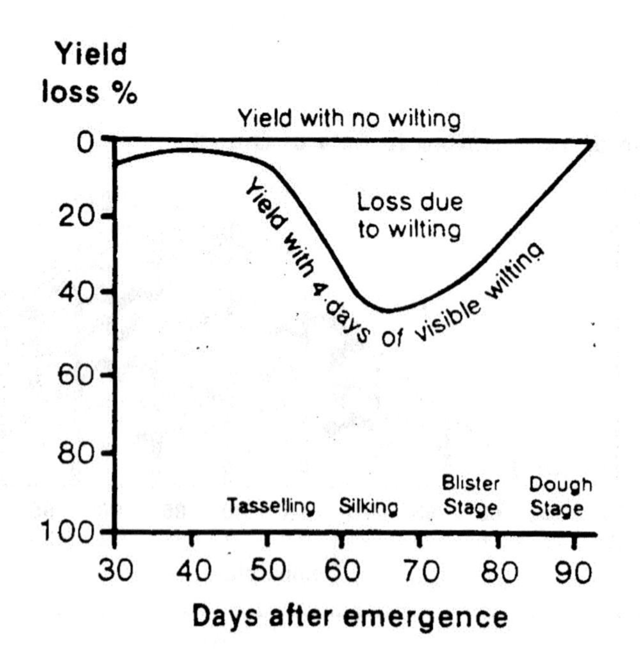

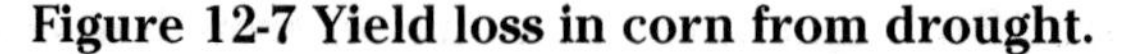
Figure 12-7 Yield loss in corn from drought.

Other crops, such as soybean, may have an **indeterminate** flowering pattern. In these crops, flowering occurs over a longer period of time; not all flowers develop together. Drought stress during flowering in indeterminate crops is not as likely to cause damage as with determinate crops. With indeterminate soybeans, drought stress during flowering may only delay flowering and if subsequent rainfall occurs, the plant may recover with only small losses in yield. Most soybean varieties grown in the Midwest are indeterminate. However, there are some determinate varieties and they can be severely damaged by drought stress at flowering just like corn, sorghum, and wheat.

Prolonged drought stress during grain or seed development can result in aborted seeds or pods. In corn, sorghum, or wheat the result is incompletely filled ears or heads. In soybeans there will be fewer pods that may not be completely filled with seeds. The plant tries to fully develop healthy seed and will reduce the number of seeds in an effort to compensate for reduced dry matter production from lower rates of photosynthesis.

Excess Water

Too much water can be just as detrimental to crop growth as a water deficit. For example, every day that the root zone is saturated corn yield is reduced 10-15%. When the soil is saturated with water from poor drainage or flooding there is little air for the roots and soil microorganisms to conduct respiration. Although there are some plants that can survive in saturated soil, such as rice, most crop plants cannot experience saturated soil for more than a few hours without damage occurring to the roots. As mentioned in previous chapters, roots need air for respiration that provides the energy needed for cell growth and maintenance and for absorbing water and nutrients. Prolonged saturation results in an anaerobic condition that causes disrupted metabolism and the formation of toxic products in the roots.

Soil microorganisms are also affected by poor aeration from saturated soil conditions. Like the roots, beneficial organisms need adequate aeration for growth and maintenance. Prolonged saturation causes the loss of beneficial organisms and the increase of harmful ones. Toxic products can accumulate which can further damage roots and beneficial organisms. Loss of soil nitrogen can also occur during prolonged soil saturation through a process called **denitrification** as discussed in Chapter 5. The result is stunted, nitrogen deficient plants that will die unless the condition is corrected.

Air Pressure and Wind

Information on the direct effects of air pressure on crop growth is limited and often controversial. However, the indirect effect of air pressure on wind is very apparent in the movement of moist or dry air masses over land areas as discussed previously.

Moderate winds apparently do not affect crop growth and may be beneficial through the mixing of air in the crop canopy to improve the carbon dioxide concentration at the leaf surface. Studies have also shown that plants subjected to moderate winds have stronger stems than those grown in greenhouses with little or no wind. However, hot dry winds increase the transpiration rate and can quickly deplete soil water causing drought stress. Persistent hot drying winds can increase the water deficit in crop leaves even though there is adequate soil water because the plant cannot take up water fast enough to keep up with transpirational demand.

Winds of extreme high velocity can cause mechanical damage to crops especially if accompanied by heavy rains or hail. Plants may be defoliated and, if winds are strong enough, lodging will occur. Corn is most susceptible to wind damage during rapid stem elongation. The internodes that are most rapidly elongating are the most brittle. Strong winds with sleet and/or snow increase crop losses even if they occur after the crop has matured.

Winds can also transport fine particles of soil, primarily very fine sands, which can cause severe leaf damage to young crop seedlings. If severe wind erosion occurs, the crop may be buried from the deposition of soil.

REVIEW QUESTIONS

1. How can a crop producer adapt to the limitations of temperature, rainfall, growing season, and wind?
2. What is the difference between weather and climate. How does each affect the crops growing in a region?
3. How do minute air particles affect precipitation?
4. Which weather component has the greatest effect on crops on a daily basis? Is any component more important than others in deciding crop adaptability in a region?
5. What effects do large bodies of water have on weather components in a region?
6. How do altitude and latitude affect temperature?
7. How do precipitation distribution and intensity affect cropping practices in a region? What precipitation distribution and intensity are found in this region?
8. What factors affect precipitation effectiveness and what measures can a crop producer take to maximize it?
9. Wheat has a higher transpiration ratio than corn but wheat does well in semiarid areas while corn does not. Why?
10. How does crop residue reduce soil damage from high intensity storms?
11. The U.S. is in the Prevailing Westerlies. How does that affect annual precipitation in this region?
12. Using Equation 12-1, if the high temperature is 85° and the low is 63°, what is the GDD that day? What if the high is 94° and the low is 76°? On which day will the crop grow faster? Why?
13. Why are grass seedlings such as corn less susceptible to freeze damage than bean seedlings? What effect does the type of emergence have on this?
14. How does an early freeze in the fall damage crop quality?
15. At what growth stage is drought stress most damaging to most crops? Why?
16. How does drought stress affect the plant's metabolism?
17. How does excess soil water affect the plant, and why?
18. What are the beneficial effects of wind on crops? What are the damaging effects?

CHAPTER THIRTEEN

Crop Pests and Control

INSTRUCTIONAL OBJECTIVES

Upon completion of this chapter, you should be able to:

1. Discuss the total losses incurred from crop pests and their importance in world agriculture and food supply.
2. Define integrated pest management. Discuss its importance in pest control.
3. List five general types of damages to crops from insects and mites.
4. Discuss the natural advantages of insects and mites that make them difficult to control.
5. List five types of crop losses from insects and mites. Give an example of each.
6. List four beneficial effects of insects. Give an example of each.
7. List three methods of insect classification. Discuss the importance of each.
8. Define metamorphosis, ametamorphosis, gradual metamorphosis, incomplete metamorphosis, and complete metamorphosis.
9. Define viviparous and parthenogenesis. Discuss how pests with these traits could cause increased damage compared to other types.
10. Define chewing insect, piercing-sucking insect or mite, and boring or tunneling insect. Give an example of each.
11. Define action threshold and discuss its importance to pest control.
12. Define genetic, cultural, biological, and chemical pest control. Give at least two examples of each.
13. Define nonpreference, antibiosis, and tolerance. Give an example of each.
14. Define pesticide. List at least four general types of pesticides.

INTRODUCTION

The best combination of genetic, climatic, soil, and cultural factors does not always guarantee successful crop production. Pests, diseases, and weeds can destroy a crop or reduce yields and crop quality enough that the producer suffers economic loss. It is estimated that crop and animal pests destroy one-half of the world's food production during growth, harvest, transport, or storage. In many tropical and subtropical climates where relative humidity and temperature are high, losses can be 60-70% of total production. In the United States alone, the combined costs of losses and control range from 25-30 billion dollars annually. Approximately 50% of this total is attributed to weeds, while insects, mites, and diseases are responsible for the rest.

Through the research and educational efforts of private industry and public agencies, crop pests have been investigated and control methods carried out so that many catastrophic losses are now averted. However, many of the control methods require the extensive and intensive use of agricultural chemicals such as insecticides, fungicides, and herbicides. The concept of **integrated pest management (IPM)** has developed many methods to control a pest instead of using one single technique. IPM is discussed later in this chapter.

Although the effects of specific insects, mites, diseases, and weeds on crops may be quite different, there are several common types of damage. For example, crop pests can cause the loss or destruction of leaf area through **defoliation** or **necrosis** of leaf tissue. Weeds compete with crop plants for water, nutrients and light. Pests can cause the destruction or dysfunction of the plant conductive tissue that disrupts the movement of water, nutrients, and photosynthate within the plant. Pests can weaken plant stems resulting in lodging and a decrease in the quantity of harvestable produce. Pests can also cause damage or replacement of the grain or fruit that reduces both quantity and quality of yield.

This chapter will discuss insects and other pests such as mites, their effects on crop growth, and possible control measures. The two succeeding chapters will cover disease and weed pests.

INSECTS AND MITES

There are approximately five to ten million kinds of insects and mites on the earth, of which fewer than one million have been given species names. The world population of insects alone has been estimated at 1.0×10^{18} and if each insect weighs only 2.5 mg (less than 1/10,000 oz), insects would weigh twelve times more than humans! More than 85,000 insect species occur in North America plus more than 2,600 species of ticks and mites. However, only about 10,000 species are designated as economically harmful to humans, livestock, and plants. Many of the insects and mites that cause economic crop damage in the U.S. have been introduced from other parts of the world. Of those introduced accidentally, the European corn borer and the Hessian fly reduce corn and wheat yields every year; the cotton boll weevil and the pink boll worm are the worst insect pests of cotton; and the alfalfa weevil and pea weevil cause economic damage to alfalfa.

In the United States, crop losses due to insects and mites range from 5-15% annually, or nearly four billion dollars. In addition, there are approximately one billion dollars spent on control measures of various kinds. Annual losses from insects and mites are about 12% for corn, 9% for sorghum, 6% for wheat, and 3% for soybeans.

Insects and mites have many natural advantages that increase their probability of survival and the likelihood of crop damage. Most insects and mites are small and often difficult to see. This characteristic, along with their extremely rapid reproduction rate, allows insects and mites to reach large population levels without detection. Many insects and mites have great mobility; they are able to fly and some are even carried by winds. Although some insects and mites are host specific, most can feed on a wide range of crop and weed species. The specialized structural adaptations of some insects and mites allow them to adapt to environmental changes. Some insects have specialized structural forms at different stages in their life cycle with periods for feeding, resting, and reproduction. This is a tremendous advantage when these life stages correspond to the same general growth stages of the host plant.

EFFECTS OF INSECTS AND MITES ON CROPS

The most apparent crop damage caused by insects and mites is reduction in yield and quality. Insects and mites reduce leaf area by direct defoliation (eating the leaves), or through the development of necrotic areas in leaves caused by sucking cell sap from the leaves. Boring and tunneling insects invade stems and roots and disrupt the translocation of water, nutrients, and photosynthetic products. **Frass**, or insect excrement and skins left from previous developmental stages, and other debris lower the quality of grain and forage. Grain or hay may be discounted at the market due to the presence of foreign material because it is less palatable as livestock feed or requires additional processing before it can be used as human food. Also, infestations generally reduce grain and forage yields because plants are stunted in growth since the normal metabolic functions of nutrient uptake and plant synthesis are disrupted.

Crops attacked by insects and mites usually have higher harvest losses than noninfested crops. Stems of many crops are weakened by feeding and frequently lodge. Plants suffering an infestation are generally less sturdy and thrifty in growth and consequently are more subject to damage from wind, rain, and other inclement weather. Finally, plants infested with insects and mites are not as productive, and are more likely to produce light weight, chaffy grain that either lowers the test weight or is lost in the separation process of harvesting.

Insects and mites commonly transmit diseases to plants, primarily the virus diseases by aphids, mites, and leafhoppers. Plants weakened by feeding are frequently more susceptible to disease infection by fungi, bacteria and viruses.

Insects and mites can also cause indirect damage to crops. Fields, pastures, and range that have been damaged by insects and mites produce less dry matter, resulting in less residue for ground cover and an increased risk of wind and water erosion. Grasshoppers can leave rangeland so bare that serious soil and water losses can occur. Corn stalk residue is often plowed under in the fall to control European corn borer leaving the soil exposed during the winter and early spring.

Not all the activities of insects and mites are destructive. Most insects and mites are beneficial and some are essential. Honey bees and other insects transfer pollen from flower to flower, a process absolutely necessary for seed development of alfalfa, clover, and some fruit trees. Ladybug beetles and larvae, and preying mantis feed on aphids. The Klamath beetle feeds on the Klamath weed, a serious weed in West Coast ranges, and has significantly reduced this weed pest. Insects improve the physical condition of the soil, primarily through increased aeration by the tunneling of insects into the soil for egg laying and overwintering. Insects also increase the organic matter of the soil; first by feeding on crop residue that hastens decay, and secondly, the bodies and other remains add organic matter directly to the soil.

CLASSIFICATION OF INSECTS AND MITES

Within the binomial system of scientific classification, insects and mites fall within the animal kingdom and the phylum Arthropoda. Insects belong to the class *Insecta*. Some common insect orders are *Coleoptera* (beetles), *Diptera* (true flies), *Hymenoptera* (ants, bees, and wasps), *Orthoptera* (grasshoppers, crickets, and locusts), and *Lepidoptera* (butterflies and moths). Mites belong to the order *Acarina*.

Life Cycles

Most insects and mites reproduce sexually but differ in their development or **metamorphosis** from the young to the adult stage. The types of metamorphosis which insects and mites exhibit are a classification system useful in insect identification. Some insects and mites are **ametamorphic** (without metamorphosis). The insect or mite hatches from the egg as a very small replica of the full-grown adult. Ametamorphic insects and mites can also be **viviparous**, giving birth to live young with no egg stage; and **parthenogenic**, reproducing asexually. These two traits may help the pest to reproduce very rapidly resulting in increased infestation levels. Many aphids and mites that damage crops are viviparous and parthenogenic.

Grasshoppers, leafhoppers, and true bugs gradually change size and shape, a growth pattern called **gradual metamorphosis** (Figure 13-1). The young are called **nymphs** and proceed through several molting stages (instars) to reach the adult reproductive stage.

Some insects develop from egg to the adult stage through **incomplete metamorphosis**. Although the young (**naiads**) change size and shape gradually, they do not look like, or function like, the adults until the shedding of the final exoskeleton at which time there is a rapid and distinctive change.

Insects with **complete metamorphosis** go through four distinct stages; **egg, larvae, pupae**, and **adult** as shown in Figure 13-2. None of the juvenile stages look like the adult, and, like incomplete metamorphosis, there is a great change in size and shape when the adult emerges from the pupa stage. As mentioned previously, distinct changes in shape and purpose during the life cycle are a tremendous advantage for these insects that allow them to adapt to changing environmental conditions. Beetles, butterflies, flies, moths, and wasps undergo complete metamorphosis.

The larvae, naiad, and nymph stages are commonly the most destructive to crop plants, primarily because they are growing and developing rapidly during these stages that require an abundant food supply. However, in some instances, both the larvae and the adult can damage crops. For example, both the

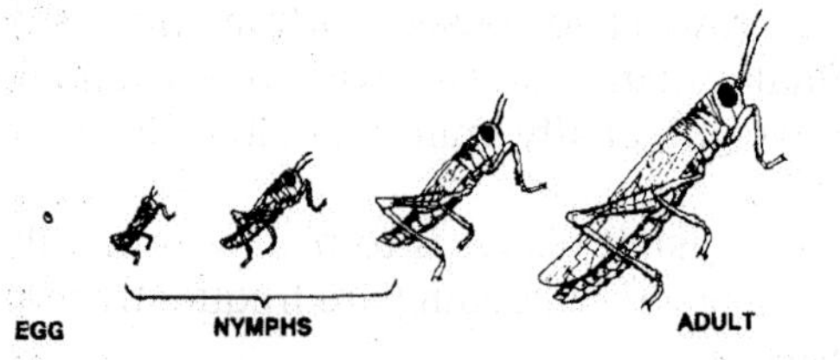

Figure 13-1 Gradual metamorphosis.

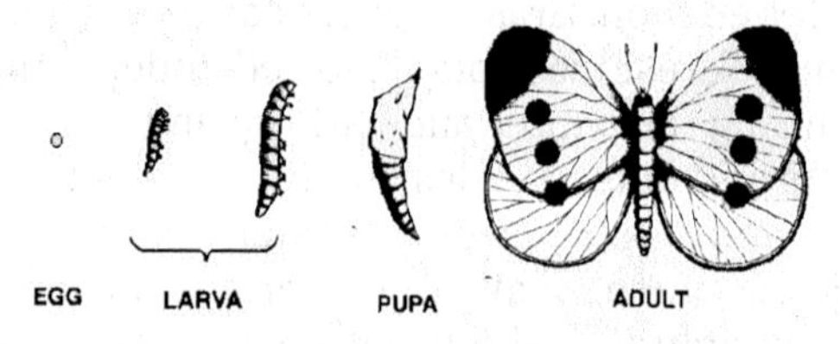

Figure 13-2 Complete metamorphosis.

larvae and adult of the alfalfa weevil can inflict economic damage; and the larvae of the corn rootworm feed on and damage the roots, while the adult beetles feed on the silks and can reduce fertilization.

Mouth Parts

As shown in Figure 13-3, insects and mites have two types of mouth parts that affect how they damage the plant. These differences can used as a basis for classification and are important in determining effective control measures. **Chewing insects** bite, tear, or lap out portions of the plant while **piercing-sucking insects and mites** pierce or rasp the plant and suck out or siphon up the sap of the plant. Aphids, mites, leafhoppers, and chinch bugs are examples of piercing-sucking pests. Sometimes the major damage by sucking insects and mites is caused by the injection of toxic substances into the leaf during feeding. Greenbugs and chinch bugs cause serious damage to sorghum and wheat in this manner. Chewing insects eat the leaves, leaf sheath, flowers, and, in some instances, the stem and roots of crop plants. Grasshoppers and armyworms may completely skeletonize corn plants leaving only the midribs and stalks. Some chewing insects, especially cutworms, feed on young seedlings near or below ground level. Still other chewing insects enter the stems, roots, ears, or heads of grain crops, feeding internally and are subclassified as **boring or tunneling insects**. Corn rootworm, European corn borer, corn earworm, and Hessian fly are examples of this type of chewing insect. Control of these insects is extremely difficult once they have moved into the plant.

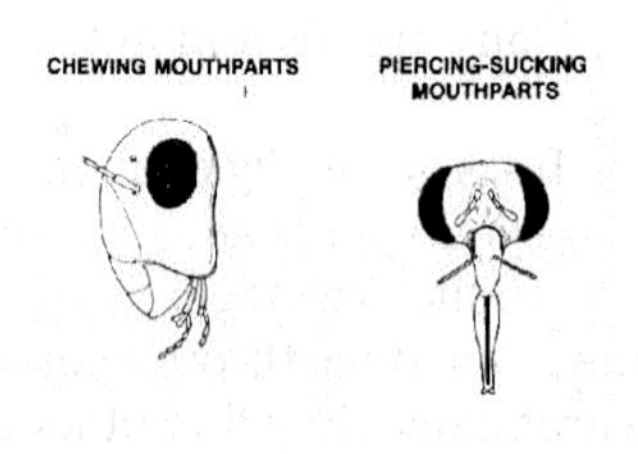

Figure 13-3 Insect mouth parts.

PEST CONTROL

Any effective and environmentally sound pest control program is based on three basic principles. First, the pest must be carefully and accurately identified. Mistaken identification can be costly economically and environmentally if the wrong control technique is used. Second, there must be an accurate estimate of the pest population and the potential crop damage that would occur at the present level or at a potentially higher level. Third, the most effective, economical, and environmentally safe control measure or measures must be decided.

Integrated Pest Management

The system that decides the need for control and type of control to use is called **Integrated Pest Management (IPM)**. Integrated pest management uses a complete system of pest control including the population dynamics of the pest, its biology and life cycle, and the effects of cultural practices and weather on reproduction and growth. It then employs the control methods that will control the pest, or reduce the population below economic damage levels, while minimizing detrimental effects on the environment. The principles of IPM are used with other types of "pests" such as diseases and weeds, for example, Integrated Weed Management. Considerable progress has been made in the use of nonchemical controls in IPM programs. However, it appears that the judicious use of pest control agrichemicals will be necessary in the immediate future to supply the quantity and quality of food needed for the world's growing population.

The pest level at which control measures are warranted is called the **action threshold (AT)**. The action threshold is also called the **control threshold** or **economic threshold**. The AT assumes that when the economic damage to the crop exceeds the cost of control, the use of some control method is justified. The AT for a high value crop such as potatoes or sugar beets may be lower than a lower value crop like hay or pasture. Conversely, the AT may be quite low for an inexpensive control measure but higher for a more costly method. Finally, the control program must correlate with the life cycle of the pest and the crop to decide when the crop is most likely to suffer damage and when the pest is most vulnerable to the control measure.

The tactics for pest control using IPM fall into the general categories of genetic control, cultural control, biological control, and chemical control.

Genetic Control

Plant breeders are constantly working to develop varieties and hybrids that are resistant to, or tolerant of, pest feeding. The biggest advantage of genetic control is that the development of resistant varieties and hybrids is a more stable method since genetic resistance lasts for several years. Most other types of control need to be repeated each year. Genetic control of insects and mites can be divided into three groups: nonpreference, antibiosis, and tolerance.

With **nonpreference control**, plant breeders are able to alter the plant's biochemistry or constituents so that a particular variety or hybrid is less palatable to the pest. The taste, aroma, color, or texture of the crop plant is undesirable to the pests and they move to a different variety or a different host plant such as another crop or a weed.

With **antibiosis**, the crop plant components have a harmful effect on the growth or reproduction of the pest when it feeds on it. One type of sorghum resistance to greenbug aphid and hard red winter wheat resistance to Hessian fly are examples of antibiosis.

With **tolerance**, the host plant does not suffer economic damage even though it may be heavily infested with the pest. Inbred lines of corn have been developed which have the capacity for prolific root regeneration after feeding by corn rootworms. Another type of resistance to sorghum greenbug feeding is a plant tolerance to the toxin injected by the greenbug.

As a result of genetic engineering there are additional opportunities to develop control methods. Recently released hybrids of *Bt* corn and cotton contain genes from *Bacillus thuringiensis* that make the plants toxic to certain larval pests (antibiosis). There will undoubtedly be more in the future.

Cultural Control

Monocultures of crops greatly increase the incidence of pest infestations as discussed in Chapter 3. **Cultural control** involves any change in the cropping sequence, tillage, cultivation, weed control, fertilization, or water management that will control or reduce the level of infestation or damage by insects and mites. Crop rotations were undoubtedly first initiated to help control all types of pests, weeds, and diseases. The changing of crops in a sequence tends to decrease the population levels, usually in the soil or the crop residue, of specific pests. Conversely, continuous cropping increases the levels of insects and mites that attack that crop. For example, corn rootworm beetles lay their eggs primarily in corn fields and not in soybean, sorghum, or small grain stubble fields. Corn that follows a different crop is not economically affected by corn rootworm.

Changing the seeding date, either earlier or later, is a cultural practice used to avoid pest infestation or reduce the damage level. Planting winter wheat after the fly-free date, the average date when the female adult has completed egg laying, is a traditional method to avoid Hessian fly damage. Early planting of corn allows the plant to develop a larger root system before the mid to late June feeding of the corn rootworm.

Control measures such as plowing, disking, or even burning crop residue heavily infested with insects and mites have been used sparingly to reduce infestation levels. The effects of these measures on possible water runoff and erosion damage must be carefully considered before they are used. Weeds along roadsides and in waste areas can be the breeding areas and the alternate hosts for insects and mites that attack crop plants. Grasshoppers commonly move into the field from weedy and grassy roadsides and waterways. Weeds in a field, particularly grassy weeds, attract egg laying adults of the armyworm. Severe infestation can develop from a small weedy area in a field. Control of weeds can reduce the overwintering and early feeding areas of many insects and mites.

Biological Control

The use of other living organisms, such as insects and disease pathogens, and the use of natural hormones and pheromones to reduce pest populations are called **biological control**. The breeding and introduction of predaceous and parasitic insects and animals have been used to control certain insects and mites. The adult and larvae of the ladybug beetle can destroy many aphids in a day. The use of birds is a traditional method employed for centuries as a biological control method. Vedalia beetle controls cottony-cushion scale insect on citrus in California.

The introduction of pest-specific disease organisms has sometimes been effective in pest control. For example, there is a virus strain that attacks only the cotton boll weevil, and the bacteria *Bacillus thuringiensis* has been found to be effective against some *Lepidoptera* and *Coleoptera* larvae including the European corn borer.

Other biological control methods include the use of natural hormones and pheromones. Hormonal chemicals found within the pests themselves can be used to disrupt their life cycle. Pheromones, which are sex attractants, can be used to attract the males to traps or can be distributed widely so that the males are confused and unable to find and mate with the females.

Chemical Control

Chemical control is the use of insecticides, miticides, and nematicides, collectively called **pesticides**, to control insects, mites, and nematodes. About 145,000 metric tons of pesticides were applied to crops in the U.S. in 1999. About 75% of all pesticides used in the U.S. were applied to agricultural land. Chemical control should be used only when other methods fail or are ineffective because it is usually the most costly control measure and has the greatest potential effect on the environment.

Pesticides are manufactured and may be applied as liquids, powders, or granules. Liquids and powders can be sprayed on crops as true solutions, emulsions, or suspensions. Powders may be dusted on plants using special equipment. Granules impregnated with a pesticide can be applied to the foliage and retained in the leaf whorl, or incorporated into the soil. Some volatile pesticides are applied to the soil as fumigants. With fumigants, the soil surface must usually be compacted or covered with plastic film or a layer of water to keep the fumigant from escaping. Regardless of the physical form or the method of application, a pesticide is applied to kill the pest directly on contact, or to coat the foliage so the pest is killed when the foliage is eaten.

Pesticides are classified by their mode of action; either as contact, stomach, or systemic poisons. **Contact poisons** are used to control piercing-sucking insects and mites since they do not ingest any external plant tissue. Contact poisons kill by direct body penetration or by entering the breathing or sensory pores. Contact poisons, commonly impregnated in granules, are also used to control boring or tunneling insects. Although these are chewing insects, usually in the larval stage, the insecticide must kill the insect before it enters the stalk, ear, or root. **Stomach poisons** are used to control chewing insects. The insect is killed when the insecticide reaches the stomach with ingested plant material. **Systemic poisons** are absorbed into the plant and then ingested by the pest during feeding. Systemic poisons can control any type of pest that it can kill. Care should be taken when using a systemic to insure that the chemical does not accumulate in plant parts that will be harvested for food or feed.

Pesticides can also be categorized into three broad groups based on chemical form. The first group is the **chlorinated hydrocarbons** of which DDT, chlordane, dieldrin, and lindane are examples. These are inexpensive and effective pesticides that can be safely applied using reasonable precautions. However, the chlorinated hydrocarbons are not readily degradable into nontoxic compounds. They have a long residual soil life and are transmitted from the soil to plants to livestock and into such foods as meat, milk, and eggs. Also, these kinds of pesticides tend to accumulate in fatty animal tissue, including humans. Finally, pest populations commonly develop a resistance to chlorinated hydrocarbons so that new formulations or higher application rates are necessary to achieve control. The combination of these factors has lead to the restricted use, or a complete ban, of many chlorinated hydrocarbon pesticides.

The **organic phosphates**, developed in Germany near the close of World War II, were found to have excellent pesticidal properties. They are effective in controlling a wide range of insects and mites, decompose more rapidly than the chlorinated hydrocarbons, and leave essentially no toxic residues. Ethyl and methyl parathion and malathion are examples of organic phosphate pesticides. Unfortunately, some of these compounds are extremely toxic to mammals and must be used with great caution. Sometimes, treated fields must be posted and cannot be entered for two to three days after treatment.

The **carbamate** pesticides is a recent group of compounds that are very effective, lower in mammal toxicity than the organic phosphates, while degrading more rapidly to less toxic forms than the chlorinated hydrocarbons. They are much safer to apply than the highly toxic phosphate pesticides. Examples of carbamate pesticides are dimetilan, metacil, and sevin.

Pesticides are highly effective in reducing crop losses from insects and mites. It is estimated that without pesticides, crop production would decrease about 30% and livestock production would decrease about 25%. Food prices would increase, food quality would decrease, and our overall standard of living would decrease. These benefits aside, pesticides are hazardous; and, if not used cautiously and judiciously, can cause severe problems to humans, animals, crops, and the total environment. Emphasis must be continually placed on alternative, less hazardous methods of pest control, and pesticides must be used only when other methods are ineffective.

REVIEW QUESTIONS

1. Why is it necessary to have a program like integrated pest management? What are its advantages?
2. How can insects and mites reduce the yield and quality of crops? How can they increase harvest losses? What other damaging effects can they have on crops?
3. What is a predaceous insect? Can you give an example?
4. How can the type of growth and development of an insect or mite affect its control? Is any type of metamorphosis any more advantageous than another? Why or why not?
5. Which growth stages are usually the most destructive to crops? Why?
6. How do the eating habits of a pest affect the type of pesticide that must be used to control it? Why?
7. How does each type of eating habit damage the plant?
8. What are the advantages and disadvantages of each type or method of pest control? Is one any better than another?
9. Can a crop producer rely only on one type or method of pest control? Why?
10. Why are insecticides and miticides more hazardous to humans than other pesticides such as fungicides and herbicides?

CHAPTER FOURTEEN

Crop Diseases and Control

INSTRUCTIONAL OBJECTIVES

Upon completion of this chapter, you should be able to:

1. Define disease, symptoms, pathogen, parasite, saprophyte, and obligate parasite.
2. List at least four types of crop losses from diseases. Give an example of each.
3. Define infectious disease and noninfectious disease. Give two examples of each.
4. List and describe four classes of pathogens that cause crop diseases. Name the type of pathogen that causes the most crop diseases.
5. List at least two unique characteristics of each type of pathogen. Compare them with each other.
6. Discuss the methods by which pathogens enter plants.
7. List at least two characteristic symptoms for each type of pathogen.
8. Discuss the conditions that are necessary for disease development. Define disease complex.
9. Define epidemic disease, endemic disease, monocyclic disease, and polycyclic disease.
10. List the three stages of disease development. Describe what happens at each stage.
11. Define primary inoculum and secondary inoculum.
12. List four types of disease control. Give an example of each.
13. Define disease resistance, disease escape, disease tolerance, and true disease resistance. Give an example of each.

INTRODUCTION

It is estimated that annual crop losses from diseases range from 10–20% of total production. The total costs from diseases, including costs of control, account for about 27% of the total losses and costs from all crop pests. Diseases reduce yields of wheat and soybeans about 14% annually, while corn and sorghum losses are estimated at about 11% and 9% respectively. Losses for cotton average about 15% each year and potato growers suffer losses from 20–25% annually.

A diseased plant is one that has become altered in its morphology, anatomy, or physiology to a degree that the effects are obvious. These effects or external changes in plant appearance that are identifiable with a specific disease are called **symptoms**. The infectious agent that invades and lives in the host plant and causes the development of symptoms is the **pathogen** or **causal organism**. Most plant pathogens are **parasites**; that is, they derive their nourishment from the plant that they have invaded.

Some are **obligate parasites** and can survive only on the living tissue of the host plants. Conversely, some are **nonobligate parasites** and can live as **saprophytes** on dead organic matter or inorganic materials when no host plants are available. Many pathogens have obligate and nonobligate phases in their life cycle that allow them to survive on dead tissue until a suitable host plant or environment develops which eases infection of living tissue again.

Types of Losses

Since plant pathogens are parasites, they rob the plant of water, nutrients and photosynthetic products for their own growth and development. They can cause chlorosis or lack of chlorophyll, and even necrosis or death of plant tissue. The pathogen may grow into and block the vascular system that disrupts translocation. All of these injuries interfere with normal photosynthesis and plant metabolism that subsequently reduces yield. Disease infection increases plant respiration rate and reduces water use efficiency resulting in an overall decrease in plant growth. The fruiting bodies of ergot and smut in cereal grains replace the kernels of grain causing a yield reduction.

In addition to yield losses, crop quality is also affected. Diseases cause premature ripening of the crop so that grain fill is never fully completed. The result is shrunken and shriveled grain of lower test weight. Grain containing smut balls or ergot is severely discounted in the market. Scabby potatoes are unacceptable for table use. In addition to the direct losses in yield and quality, there are indirect harvest losses since many diseases cause lodging, ear and fruit dropping, and leaf loss. Finally, some diseases are hazardous to humans and livestock. Some people are allergic to wheat rust, smut, and other types of spores. Disease contaminants in grain or forage crops reduce palatability and can cause sickness or death. The presence of ergot or the residue of small grains scab can cause sickness or even death when fed to livestock or used for human consumption.

CLASSIFICATION OF DISEASES

Diseases are commonly classified into two major categories, infectious and noninfectious. **Infectious diseases** are caused by some living organism that invades the host plant and disrupts normal growth enough that symptoms are visible or productivity is reduced. **Noninfectious diseases** are the result of conditions or circumstances that cause crop injury but which involves no pathogen. Examples of noninfectious diseases are weather damage, soil factors such as nutrient excesses or deficiencies, or chemical injury from pesticides. These are discussed later in this chapter.

Pathogens Causing Infectious Disease

Fungi

Fungi are responsible for the greatest number and diversity of crop losses. They are multicellular branched microorganisms that usually spread by spores. They may be parasitic or saprophytic and many are both. Fungi are generally very adaptable to the host and the environment and may live in the soil or crop residue, on an alternate weedy host, as well as in the host plant.

Reproduction in fungi is mostly sexual. The reproductive structures (spores) are very resistant to adverse environmental conditions and are capable of being carried great distances by wind. Once a spore lands on a susceptible host plant, the spore germinates and enters the plant by penetrating the epidermis. Since they commonly reproduce sexually, they can readily adapt to, and overcome, genetic resistance in crop varieties. Some examples of crop diseases caused by fungi are the rusts, smuts, stalk rots, and downy mildews.

Bacteria

Bacteria, the smallest living entities, are primitive unicellular organisms that reproduce primarily by division. Like fungi, they can be parasitic or saprophytic. Most bacteria do not form spores but do have the capacity to develop resistant protective coverings under adverse conditions. Bacteria are able to enter plants only through natural openings such as the stomata or through wounds in the plant tissue. Symptoms of bacterial diseases are soft rots, leaf spots, bacterial galls, extensive blights, or vascular diseases. Diseases caused by bacteria include bacterial blight of beans, Goss' bacterial wilt of corn, common scab of potatoes, and bacterial blight of alfalfa.

Virus

A **virus** is a submicroscopic pathogenic particle consisting of an inner nucleic acid core surrounded by an outer sheath of protein. An example is shown in Figure 14-1. Viruses do not carry on the metabolic processes of digestion and respiration and are not considered to be living organisms in themselves. Most are obligate parasites and cannot grow and multiply except within a living host, either a plant, insect, or fungus.

The most common evidence of a virus infection is an alteration of the leaf chlorophyll into a characteristic mosaic, mottled, or striped pattern. These symptoms have given rise to the use of terms like mosaic, yellows, mottle, or stripe with reference to virus diseases. In many cases, growth is malformed into leaf curling, edge crinkling, or excessive branching with diseases named after the symptom such as curly top of sugar beets and witches broom of potatoes. Virus diseases can also cause stunting and premature death; for example, maize dwarf mosaic and yellow dwarf of potato.

Nematode

Nematodes are microscopic, eel-like worms that can attach to the roots, stems, leaves, and even inflorescences of plants. They are roundworms and can live in soil or water; but are not closely related to earthworms or wireworms. The majority of nematodes that damage crop plants feed on, or damage, the roots. They cause root knots or galls, root lesions, stunted root systems, or injured root tips. The aboveground symptoms are lighter colored foliage, stunted growth, and wilting even when soil water is adequate. Examples of nematodes that infect crops are the soybean cyst nematode, the sugar beet cyst nematode, and the corn cyst nematode.

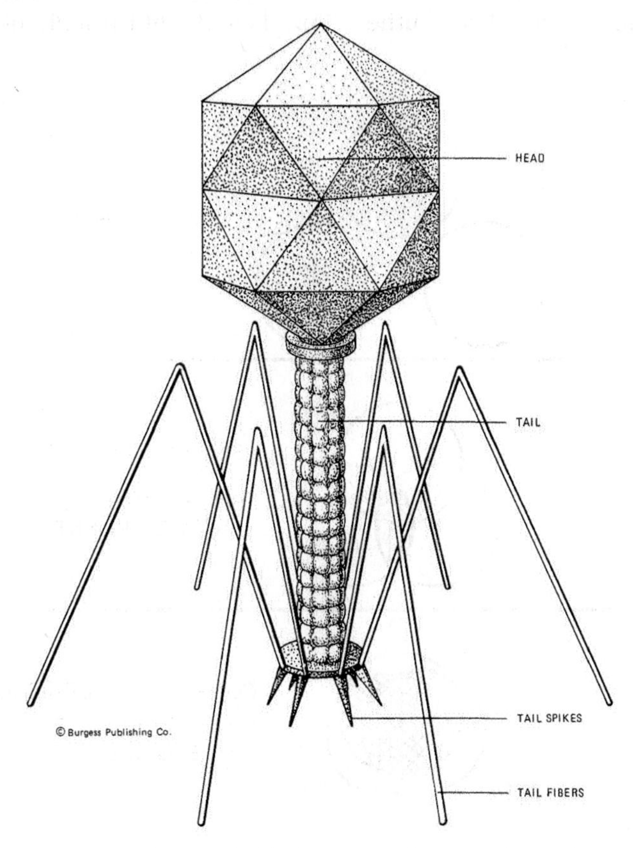

Figure 14-1 An example of a virus.

REQUIREMENTS FOR DISEASE OCCURRENCE

Certain conditions must be met before a disease will occur. First, there must be an organism capable of causing the disease; and secondly, there must be a susceptible host plant. Thirdly, the environment must be favorable for the development of the disease. This assumes that there is a suitable vector for the transmission and spread of the causal organism from susceptible host to susceptible host. If these conditions do not occur simultaneously, the disease probably will not occur. The interaction of these three conditions is known as the disease complex and the severity of the disease is frequently decided by the level of interaction as shown in Figure 14-2.

Causal agent, pathogen, or **incitant** are all terms used to designate the organism capable of infecting the host plant under the right environmental conditions. Of the multitude of nematodes, fungi, bacteria, and viruses that exist in nature, only a limited number are pathogenic to field crop plants. Most plants are immune to most pathogens. Some pathogens may infect plants but the level of infection and damage is below economic levels of concern. However, the majority of cultivated plants are subject to infection at economic levels by at least one pathogen, and some crop plants are susceptible to many.

The crop plant host must be susceptible to invasion, establishment, and development of the pathogen within the cells and tissue of the plant or the disease will not develop. Some plants have a mechanical resistance to invasion or a biochemical immunity that prevents disease establishment or development. A few crop species or varieties have disease tolerance; that is, the host plant has the ability to survive and produce acceptable yields even though disease infected.

The environment is often the determining factor in disease development. The fields may be planted to a susceptible crop variety and the pathogen may be present; but, without a favorable environment, the disease does not occur or fails to develop to economically destructive levels. The temperature, relative humidity, wind direction and velocity, soil pH, soil water, and soil fertility are all important environmental factors. Sometimes, poor cultural practices and/or inadequate weed and insect control will predispose the crop plant to disease invasion and infection.

When all of these factors are present and in balance, a plant disease will begin, develop, and reach epidemic levels. An **epidemic disease** is one that occurs frequently over a wide crop area but only periodically. The 1971 occurrence of Race T of Southern Corn Leaf Blight throughout the Corn Belt was an

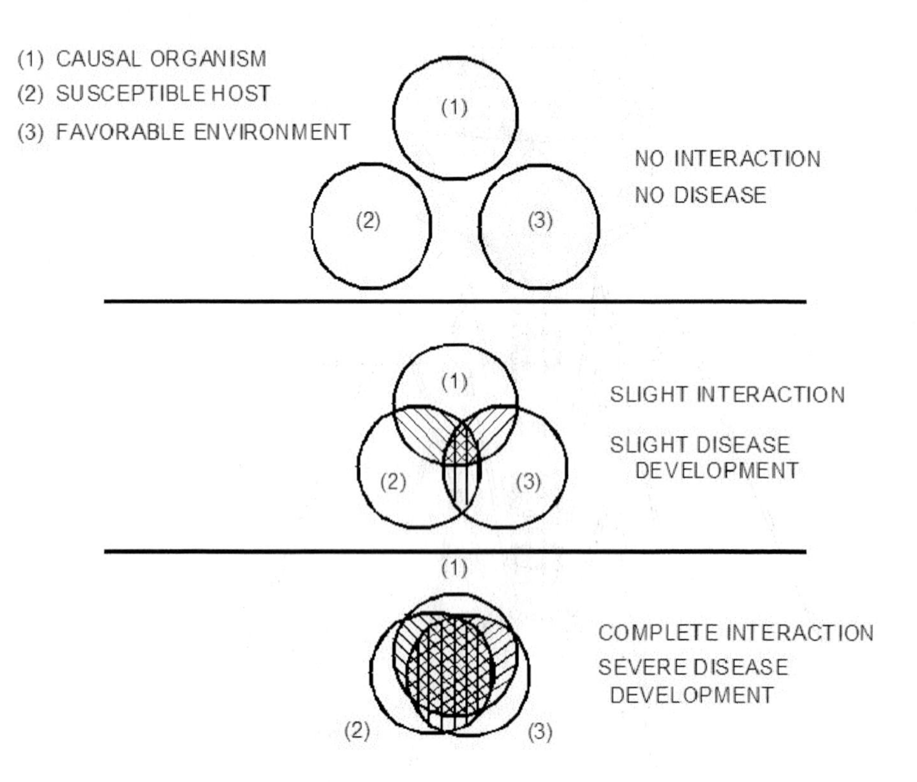

Figure 14-2 Diagram of interactions in a disease complex.

epidemic disease that caused a 14% nationwide yield loss. An **endemic disease** is usually present from year to year in a crop production area, causing moderate to severe damage every year. Various kinds of corn stalk and root rots are examples of an endemic disease.

STAGES OF DISEASE DEVELOPMENT

There are three pathogenic stages in the development of a disease that may occur only once or many times in the seasonal cycle of the disease. These stages are illustrated in Figure 14-3.

Inoculation

The first stage is **inoculation**. Some form of the pathogen, called the **inoculum,** is transferred to the plant and penetrates or enters it. Many pathogenic and nonpathogenic organisms may enter the plant through stomata or wounds and even penetrate cells but die without any further sign of disease development. Thus there are many cases of inoculation, the initial stage, without the progression to subsequent stages.

Primary inoculum is the first inoculum present and can either come from forms of the pathogen that overwintered in the area or is brought into the area by wind, insects, or another vector. After the disease becomes established in the first plants infected, secondary inoculum can be produced which further spreads the disease to nearby plants. There may be many cycles of secondary inoculum production during a disease epidemic.

Insects may inject the inoculum directly into the plant during feeding. Fungal pathogens living in the soil or on decaying soil organic matter may encounter root hairs and/or small branch roots and enter

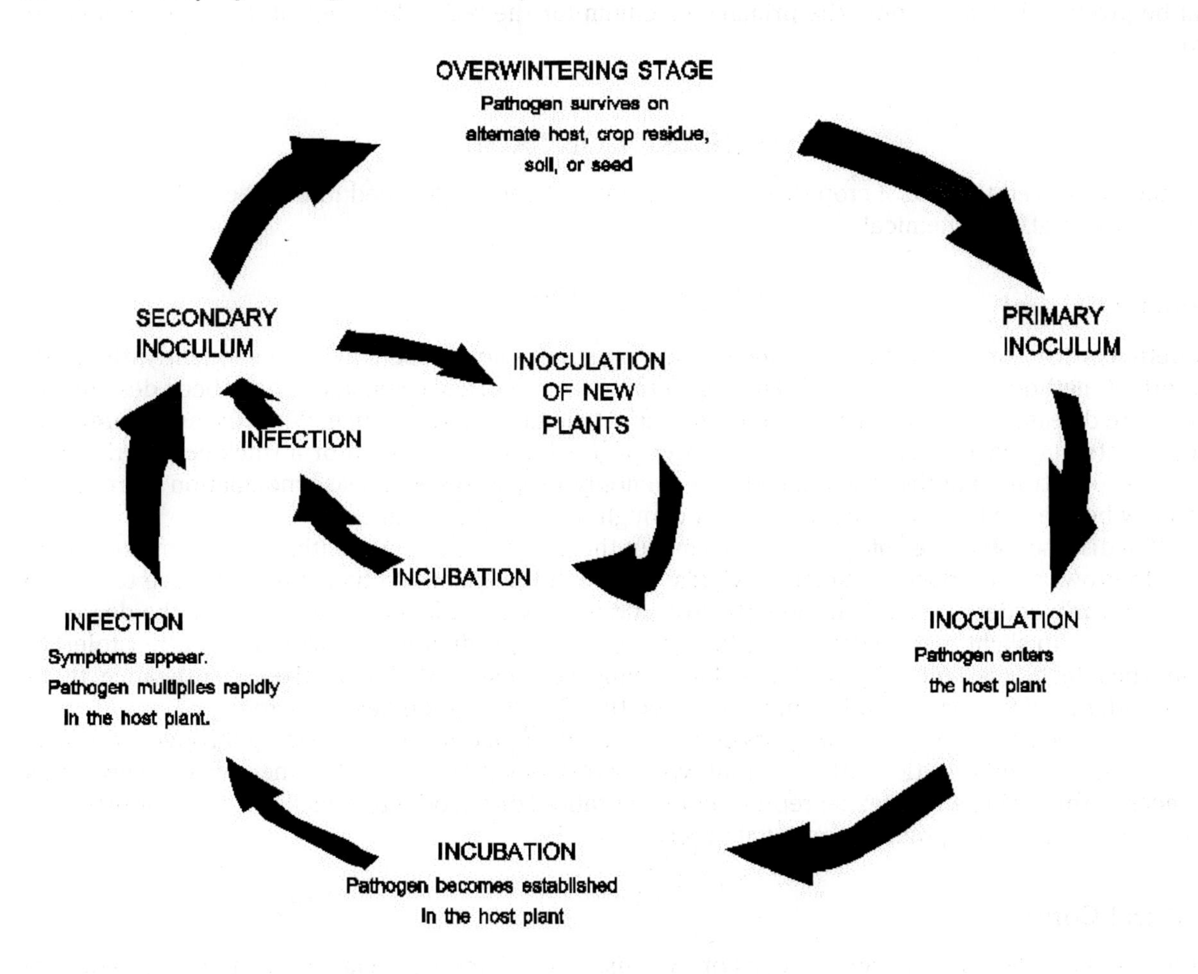

Figure 14-3 Diagram of a typical disease cycle.

root tissue. With many diseases the spores or other reproductive bodies may fall on the leaves or the axils of the leaves, germinate, and penetrate leaf or stem tissue. Whatever the method, the inoculum must enter the plant for subsequent disease stages to develop.

Incubation

During the second stage, **incubation**, the pathogen becomes established in the host plant. There is no visible change in the appearance or the apparent metabolism of the host plant. The incubation period begins immediately after the pathogen enters the plant and ends when the plant reacts to the disease and symptoms appear. This may be a relatively long period, delayed by unfavorable environmental conditions; but with most plant diseases it is quite short, usually two to three days.

Infection

The **infection** stage is the final stage in disease development. During this stage, disease symptoms appear and the producer becomes aware of the disease. The pathogen is multiplying rapidly within the plant. Tissues are ruptured, xylem and phloem vessels may become plugged, and chlorosis and/or necrosis may occur on leaves, stems, or fruits. Economic crop losses are the result of this disease stage.

An important factor in the economic impact of a disease is whether it is monocyclic or polycyclic. A **monocyclic** disease goes only through one cycle as illustrated in Figure 14-3. It produces little secondary inoculum. If the disease is **polycyclic**, secondary inoculum is produced during the infection stage of disease development and the disease may go through several cycles as shown in the inner path in Figure 14-3. Also, if the disease overwinters in the area, special forms or spores of the pathogen will be produced which supply the primary inoculum for the redevelopment of the disease the next growing season.

DISEASE CONTROL

The methods used to control crop diseases are many but can be grouped into three main categories: genetic, cultural, and chemical.

Genetic Control

Genetic control, or host plant resistance, is the most commonly used, and consequently, the most important method of disease control. Three general classes of disease resistance have been designated. These are disease avoidance, disease tolerance, and true disease resistance. With **disease avoidance,** crop plants may have morphological structures such as sunken stomata or a thick cuticle that discourages penetration of the inoculum. More commonly, crops escape disease inoculation by reaching maturity before the level of inoculum is great enough to infect a large area.

With **disease tolerance,** infection may occur, but the crop host is able to withstand the invasion, continue to grow, and produce satisfactory yields. Although the exact mechanisms of disease tolerance are not clearly understood, it is frequently crop and variety specific. In some instances, the degree of parasitism is so slight that no detectable loss of dry matter production occurs while in other tolerant crops the infection is limited by mechanical or chemical exclusion. With **true disease resistance,** there is no evidence of incubation and/or infection even though inoculation has occurred.

Plant breeders and plant pathologists cooperate to introduce, select, or develop through hybridization new strains and varieties with one or all types of resistance. However, the emergence of new forms or races of the pathogen by sexual reproduction, mutation, or introduction makes the development of resistant varieties and strains a continual process.

Cultural Control

With or without disease resistant varieties or hybrids, cultural practices can be effectively used to control some plant diseases. **Cultural disease control** involves any change or manipulation of field operations that alters either the life cycle of the pathogen or the host so that inoculation or infection does not occur. Crop rotations change the crop host and also reduce the level of the primary inoculum. The use of a crop rotation reduces the buildup of diseases that frequently develops in continuous cropping. A change in planting date, either earlier or later, often helps a crop escape or withstand disease infection. Later planting of winter wheat reduces the amount of fall growth, leaving plants less susceptible to crown and root rots. Conversely, earlier planting of spring oats allows the crop to mature ahead of infectious levels of leaf rust.

Proper crop management is very effective in reducing infection or the severity of the disease if infection does occur. Optimum plant density; timely planting, cultivation, and fertilization; and proper water management are all practices that promote vigorous crop growth. Healthy plants are less disposed to disease while stressed plants are more likely to suffer disease infection and damage. For example, the combination of too much nitrogen fertilizer and very high plant population increases the chance of stalk rot infection in corn.

Tillage and cultivation reduce the source of primary inoculum of some diseases by burying crop residues. Weed control of all kinds often eliminates the alternate disease host but always reduces the competition of weeds for light, water, and nutrients. Soil and water management can be effective tools in disease control. Drainage to correct cool, waterlogged soil conditions can reduce the incidence of seedling blights and promote vigorous early crop growth. Correction of soil acidity can improve nutrient availability and improve crop growth. Choice of the type of nitrogen fertilizer and the time of application have been shown to increase or decrease the incidence of stalk, crown, and root rots.

The use of disease free seeds or propagation materials helps to eliminate diseases transmitted in this manner. Certified seed, sod, or transplants are grown under disease free conditions and are inspected before sale to insure freedom from disease.

Chemical Control

Chemical control of crop diseases is not as common as chemical control of pests and weeds. Fungicides accounted for only about 11% of all chemical pesticides used in the U.S. in 1999. In most cases, chemical control must be applied before inoculation and remain on the plant or be reapplied as long as there is a threat of inoculation. Once the pathogen has entered the plant, chemical control is difficult if not impossible. Chemical control of plant diseases is strictly preventative. There are no treatments to reduce the symptoms once a plant has the disease.

Chemical disease control may be effective but is not always feasible on field crops. Leaf rusts of cereal grains can be controlled by fungicide application, but the costs are often greater than the economic gain. However, with such high value crops as vegetables, tobacco, or sugarbeet, chemical control can be economically effective.

Seed treatment to protect seedlings from **damping-off** diseases is a commonly used method of chemical control. It is also effective in the control of such seed-borne diseases as covered smut of cereals. Seed treatment, however, is ineffective in the control of diseases that are transmitted internally through the seed unless a systemic fungicide is used which is taken up by the seed. Systemic seed treatment should only be used if necessary as it is more expensive and there are limitations on the use of the crop for grazing and haying.

NONINFECTIOUS DISEASES

Initially, we described a disease as any abnormality in the plant structure or function. Consequently, any detrimental effects of the environment can produce symptoms and permanent damage that could be designated as a disease. However, in these cases there is no living organism, no virulent pathogen, and no parasitic agent that caused internal plant damage. Noninfectious, nonparasitic diseases are the result of such external forces as weather, soil condition, or misapplication of chemicals.

Weather Damage

Crop injury from low and high temperatures is common every season. Late frosts in the spring and early frosts in the fall can result in necrosis of all or some of the vegetative tissue. Extremely low winter temperatures can cause freezing injury or even winter killing to perennial and/or winter annuals. The damage is usually to the crown area and the upper root system. High temperatures commonly occur with low relative humidity and high wind velocity. This combination causes soil moisture depletion and the usual drought symptoms of wilting, stunted growth, and desiccation and death of tissue. Succulent leaves of many plants suffer from sun scald. The damage usually occurs when a period of favorable moisture and temperature, usually with cool and cloudy weather, has promoted rapid vegetative growth. When this is followed by hot, very sunny days, sun scald occurs. The injured tissue loses turgidity, followed by rapid desiccation, resulting in brown leaf spots which may bleach to light brown to white. Corn, potatoes, and the fruits of tomatoes most commonly exhibit sun scald.

Weather damage can also be caused by high winds, hail, or lightning. High winds and hail can damage leaves, and, in severe cases, completely defoliate plants. Lightning damage to field crops is rare and is often overlooked until two or three weeks after the incident. The damage is observed as a rough circular pattern with dead plants in the center and various stages of stunted plants around the periphery. The damage resembles a rapidly developing disease that is spreading outward from the point of initial infection.

Soil Problems

Soil related nonparasitic diseases include low and high pH, plant nutrient deficiencies, and sufficient high soil water level to develop anaerobic conditions. Acid soils can be directly toxic to some crop species but growth depression is the most common symptom. Increased solubility of iron, aluminum, and manganese at low soil pH often allows these elements to reach yield depressing levels. Conversely, high pH can also suppress crop growth and yields. An accumulation of soluble salts often occurs in association with high soil pH that reduces water and nutrient availability to the crop. The presence of free lime in calcareous soils can increase soil pH that can cause induced iron chlorosis in sorghum and other sensitive crops.

Chemical Damage

Over application of a pesticide, or the application of the wrong pesticide, can cause deformed growth or death of crop plants. Symptoms often strongly resemble the disease symptoms of fungus, bacteria, virus, or nematode infections. Herbicide carry-over is a common cause of chemical damage. Injurious gases, such as anhydrous ammonia, ozone, and industrial pollutants can reduce growth and cause bleaching, dehydration, or scorching of leaf tissue, and, sometimes, death of the crop. These types of injury, fortunately, are very rare.

REVIEW QUESTIONS

1. What are the differences between a parasite and a saprophyte? Which type of pathogen is rarely saprophytic?
2. What are some ways that a pathogen can affect the plant's metabolism? How do these affect crop yield and quality?
3. Why do fungi cause the greatest number of diseases? Why are they so adaptable to genetic plant resistance?
4. What are some ways that inoculum can be spread to plants?
5. What are some characteristic symptoms for each type of pathogen? How can you decide what type of pathogen is causing a disease just by the disease name?
6. What conditions are necessary for disease development? How does their interaction with each other decide the severity of the disease?
7. What is the difference between an epidemic and an endemic disease? Is one any easier for a crop producer to control through management? Why?
8. At what stage of development do symptoms appear? What can be done about the disease at this time? Why?

9. Why is genetic resistance the most common method of disease control? What are its advantages and disadvantages?
10. How do cultural cropping practices affect disease development? How do they affect the three requirements for disease development?
11. Why is chemical disease control usually not very effective?
12. Is it possible to breed resistance to noninfectious diseases? Can cultural or chemical control be used on noninfectious diseases? How?

CHAPTER FIFTEEN

Weeds and Weed Control

INSTRUCTIONAL OBJECTIVES

Upon completion of this chapter, you should be able to:

1. Define a weed. Discuss the importance of weed control to crop production.
2. List the growth habits of weeds that make them hard to control.
3. List three classification systems for weeds. List at least two examples for each one.
4. List six types of losses from weeds. Give at least one example of each.
5. List four general methods of weed control. Give at least two examples of each one. Discuss how each example disrupts a weed's life cycle.
6. Define herbicide. Define selective, nonselective, translocated, and nontranslocated herbicides, and soil sterilant. Give an example of each.
7. Define preplant, preemergent, and postemergent applications. Discuss the advantages and disadvantages of each application method.
8. Define residual and nonresidual herbicides. Discuss the advantages and disadvantages of each.

INTRODUCTION

A weed is any plant that is growing where it is not wanted. Most of the time we think of certain types of plants as weeds or undesirables, and other types of plants as crops or desirables. However, a corn plant growing in a soybean field is just as much a weed as any other unwanted plant and can cause just as much damage.

Losses from weeds and the cost of controlling them account for about 42% of the total costs and losses from all crop pests. Yield losses are greater in soybeans and grain sorghum because they are poor competitors with weeds early in the growing season. Losses are less with corn because better herbicides have been developed for that crop, and because corn is a good early competitor with weeds. Losses in wheat are lower because it is seeded in narrow rows which help it compete with weeds.

As was already discussed in Chapter 5, most tillage is done for the purpose of controlling weeds. The cost of chemical control of weeds is a major expense with most crops and is getting more expensive each year. Weed pests can sometimes dictate the types of crops that can be grown on a particular farm and may force the use of crop rotations or other special practices. These and other problems make weed control one of the most important factors in crop production.

Notice that in this chapter we only discuss controlling weeds; we never discuss eradicating them. While it is possible to control weeds so as to minimize economic losses from them, it is virtually impossible to completely eradicate them.

Many weeds have high rates of seed production. For example, rough pigweed can produce nearly 200,000 seeds per plant, lambsquarter almost 75,000, mustard over 500,000, and mullein almost 225,000. In contrast, corn produces only 500 to 600 seeds per plant. All annual weeds reproduce only by seeds. Many weeds can regenerate key plant parts. For instance, dandelion can regenerate the crown. Many weeds use vegetative reproduction such as bulbs, rhizomes, and stolons. These make control by tillage very difficult. Most weeds produce dormant seeds that can delay germination for many years, thus perpetuating the species for many years. One study showed that after almost 40 years, 91 percent of jimsonweed, 38 percent of velvetleaf, and 7 percent of lambsquarter seed still germinated.

Most weeds are very adaptable to our control measures, and, in some cases, have developed resistance to chemicals. However, it is possible to effectively control weeds if proper management practices are used in the production of a crop.

CLASSIFICATIONS OF WEEDS

The types of weeds present in any crop field or pasture can greatly influence the control practices that will be needed to control them. Knowledge of the life cycle, botany, and growth habit of various weeds is essential in formulating control measures that are both effective and economical.

Life Cycle

Like all plants, weeds are either summer annuals, winter annuals, biennials, or perennials. These classifications based on life cycle are discussed in detail in Chapter 2. Weeds tend to be a problem in crops that most closely fit their life cycle. Winter annual weeds will most likely be found in winter annual crops, and summer annual weeds will most likely be found in summer annual crops. Biennial and perennial weeds will be most troublesome in perennial crops such as forages and pastures. Some perennial and biennial weeds, however, can be serious in annual crops if the cultural practices for that crop fit their life cycle.

Biennial and perennial weeds are much harder to kill than annual weeds because they usually have extensive underground storage and reproductive organs such as fleshy taproots or rhizomes. Normal tillage operations will not kill these plants since they do not affect the underground parts and the weeds simply reproduce themselves. Repeated tillage operations can eventually kill these weeds by depleting their food reserves but such practices are not always practical with field crops. Chemical control of biennial and perennial weeds is possible but the timing of the application of the chemical is important to insure that it is translocated to the underground parts thus killing the entire plant. For example, early fall is usually the best time for chemical control of perennials and biennials.

Although annual plants are easier to kill than biennials or perennials, they are by no means easier to control. Annual weeds can produce thousands of seeds which can survive for years in the soil. These seeds are easily transported by wind, water, animals, and machinery. The difficulty in controlling annual weeds is shown by the fact that most common weeds that infest crop fields are annuals.

Broadleaves or Grasses

Another common classification system for weeds refers to whether they are broadleaves or grasses. **Broadleaf** is a common term used to indicate plants which belong to the dicot subclass. **Grasses**, of course, are those plants which belong to the grass family of the monocot subclass. In some cases, other monocots such as sedges are included in this classification, although they are not technically grasses.

There are some distinct differences between monocots and dicots in growth habit, as you already know. There are also distinct differences between these subclasses in their physiology which affect the toxicity of many herbicides. Many herbicides are selectively toxic to either broadleaves or grasses but have little effect on the other type of plants. For example, 2,4-D is very toxic to all broadleaves but has little effect on grasses. These physiological differences form the basis for herbicide selectivity which will be discussed in more detail later in this chapter.

Common or Noxious

Many states have laws which govern the control and movement of plants and seeds of serious weed pests. Such weeds are designated as **noxious weeds.** In many states, landowners can be fined for allowing noxious weeds to grow on their land; and crop seed, which contains noxious weed seeds, cannot be sold for seed. In some states, noxious weeds are further classified as primary or prohibited noxious weeds, or as secondary or restricted noxious weeds. **Prohibited noxious weeds** are those which are extremely difficult to control. The sale of crop seed which contains the seed from any of these weeds is unlawful. **Restricted noxious weeds** are those which are hard to control, but not as difficult as prohibited weeds. Sale of crop seeds with restricted weed seeds is permitted, but strict tolerances are usually designated. **Common weeds** are all other weeds. There are no restrictions to the sale of crop seeds with common weed seeds, but every effort should be made to remove them. Weed control is difficult enough without introducing more seeds into the field!

While the designation of a weed as common or noxious can be somewhat arbitrary and does vary from state to state, the concept of noxious weeds is valuable because it calls attention to weeds that are especially costly or troublesome to control; and, in some states, funds are made available to aid the farmer in the control of these weeds.

LOSSES FROM WEEDS

Weeds compete for the use of the limited resources and space in and on our land. They compete directly and indirectly with our food supply, and can even affect our health by producing large quantities of pollen. We will now look at some of the ways that weeds cause losses for agriculture.

Lower Crop Quality

Since weeds compete directly with crop plants, severe weed infestations can cause crop plants to produce grain which is shriveled and low in test weight. Presence of weed seeds can cause grain to be unpalatable to livestock, and they may also cause the grain to be rejected for market. The presence of weeds in a mature crop field can cause harvest problems by increasing threshing losses and the amount of trash in the harvested grain. Green weeds at harvest can raise the moisture content of the grain if green, wet pieces of weed plants are harvested with the grain. This is a problem mainly with winter wheat and other crops harvested during the summer. Weeds in forages can reduce the palatability and feeding value of the crop.

Decreased Crop Yield

Weeds compete with the crop for space, nutrients, light, and water. The average yield losses from weeds for several crops were given earlier in this chapter and ranged from 10 to 17% yearly. Severe infestations of weeds can easily reduce crop yields by one-half to two-thirds, and can many times result in total crop loss. Weeds, by their very nature, are fierce competitors, and usually can gain a competitive advantage over crop plants if given the opportunity. Proper crop management, however, is designed to favor the crop at the expense of the weeds.

Harbor Insects and Diseases

Some weeds act as alternate hosts for certain diseases. For example, black stem rust of wheat will use European barberry, quackgrass, or wild oats as alternate hosts until the next crop of wheat is established. Curly top of sugarbeet can infest many weeds in wastelands near the sugarbeet fields. Control of these alternate hosts disrupts the life cycle of the disease and greatly aids in its control.

Weeds can also serve as alternate hosts to crop insects or they can provide protection for harmful insects. Onion thrips can live on ragweed and wild mustard when onion plants are not present. Weeds along roadsides and in waste areas provide shelter for grasshoppers during the day, and the grasshoppers move into the field during the night to feed on young, tender crop plants. This can be a problem with winter wheat in the fall.

Increased Irrigation Costs

Weeds increase the total water use in a crop field. This can be detrimental in both dryland and irrigated crops. In irrigated crops, weeds can be extremely costly because the costs of production are very high and the cost of pumping or purchasing water is steadily increasing. Weeds can also clog irrigation ditches and canals, making it difficult to move water to the crop and stealing valuable water before it reaches the crop. Also, weed seeds can be carried in the water to the crop fields

Weed control in irrigation ditches and canals is difficult. The ready supply of water favors weed growth. Mechanical weed control is almost impossible because of the muddy conditions that exist along the canal or ditch. Chemical control is possible but the selection of herbicides is limited because many herbicides can be carried in the irrigation water and cause considerable damage to the crop.

Livestock Injury

Certain weeds in fields or pastures used for hay or grazing can be very harmful to livestock. Some weeds have barbs, awns, or other structures which can cause physical injury to livestock. These plant parts can get stuck in the animal's coat, hooves, or face and cause physical discomfort which lowers productivity. Sometimes they can get lodged in the animal's mouth, throat, or digestive tract and can cause serious injury or even death.

There are many weeds which are poisonous to livestock. Some contain prussic acid, which, when eaten by ruminant animals, produces hydrocyanide which can cause death. Some weeds are high in nitrate. Ingestion of nitrates interferes with the blood's ability to transport oxygen. Nitrate poisoning can be deadly to young livestock, and can cause abortions in pregnant livestock.

Isolation and removal of poisonous or injurious weeds from hay or grazing land are necessary to prevent losses. However, caution must be exercised in the control method used. Treatment of poisonous weeds with herbicides can sometimes increase their palatability for a short time. Careful attention to label directions is necessary to avoid serious livestock losses. Also, injurious plants should be physically removed after they have been killed to prevent possible injury to livestock that may move through the area.

Decreased Land Values

The presence of severe infestations of weeds, especially noxious weeds, can decrease the value of land. Lower crop yields and increased costs of weed control will make the land less profitable for crop production. In addition, the presence of certain weeds can restrict the choice of crops that can be grown. These weeds could be heavy competitors with certain crops, or the herbicides necessary to control the weeds may be used only on certain crops.

TYPES OF WEED CONTROL

As you have learned in the discussion so far, there are many ways or methods to control weeds. Most examples have been with mechanical or chemical control. While these two methods are the most common, there are other alternatives which can be used alone, or in conjunction with other methods. In this section, we want to look in detail at the major ways to control weeds. As discussed in Chapter 13, **integrated weed management** provides a systematic program for determining the best management practices for weed control.

Mechanical Control

Mechanical weed control involves the physical destruction or removal of weeds. Mechanical weed control is mostly tillage but can also involve mowing, burning, or some other method.

Tillage

As already stated, most tillage is for weed control. Tillage is designed to bury weeds or weed seeds, or to cut the roots of weeds. Burial is very effective with small annual weeds. It is important to bury all the growing points in order to kill the plant. Burial is not as effective on biennial or perennial plants which have underground stems or roots that can produce new vegetative growth such as field bindweed, Canada thistle, or hemp dogbane. Repeated tillage of these plants is necessary to deplete the food reserves of the underground parts.

Subsurface tillage implements are designed to cut the root system of the weed causing it to die from desiccation. These tillage practices have the added advantage of not incorporating much of the surface residue in the field. However, weed control with these implements is more difficult and may not be successful if the soil moisture is high or rain occurs before the weeds die. When this happens, the weeds will regenerate their root systems and the only thing accomplished by tillage was to transplant them in the field with little or no injury.

The best method or implement for weed control depends on the type and size of weeds present, the soil moisture level, the amount of residue on the soil surface, and the erodibility of the soil. As discussed previously with tillage in Chapter 5, the best tillage method varies from field to field, and from year to year. Thorough knowledge of one's soils and weeds is necessary to maximize weed control.

Mowing

Mowing is another mechanical method of weed control. It can be very effective on tall annual weeds. Repeated mowing of perennial weeds can be effective in depleting the food reserves of the underground stems or roots. With tall annual weeds, mowing can sometimes kill the plants if they are nearing maturity, but it is important to mow before the plants produce seed. Mowing is mainly used in noncrop areas and is not practical with most field crops. Mowing of small areas in pastures is sometimes used to control annual or perennial weeds.

Burning

A third method of mechanical weed control is burning. Burning can kill small weeds, but may have little effect on older weeds or perennial weeds. In most cases, complete burning of cropland is not recommended because the removal of plant residues can increase soil erosion.

Burning rangeland which is mostly comprised of warm-season plants in the early spring can help to control undesirable cool-season plants and woody perennials. If the cool-season plants and perennials have already begun growth they will be harmed by the fire; while the warm-season plants, which are still dormant will hot be harmed. An example of a region where burning of rangeland has been shown to be beneficial is the Flint Hills of Kansas and Oklahoma. Rangeland which has a high erosion hazard, such as the Sand Hills of Nebraska, should never be burned.

Cultural Control

Cultural weed control involves the use of crop management practices which give the crop a competitive edge over the weeds, or which disrupt the weed's life cycle.

Crop Rotation

Crop rotation can be very effective in helping to control weeds. As mentioned earlier, weeds tend to infest those crops which most closely match their life cycle. By seeding different crops in the same field in different years, the timing of tillage operations, seeding and harvest dates, type of herbicide used, and other practices will vary from year to year. This variation in cultural practices disrupts the life cycle of most weeds, thus aiding in their control. Research has shown that a crop rotation that involves several crops including a perennial, together with proper tillage, can effectively control all but a few weeds. In some cases, it is necessary to use crop rotation to control weeds that are especially competitive in a certain crop. For instance, chemical control of shattercane in corn or sorghum can be very difficult because of the lack of effective herbicides that can be used in those crops. Rotating to soybeans or some other crop, however, enables the producer to use herbicides that are very effective on shattercane.

Seeding Date and Rate

Seeding the crop at the optimum date and rate or population will increase its chances at gaining a competitive advantage over weeds that will occur in the field. A crop that is seeded as soon in the spring as the soil temperature reaches the minimum for germination will have a good chance of getting a jump on the weeds. This is assuming, of course, that good quality seed was used that will germinate into vigorous seedlings.

Every location in a field can support a certain amount and type of plant growth depending on the supply of water, nutrients, and other factors. Proper seeding rate will result in a population of crop plants that will fill most of the ecological niches in the field and leave little room for weeds, without providing too much competition of the crop plants with each other. Seeding rates that are too high will cause too much competition between crop plants, and weeds will easily gain the competitive advantage once the crop begins to fail.

Row Spacing

Matching the row spacing to the growth habit of the crop is valuable in controlling weeds because it maximizes shading of the soil surface and prevents weeds from getting established. The proper row spacing is determined by the leaf area produced by the plant and also by the width of spread of the leaves between the rows. For instance, erect leafed hybrids of corn should be seeded in narrower rows than hybrids which have more lax leaves because the spread of the erect leaves is less, resulting in more light reaching the soil at the wider row spacing. Soybean varieties which branch profusely can be planted in wider rows than other varieties with minimal branching.

Biological Control

Biological weed control uses insects, diseases, predators, or other plants that are harmful to weeds without causing damage to crops. The primary purpose of biological control is to put the weed at a competitive disadvantage with the other plants it is associated with, including crop plants. The progress and success of biological weed control has been mixed, but the potential is great enough to warrant continued research and effort in this area.

Biological control has been used to help control musk thistle in the Plains States, St. Johnswort in California, klamath weed in rangeland in the Pacific Northwest, and prickly pear in Australia. In these cases a predator was introduced into the region that occurred naturally in other parts of the world. This illustrates one of the basic concepts of biological control. In many cases, a weed becomes a serious pest because it was introduced into a region without its naturally occurring enemies that keep it under control.

Another area of biological weed control that is receiving attention is the use of other plants to control weeds using **allelopathy** which was discussed in Chapter 3. For example, sorghum has shown the ability to suppress broadleaf weeds. Compounds given off by sorghum roots inhibit weed seed germination and early growth. Genetic engineering may be able to enhance this process in sorghum and even transfer it to other grassy crops.

There are certain concepts that must be followed if biological control of weeds or any pest is to be successful. First, if an insect or disease is being introduced it must be host specific for the weed, and should not easily adapt to other plants when the weed population has been reduced. Failure to assure this could result in a bigger problem than the weed represented. For example, flowerhead weevil was introduced to control musk thistle in Nebraska but it now is threatening a native non-weedy prairie species, the Platte thistle.

A second concept is the ease by which the insect or disease can become established in the area. This has been a major stumbling block in establishing biological control in many areas. Sometimes the weed is better adapted to the region than the predator is, and repeated attempts at introducing the biological control are necessary before success, if any, is achieved.

A third concept of biological weed control is that it must not completely eliminate the weed; instead, it must simply control it below economic thresholds. If the insect or disease completely eliminated its host plants, it would be eliminated itself as well. Then, if the weed becomes established again in the area by seed or other means, there would be no biological control to keep it from once again becoming a serious pest.

Chemical Control

Chemical weed control involves the use of chemicals, called **herbicides**, for the control of weeds. Chemical weed control is a very common method of weed control in industrialized agriculture such as found in the United States. Herbicides are primarily used to eliminate or replace tillage operations for weed control, which partly compensates their cost. In addition, proper use of herbicides can provide better weed control than tillage. Because of its importance, chemical weed control is discussed in detail in the next section.

CHEMICAL WEED CONTROL

Herbicides account for almost 60% of all pesticides applied to crops in the U.S. Row crops account for about 75% of all herbicide use. Weeds are more likely to compete in wider rows and many acres of row crops are grown in continuous cropping systems where weeds are more likely to be a problem.

Herbicides are plant poisons, so their use in crop production can be disastrous if proper procedures and rates are not used. Careful study of the label that accompanies every herbicide is necessary to assure that it is being used with the least danger to the crop and the environment; and the maximum effectiveness for weed control. In order to safely use herbicides, the producer or applicator must be aware of the properties associated with it. It is these properties that we now want to study in detail.

Degree of Selectivity

A herbicide can be classified as selective, nonselective, or as a soil sterilant. A **selective herbicide** is one that will kill only certain types of plants. Most herbicides used on cropland are selective because they control weeds while doing little, if any, damage to the crop. For example, 2,4-D is a selective herbicide because it is toxic to broadleaves but has little effect on grasses. Atrazine will control many broadleaf and grassy weeds while doing little damage to corn. A selective herbicide is not toxic to some plants because it can be metabolized into a nontoxic substance.

A **nonselective herbicide** is one that will kill any plant it contacts. Nonselective herbicides can be used with growing crops but extreme caution must be used to prevent crop damage. It is necessary to use special applicators such as rope-wick applicators, or directed, recirculating sprayers to assure that the herbicide is applied to the weed but not the crop. For example, nonselective herbicides are sometimes used to kill volunteer corn plants in soybean fields. Since the corn plants will grow taller than the soybean plants, the herbicide can be carefully applied to the corn without touching the soybeans. Nonselective herbicides are also used in noncrop areas such as right-of-ways and around buildings where total control of vegetation is desired.

Some herbicides prevent any vegetation from growing for a period of months or even years. These are called **soil sterilants.** Soil sterilants are never used on cropland as you would guess. They are used in noncrop areas where complete control of vegetation is desired for a long period of time. The main difference between nonselective herbicides and soil sterilants is the length of time they are effective. Nonselective herbicides may have little or no residual activity after application.

Whether a herbicide is classified as selective, nonselective, or as a soil sterilant will many times depend on the rate that it is applied to the soil or plant. For example, atrazine is normally used as a selective herbicide, but increasing the rate slightly can turn it into a nonselective herbicide. If the rate is increased enough, it will become a soil sterilant that can last for two years or longer. This aptly illustrates the importance of applying a herbicide at the proper rate, especially with crops.

Site of Action

Translocated herbicides are taken up into the plant through the roots or leaves and moved to sensitive areas such as a growing point or storage organ. Most herbicides are translocated including those which are applied before germination of the crop and weed seeds. When applied preplant, a translocated herbicide may not provide good weed control if it is not moved into the soil where it can be taken up by the weed seedlings. Many times it is necessary to incorporate the herbicide to insure that it will be effective if rainfall does not occur to move it into the soil.

Translocated herbicides are important in controlling perennial weeds with underground storage organs as discussed earlier in this chapter, but timing of application is important. In most cases, the herbicide will simply be carried by the flow within the xylem or phloem until it reaches the sensitive areas of the plant, so it is important to apply the herbicide when the net flow within the plant is toward the sensitive organ such as a rhizome or fleshy taproot which cannot be destroyed by tillage.

Nontranslocated herbicides, also called **contact herbicides,** are not moved within the plant but simply kill the plant tissue it contacts. Most contact herbicides are nonselective. Thorough, even coverage of the herbicide is essential to get good weed control and can be a problem if the plant has pubescent (hairy) leaves, or leaves with a heavy waxy layer over their surface. In these cases it may be necessary to use a detergent or oil with the herbicide to assure that good coverage and penetration of the herbicide will occur.

Some contact herbicides are preplant applied to the soil before germination and emergence of weeds. When weed seedlings contact the treated soil, they are killed. This is an example of a selective contact herbicide.

Mode of Action

The method by which a herbicide kills depends on how the plant translocates and metabolizes the chemical. Some herbicides inhibit lipid or amino acid synthesis, chlorophyll formation, or photosynthetic reactions. Other herbicides act as growth regulators and interfere with normal metabolism. Others will disrupt cell membranes. Crop plants that are not affected by a particular herbicide are able to break down the chemical thus rendering it harmless. The different types of chemicals available as herbicides give the producer many choices to kill a wide variety of weeds in a wide variety of crops.

Time of Application

As mentioned before, the timing of the application of a herbicide can affect its ability to control certain weeds. However, the time that herbicides are applied can also be used as a method of classifying them, and is important in determining how they are used in the overall management of a crop.

A **preplant** herbicide is applied before the crop is seeded and is usually applied as part of seedbed preparation. An **early preplant (EPP)** herbicide is applied 10 to 30 days before seeding and may or may not be incorporated into the soil. A **preplant surface applied (PPSA)** herbicide is applied up to 10 days before seeding and is not incorporated into the soil. A **preplant and incorporated (PPI)** herbicide is also applied up to 10 days prior to seeding and immediately incorporated because it will evaporate or be broken down by sunlight.

A **preemergent (PRE)** herbicide is one that is applied before the weeds and/or the crop emerges but after the crop has been seeded. Most PRE herbicides are applied by the planter at the time the crop is seeded after the seed had been placed in the soil. The biggest advantage of a PRE application is that the herbicide can be banded over the seed row, reducing the amount of herbicide applied. A disadvantage of PRE application is that those herbicides that require immediate incorporation cannot be used.

A **postemergent (POST)** herbicide is one that is applied after the crop and weeds have emerged from the soil. POST applications are usually used to control weeds which occurred because a previous control measure failed to provide adequate weed control. Thorough, even coverage of the weeds is essential as discussed previously. The time of day that a POST herbicide is applied can also affect its ability to control weeds. Applications during the middle of the day when the weed is most active metabolically are more effective than those applied near sunrise or sunset. However, there may be more wind during mid-day, and the herbicide may drift and cause damage to nearby areas.

Length of Herbicidal Activity

How long a herbicide will remain active in the soil is important from the standpoint of weed control and for planning the cropping system for a particular field. Weed control is essential during the first 4-6 weeks after the crop has been seeded. Weeds that emerge after that time period do little or no damage because they cannot compete with the crop. Also, if a different crop is to be seeded in the field the next year, it is important that there be no carryover of herbicide from the previous crop. A crop producer has to always think ahead when planning a weed control program.

A **nonresidual** herbicide is one that remains toxic for only a very short time after application. It can be broken down by sunlight or by microorganisms in the soil. It can become attached to soil particles in a process called **adsorption**. It could also rapidly evaporate or leach from the soil. Nonresidual herbicides are mostly applied postemergent to control or correct a specific weed problem.

A **residual** herbicide is one that is more or less resistant to breakdown or other types of loss and will remain toxic for a few weeks to more than one year after application. As mentioned before, careful consideration for the succeeding crop must be exercised when using herbicides which have a long residual period. The length of time that a herbicide will remain active in the soil depends first of all on the type of herbicide. Some herbicides are inherently more residual than others. However, the amount of rainfall or irrigation water, soil pH, soil temperature, soil organic matter, and soil texture can also affect the length of residual activity of any particular herbicide. Less rainfall, higher soil pH, lower soil temperature, lower soil organic matter, and coarser soil texture can all increase herbicide carryover. When these conditions exist, or are expected to exist, it may be necessary to decrease the rate of application of the herbicide or avoid using it altogether to prevent damage to the next crop.

REVIEW QUESTIONS

1. Why are weeds almost impossible to eradicate? What is the difference between eradication and control?
2. Why do some weeds become a problem in some crops but not in others? How can a crop producer take advantage of this?
3. Why are weeds classified as broadleaf or grassy? How do these terms fit into the Botanical Classification System?
4. In what ways can weeds cause a decrease in crop yield and quality? What other types of losses are there, and how do weeds cause them?
5. How does tillage control weeds? What are the advantages and disadvantages of this method of weed control?
6. How do crop rotation, seeding date and rate, row spacing, and other cultural methods help to control weeds? What are the advantages and disadvantages of these methods?
7. What are the advantages and disadvantages of biological weed control?
8. What are the advantages and disadvantages of chemical weed control?
9. Why is a thorough knowledge of the properties of a herbicide important? What information should a producer have before applying a herbicide? How can a producer use this information?
10. Is any one method of weed control better than another? Can a producer use only one method? Why or why not?

Laboratory Exercises

Part I

Plant Structures and Organs

LABORATORY EXERCISE 1

Seeds and Germination

INSTRUCTIONAL OBJECTIVES

Upon completion of this exercise, you should be able to:

1. Define monocot and dicot.
2. List the names of the two plant families discussed in this exercise including at least three examples of each family.
3. Explain the difference between a fruit and a seed.
4. Define caryopsis.
5. Identify the parts of a corn caryopsis and its embryo.
6. Identify the parts of a bean seed and its embryo.
7. Write the definitions of seed dormancy and seed germination.
8. List the three requirements necessary for germination in all seeds.
9. List the steps involved in hypogeal emergence and the functions of the different parts of the germinating seedling.
10. List the steps involved in epigeal emergence and the functions of the different parts of the germinating seedling.
11. Identify the parts of a corn seedling and a bean seedling.

CLASSIFICATION OF FLOWERING PLANTS

Chapter 6, Seeds and Seeding, has information that you will use in completing this exercise. Read *Seed Structure and Function*, pages 64-65, that covers differences between grass and legume seeds.

The subclasses: *Monocotyledonae and Dicotyledonae*, **monocots** and **dicots,** refer to the number of **cotyledons**, or seed leaves, in the seed embryo. Monocots have one cotyledon and dicots have two cotyledons. There are many other differences besides number of cotyledons and you will learn these as you study the plants.

Monocots and dicots are further grouped into closely related plants called **families**. Agronomically, the most important monocots are in the **Grass** family (*Poaceae*), and the most important dicots are in the **Legume** family (*Fabaceae*).

Grasses include wheat, corn, sorghum, oats, barley, bromegrass, rice, sugarcane, tall fescue, bluestem, blue grama, and buffalograss. Examples of legumes are alfalfa, red clover, soybean, field bean, peanut, and pea.

Although this book is mostly concerned with these two families, keep in mind that there are many other families of both monocots and dicots that are important as sources of food, fiber, oil, and shelter, as well as competing against crops as weeds.

SEED STRUCTURE

All seeds contain an **embryo**, **seed coat**, and **food supply**. In grass seeds, the food supply is in the **endosperm**. In legumes, the food supply is found in the cotyledons. Since the endosperm is not part of the embryo, the embryo does not occupy all the space within the seed. In legumes, however, all the space within the seed is the embryo.

The Grass Seed

The grass "seed," is not a true seed. It is a special type of **fruit** called a **caryopsis**, which has its seed coat fused to the **pericarp** or mature ovary wall.

The corn kernel, because of its large size, is good for studying the various parts of the grass seed. Remember that all grass seeds, no matter how small, have basically all the parts that you will see in a corn kernel.

Obtain a soaked kernel of corn from the instructor. Locate the embryo or germ side of the kernel. It is probably recessed and a lighter color. Use a razor blade to cut the kernel through the center of the embryo from top to bottom. Ask for help, if needed.

Using Figure 6-1 on page 65 as a guide, locate the following parts of the corn seed and label Lab Figure 1-1.

- ☐ **Pericarp:** the seed coat and pericarp are fused (inseparable) which makes the fruit a caryopsis.
- ☐ **Endosperm:** carbohydrates and other energy-containing compounds that the embryo uses during germination.
- ☐ **Cotyledon:** embryonic seed leaf. Sometimes it is called the **scutellum.**
- ☐ **Coleoptile:** protective sheath covering that covers and protects the foliar or true leaves during emergence of the seedling through the soil.
- ☐ **Plumule:** contains the **embryonic foliar leaves** and the **mesocotyl**. It develops into the shoot that emerges above the ground. The mesocotyl or first internode elongates to push the coleoptile and foliar leaves to the soil surface.
- ☐ **Hypocotyl:** internode just below the cotyledonary node. It connects to the radicle.
- ☐ **Radicle:** develops into the primary root at germination.

Two other important structures found in the embryo are not labeled in Figure 6-1. However, you should include them in Lab Figure 1-1.

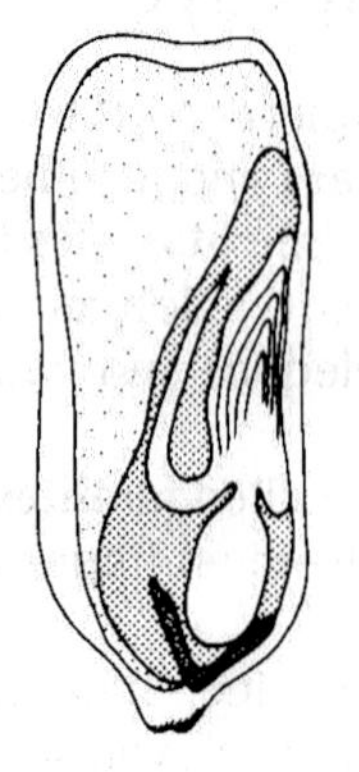

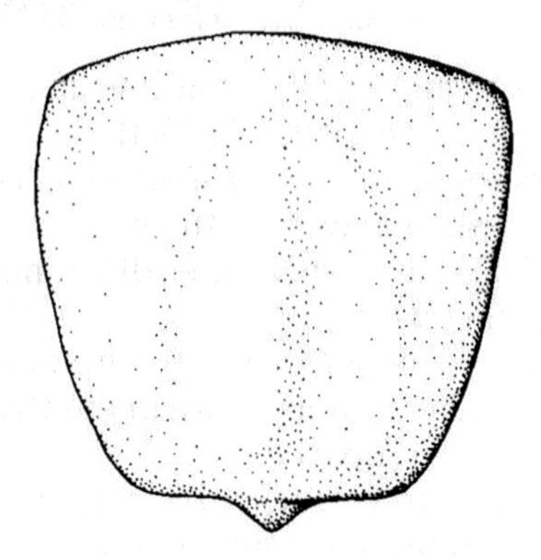

Lab Figure 1-1 Parts of the corn seed.

- □ **Coleorhiza:** protective sheath or covering which encloses the radicle. It is located at the base of the radicle.
- □ **Cotyledonary node:** point where the cotyledon attaches to the plumule and the hypocotyl.

The Bean Seed

Botanically, the bean is a true seed. The bean fruit includes the **pod**, which is the mature ovary wall, and the seeds, which each develop from an **ovule.**

Obtain a soaked bean seed from the instructor. Using Figure 6-2 on page 65 as a guide, locate the following parts of the bean seed and label Lab Figure 1-2.

- □ **Hilum:** the scar left on the seed where it attached to the ovary wall.
- □ **Micropyle:** a minute scar where the pollen tube entered the ovule before fertilization of the egg.
- □ **Raphe:** a slight ridge that runs along the edge of the seed.
- □ **Testa:** the seedcoat.

Carefully peel the seedcoat from the seed and gently separate the two halves of the embryo. All the structures you see are parts of the embryo as there is no endosperm in legume seeds. Food for the embryo is stored in the cotyledons. Locate the following parts of the bean seed and label Lab Figure 1-2.

- □ **Cotyledons:** the two halves of the embryo. They are embryonic seed leaves and contain the other food reserves needed during seed dormancy and germination.
- □ **Plumule:** contains the **embryonic foliar leaves** and the **epicotyl**. The embryonic foliar leaves develop into the "true" unifoliolate leaves seen after emergence. The epicotyl or first internode elongates and develops into the stem of the plant. The entire shoot of the plant develops from the plumule.
- □ **Hypocotyl:** elongates to pull the cotyledons and plumule out of the ground in epigeal emergence.
- □ **Radicle:** develops into the primary root at germination, and later develops into the main root or taproot of the plant.
- □ **Cotyledonary node:** where the two cotyledons attach to the rest of the embryo. It is located between the epicotyl and hypocotyl and is the first node of the stem.

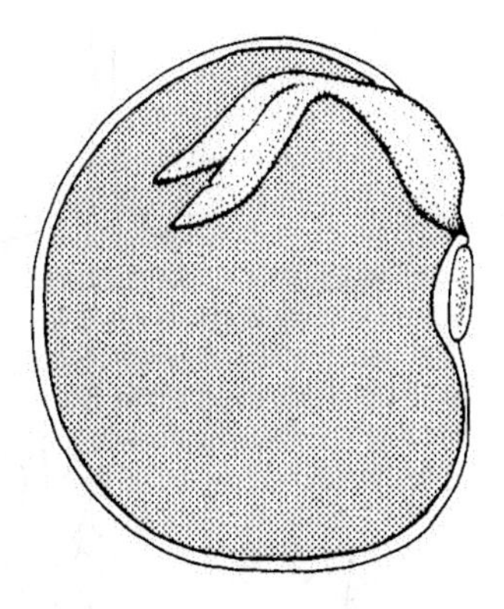

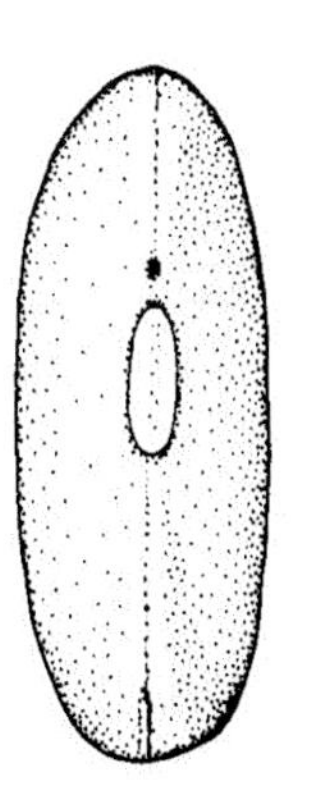

Lab Figure 1-2 Parts of the bean seed.

SEED GERMINATION

Read *Seed Viability and Germination*, pages 65-69, in Chapter 6. This section discusses the processes that seeds go through during germination and emergence.

Hypogeal Emergence

All grasses have **hypogeal emergence**, where the cotyledon remains in the soil. Also, some legumes, such as peas have hypogeal emergence, as well as other monocots and some dicots. With hypogeal emergence, the first node (cotyledonary node) is located in the soil where the seed was placed. Emergence occurs from elongation of the first one or two internodes at the base of the stem.

Lab Figure 1-3 shows the progressive steps in hypogeal emergence of corn. Obtain a corn seedling from the instructor. Using Figure 6-4 on page 67 as a guide, locate the parts of the embryo that have developed into the seedling and label Lab Figure 1-3. Understand the function of each part of the seedling and its role in germination and emergence.

- ☐ **Primary root:** develops from the radicle. This is the first structure to emerge from the seed. It supplies the seedling with moisture needed to continue emergence.
- ☐ **Mesocotyl:** elongates to push the coleoptile and the enclosed foliar leaves to the soil surface. The mesocotyl is the first internode and is sometimes called the epicotyl. In some species, such as wheat, it is mostly inactive. The hypocotyl is present but does not elongate and functions in hypogeal emergence.
- ☐ **Coleoptile:** protects the plumule until it reaches the soil surface. In monocots with an inactive mesocotyl, it elongates to allow emergence of the seedling.

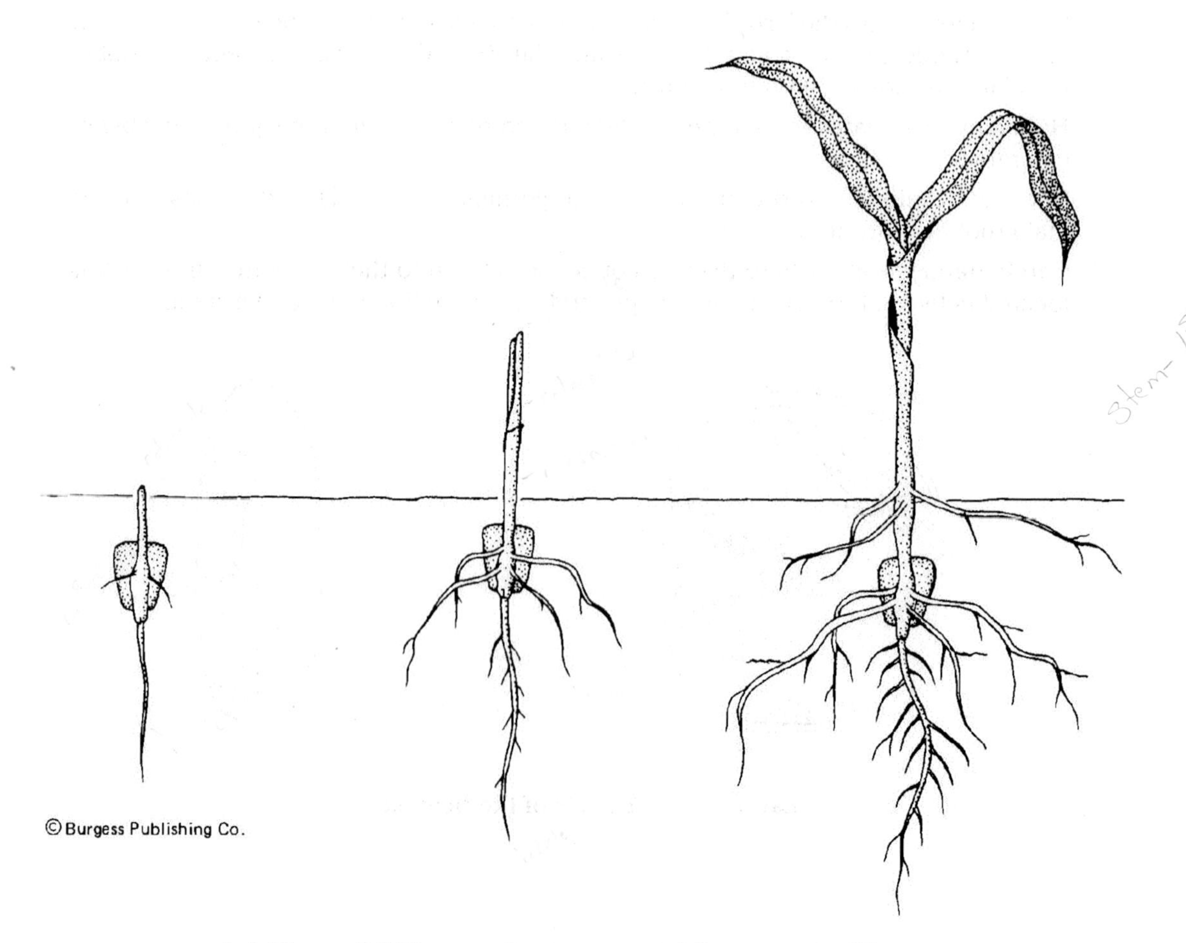

Lab Figure 1-3 Hypogeal emergence of the corn seedling.

- **Adventitious roots:** also called **secondary roots**. They develop from the second to sixth nodes of the plant and soon replace the primary roots as the major root system of the plant. Additional secondary roots sometimes emerge from nodes above the soil and are called **prop roots** or **brace roots**. Note that the depth of secondary roots depends on mesocotyl and coleoptile elongation.

There are other structures important in emergence that are not labeled in Figure 6-4 that should be labeled in Lab Figure 1-3.

- **Seminal roots:** emerge from the cotyledonary node. The seminal roots and the primary root make up the primary root system. Note that their depth in the soil depends on the planting depth as they emerge directly from seed that remains where it was placed in the soil.
- **Foliar leaves:** become the first "true" leaves of the plant.
- **Coleoptilar node:** located at the base of the coleoptile just above the mesocotyl. It develops into the main stem or crown of the plant and also develops the secondary or adventitious root system. It is the second node.

Epigeal Emergence

In epigeal emergence the cotyledons emerge from the soil instead of remaining in the soil as with hypogeal emergence. Lab Figure 1-4 shows the steps in epigeal emergence of the bean seedling. Obtain a bean seedling from the instructor. Using Figure 6-5 on page 68 as a guide, locate the parts of the embryo that have developed into the seedling and label Lab Figure 1-4. Understand the function of each part of the seedling and its role in germination and emergence.

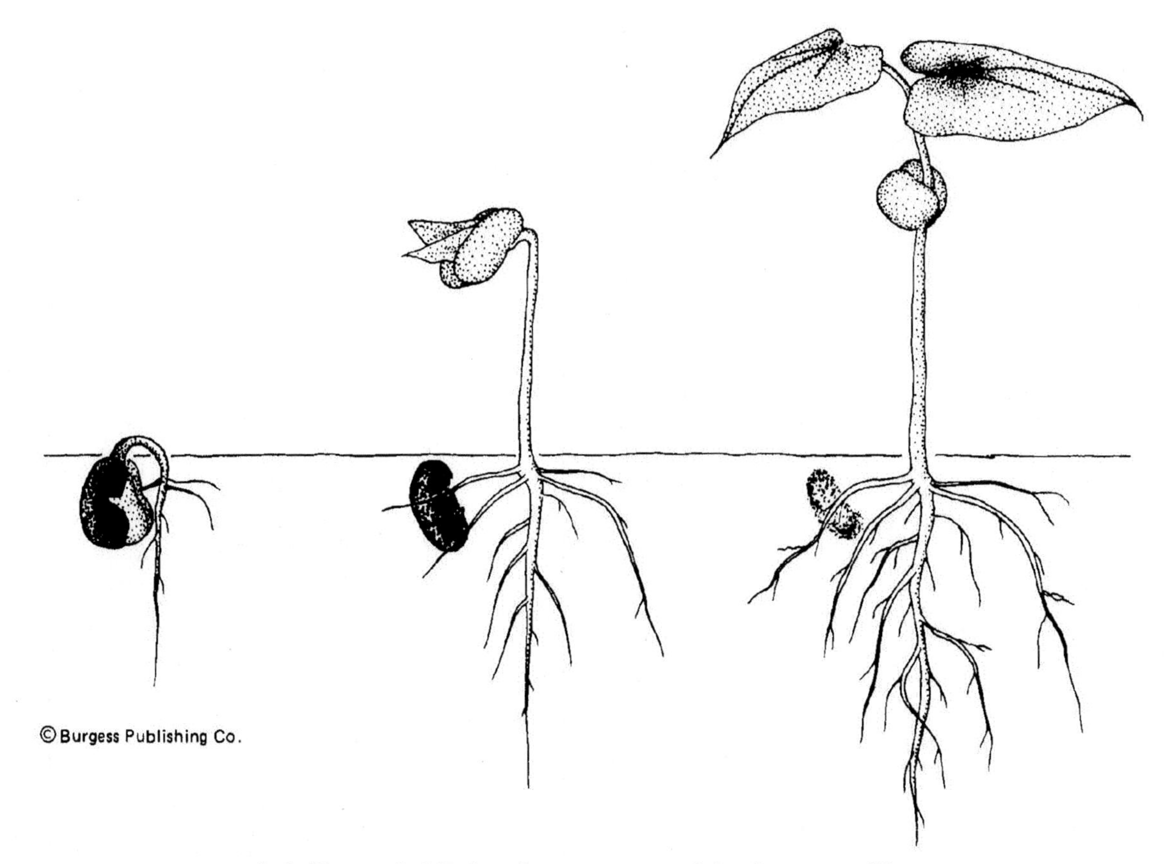

Lab Figure 1-4 Epigeal emergence of the bean seedling.

- **Primary root:** develops from the radicle. This is the first structure to emerge from the seed and it develops into the main root or taproot. It supplies the seedling with moisture needed to continue emergence.
- **Secondary roots:** branch from the primary root. The taproot and its branches develop into the root system for the plant.
- **Hypocotyl:** elongates and arches toward the soil surface to pull the cotyledons and the rest of the seedling above the soil surface. The arch of the hypocotyl is the first structure to emerge from the soil. The seedcoat is usually left in the soil.
- **Cotyledons:** after emergence from the soil, they develop chlorophyll and are useful in photosynthesis. After the foliar leaves develop, the cotyledons usually wither and drop off the plant.
- **Epicotyl:** elongates and supports the first foliar leaves. It develops into the main stem of the plant.
- **Foliar leaves:** become the first "true" leaves of the young plant.

STUDY GUIDE

Complete the following table to help you learn the material in this exercise.

	Grasses	Legumes
Number of cotyledons?	___	___
Botanical subclass?	___	___
Endosperm present?	___	___
Location of food supply?	___	___
Seed a fruit?	___	___
Type of fruit?	___	___
Seedcoat name?	___	___
Name of structure above cotyledonary node?	___	___
Name of structure below cotyledonary node?	___	___
Type of emergence?	___	___
First structure to emerge from the seed?	___	___
Seminal roots present?	___	___
Elongating structure that brings seedling up?	___	___
First structure to emerge from soil?	___	___
Location of cotyledons after emergence?	___	___

LABORATORY EXERCISE 2

Vegetative Characteristics of Grasses and Legumes

INSTRUCTIONAL OBJECTIVES

Upon completion of this exercise, you should be able to:

1. List at least three functions of roots.
2. Identify the four tissues present in a young root and the functions of each.
3. List the functions of the xylem and phloem.
4. Explain the difference between a root hair and a branch root.
5. Explain the differences between a monocot and a dicot root system.
6. List three functions of stems.
7. Define rhizome, stolon, and crown. List at least two other modified stems.
8. Define node and internode.
9. Explain the difference between a stem and a root.
10. List the tissues found in a monocot stem and the functions of each.
11. List the tissues found in a dicot stem and the functions of each.
12. Explain the differences between monocot and dicot stems.
13. Define a culm.
14. Identify the three parts of a grass leaf. List the function of each. Identify ligules and auricles.
15. Identify the three parts of a dicot leaf. List the functions of each.
16. Identify the different types of leaf venation and which type occurs with monocots or dicots.
17. Explain the differences between simple and compound leaves. Identify the structures associated with each. Identify the two types of compound leaves.
18. Define leaf shape, leaf margin, and leaf arrangement. Identify the three types of leaf arrangement.
19. Identify the three tissues found in the leaf blade. List the function of each.
20. List the characteristics and functions of leaf stomata.

Chapter 7, Crop Roots, and Chapter 8, Crop Stems and Leaves, have information that you will use in completing this exercise. You should begin this exercise by reading *Root Structure and Growth*, on page 78.

ROOTS

Roots provide mechanical support for the aboveground parts of the plant, absorb water and nutrients from the soil, and transport water and nutrients to the aboveground parts of the plant. Roots also store food in some plants. Food storage occurs mostly in perennial or biennial plants that have large fleshy roots, such as sugarbeet or carrots. The reserves keep the plant alive through the winter and provide the energy to resume growth in the spring. In a few species, roots can function in vegetative propagation to produce new plants. However, vegetative propagation by roots is not important in agronomic crops.

Root Tissues

In Exercise 1, you learned some differences in the seeds of monocots and dicots. There are also differences between monocots and dicots in root structure. Although the tissues are similar, their placement in the root is different.

Lab Figure 2-1 shows a cross section of a monocot root and Lab Figure 2-2 shows a cross section of a dicot root. Using Figures 7-2 and 7-3 on page 80 and 81, label the root tissues in Lab Figures 2-1 and 2-2.

- ☐ **Epidermis:** a single layer of cells on the root surface. Epidermal cells absorb water and nutrients from the soil. Some epidermal cells elongate outward from the root to form **root hairs**. These help absorption by increasing root surface area and contact with soil particles.
- ☐ **Cortex:** contains loosely packed cells that conduct water and nutrients to the vascular cylinder. Another function of the cortex is storage of food and nutrients.
- ☐ **Endodermis:** a single layer of cells that separates the vascular cylinder from the cortex. It functions in regulating the types of absorbed nutrients that are allowed to enter the vascular cylinder.
- ☐ **Vascular cylinder:** a group of tissues that connect the root to the rest of the plant and conduct materials to and from the root. It is also called the **stele**. The tissues found in the vascular cylinder are the:
- ☐ **Pericycle:** separates the xylem and phloem.
- ☐ **Xylem:** conducts water and nutrients from the root to the stem and leaves.
- ☐ **Phloem:** conducts sugar from photosynthesis and other compounds from the leaves to the root. These materials are used for growth and energy within the root.
- ☐ **Pith:** found in the center of a monocot.
- ☐ **Cambium:** cells can divide to produce secondary growth in perennial plants. They are found only in dicots and are not important in annual crop plants.

Comparing Monocot and Dicot Root Systems

Monocots have a **fibrous root system**. In grasses, the main root system develops from adventitious roots that originate from the coleoptilar node and above. The primary root and seminal roots that develop from the seedling are eventually crowded out. Dicots have a **taproot** system in which the primary root persists, produces branches, and continues as the main root of the plant.

Look at the examples of the monocot and dicot root systems provided and Figure 7-4 on page 82. Notice that a fibrous root system has many slender roots of about the same diameter, which originate at the base of the stem. Branch roots are produced on the main roots. The dicot has a main vertical root, called the **taproot**, which has branches originating from it.

Branch roots originate in the pericycle and contain all the root tissues listed above. In addition there may also be root hairs present. Don't confuse branch roots with root hairs. Root hairs are extensions of single epidermal cells on the surface of the root.

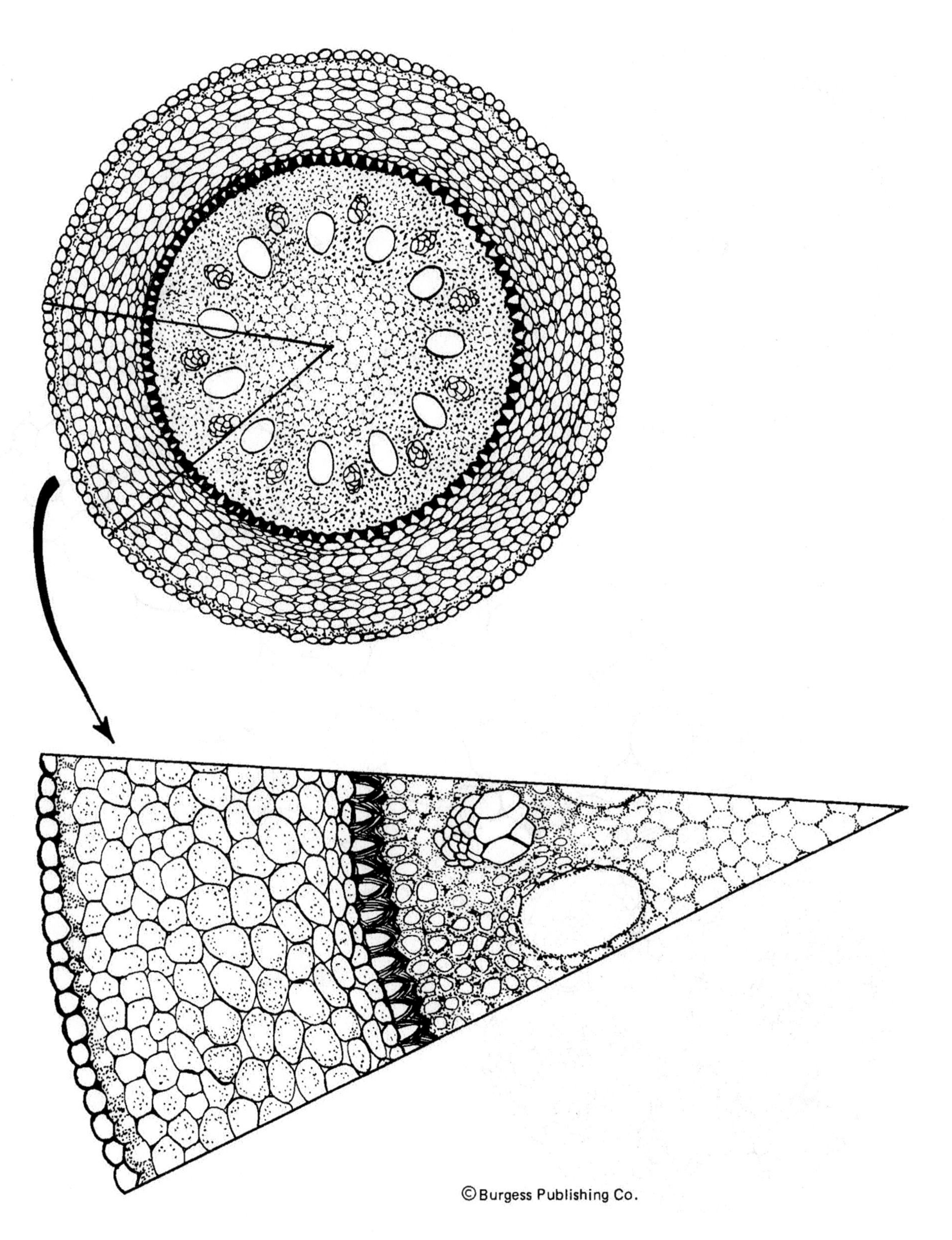

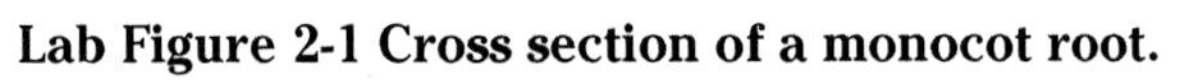

Lab Figure 2-1 Cross section of a monocot root.

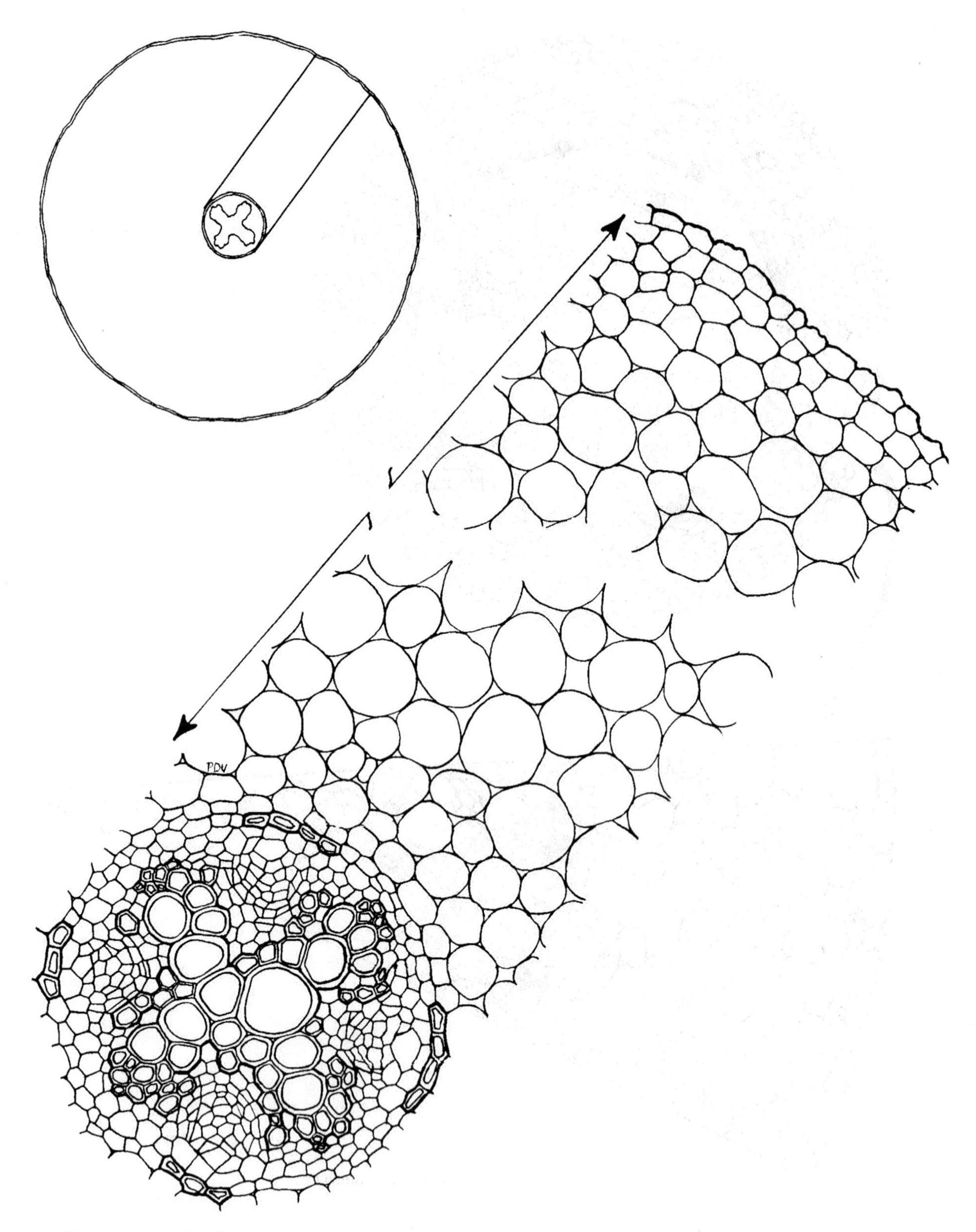

Lab Figure 2-2 Cross section of a dicot root.

STEMS

Chapter 8, Crop Stems and Leaves, has information that you will use to complete this part of the exercise. Read *Introduction, Crop Stems,* and *Stem Structure*, pages 92–99.

Stem Structure

A stem has **nodes** and internodes. Nodes are structures where one or more leaves, branches, and flowers are attached, and **internodes** are the regions between the nodes. Roots do not have nodes, which makes it easy to identify them from stems.

Monocot stems have vascular bundles that are scattered throughout the interior. Dicot stems have vascular bundles arranged around the perimeter. Some grasses, such as wheat, have hollow stems called **culms**. Solid grass stems are called **stalks**.

Lab Figure 2-3 shows a cross section of a monocot stem. Using Figure 8-1 on page 93, label the stem tissues in Lab Figure 2-3.

- □ **Epidermis:** a single layer of cells on the outside of the stem. It is covered with a waxy layer called a **cuticle**. The waxy substance is called **cutin**. The epidermis may also contain hairlike structures called **pubescence**.
- □ **Parenchyma:** large, thin-walled, and loosely arranged cells that occur between vascular bundles. They are the bulk of the stem.
- □ **Vascular bundles:** serve the same function described previously with roots. The bundles are scattered throughout the stem and contain the:
  - □ **Xylem:** located within the bundle toward the center of the stem. The xylem conducts water and nutrients from the root through the stem to the leaves, seeds, etc.
  - □ **Phloem:** located within the bundle toward the outside of the stem. The phloem conducts sugar from photosynthesis and other compounds from the leaves to the rest of the plant. Monocot stem phloem cells are also called **sieve tubes**.

Lab Figure 2-4 shows a cross section of a dicot stem. Using Figure 8-2 on page 94, label the stem tissues in Lab Figure 2-4.

- □ **Epidermis:** similar to that found in monocot stems.
- □ **Cortex:** located between the epidermis and the ring of vascular bundles.
- □ **Vascular bundles:** arranged in a ring near the outside of the stem. They are usually larger than those found in monocots and contain a cambium.
  - □ **Xylem:** located within the bundle toward the center of the stem.
  - □ **Phloem:** located within the bundle toward the outside of the stem.
  - □ **Cambium:** located between the xylem and phloem. The cambium contains meristematic tissue, which can cause secondary growth and subsequent increase in diameter of the stem in woody plants. It usually is not active in herbaceous stems.
- □ **Pith**—occupies the center of the stem and functions mainly in food storage.

LEAVES

Read *Crop Leaves* on pages 99–107. Leaves attach to the stem at a **node** in both monocots and dicots. Despite that similarity, there are many differences between monocot leaves and dicot leaves that are valuable for identifying plants.

The Grass Leaf

Lab Figure 2-5 shows the structures of a grass leaf. Using Figure 8-6 on page 100, label the parts in Lab Figure 2-5.

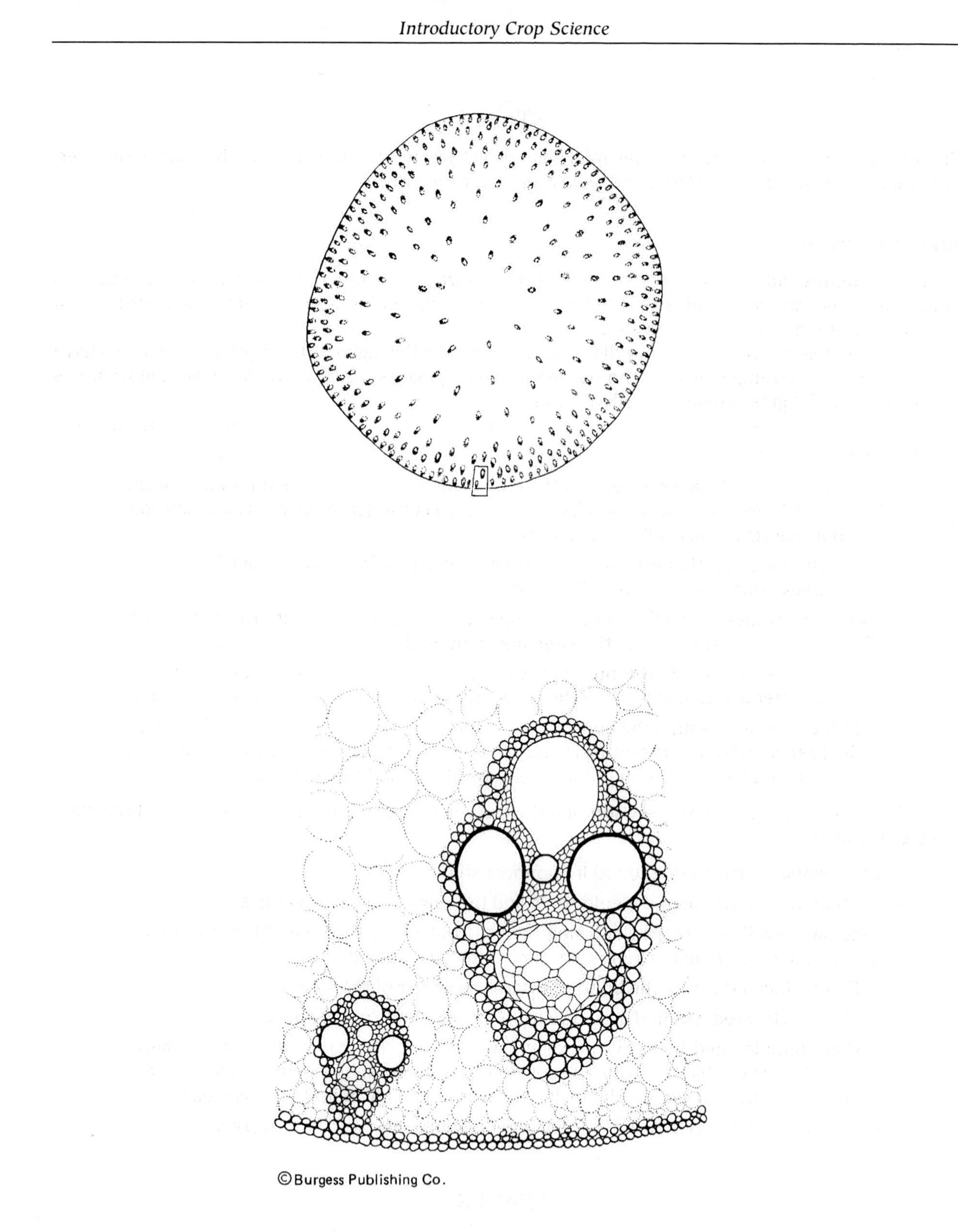

Lab Figure 2-3 Cross section of a monocot stem.

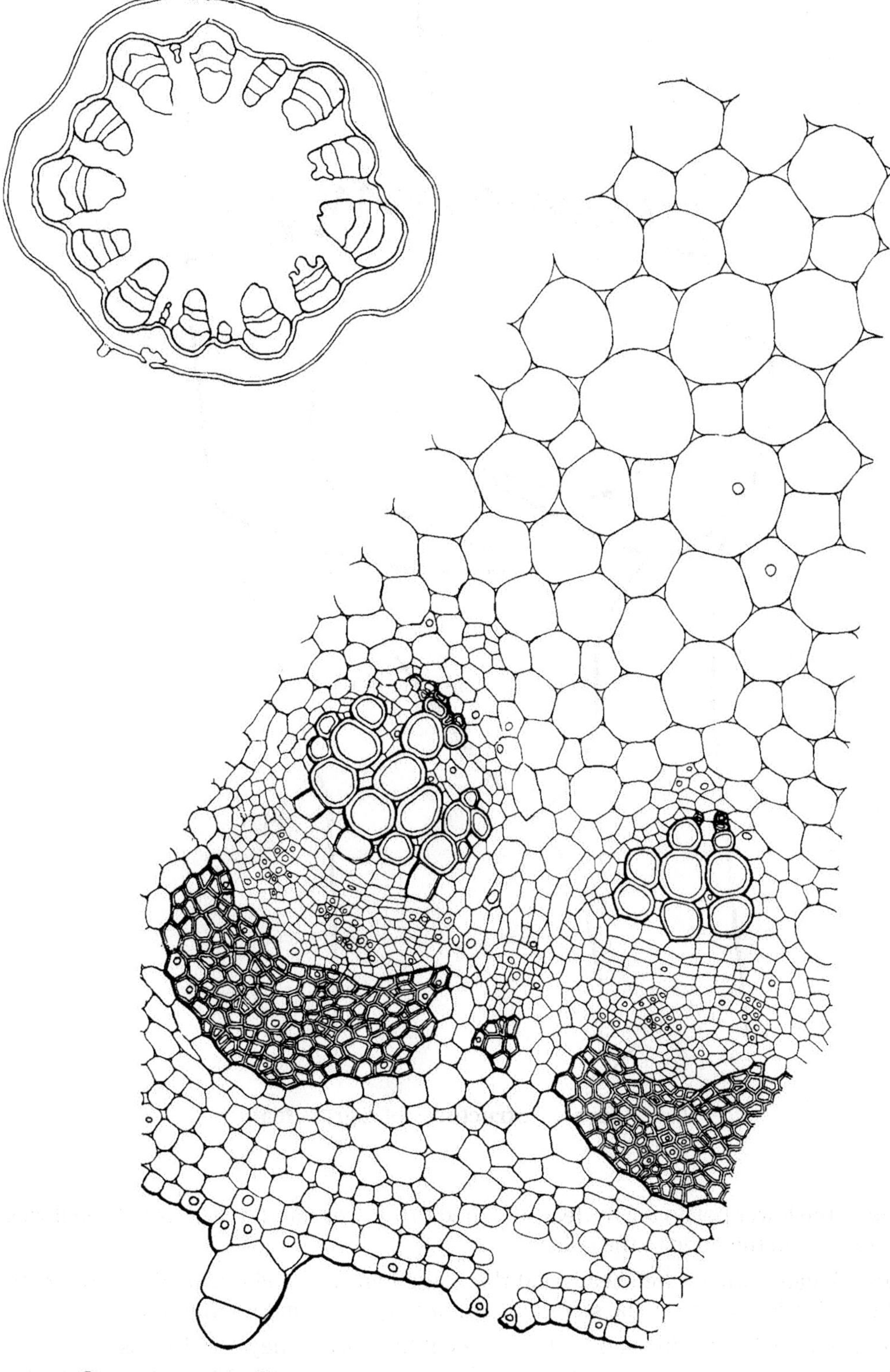

Lab Figure 2-4 Cross section of a dicot stem.

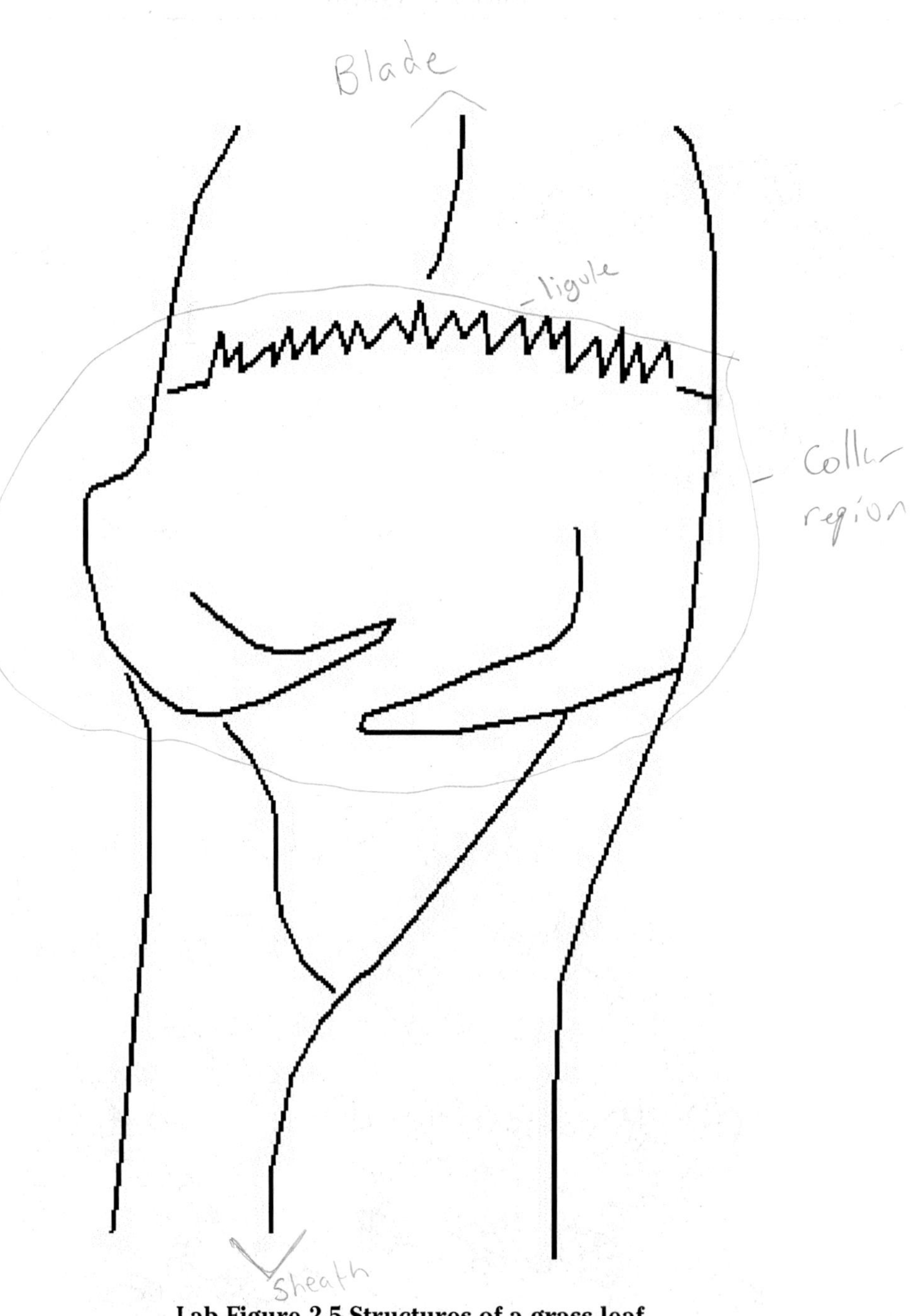

Lab Figure 2-5 Structures of a grass leaf.

- **Sheath:** the lower portion of the leaf, attached to the stem at a node. It usually encloses the stem from the node to the collar.
- **Collar:** located where the sheath and the blade join. The collar provides support to hold the blade away from the stem. The collar may also contain other structures.
- **Auricles:** clawlike appendages on the edges of the collar. They may be absent, short, long, rounded, pointed, smooth or hairy. Auricles make excellent identification tools.
- **Ligule:** located on the inside of the collar next to the stem. Ligules are of various sizes and shapes and can be long or short hairs, a membrane, or absent. The ligule is also a valuable identification tool.
- **Blade:** the flattened part of the leaf that extends outward from the collar and the stem. Blade characteristics can also aid in identification.

The Dicot Leaf

Lab Figure 2-6 shows the structures of a dicot leaf. Using Figure 8-7 on page 101, label the parts in Lab Figure 2-6.

- ☐ **Stipules:** small leaflike **bracts** or modified leaves located at the base of the petiole where it attaches to the stem, or on the side of the stem. Petioles vary in size and shape or may be absent. They are good identifying characteristics.
- ☐ **Petiole:** a stalk that connects the blade to the stem. If it is missing, the leaf is **sessile.** The angle formed by the petiole and stem is the **leaf axil.**
- ☐ **Blade:** the flattened part of the leaf where photosynthesis occurs.

Leaf Venation

A **leaf vein** contains the vascular tissues, xylem and phloem, which connect with the vascular tissues in the stem. Leaf venation forms the pattern of the leaf veins in the leaf blade. **Parallel venation** is a non-branching pattern of parallel veins that run the length of the leaf blade. All monocots have parallel venation. **Net venation** is a branch pattern where the veins form a network. All dicots have net venation. There are two types of net venation. **Pinnate venation** has a central vein, called a **midrib**, which runs the length of the blade with smaller veins branching from it. **Palmate venation** has many major veins that originate at the base of the blade with smaller veins branching from them. Figure 8-8 on page 102 shows several examples of venation. Leaf venation is another good way to identify plants.

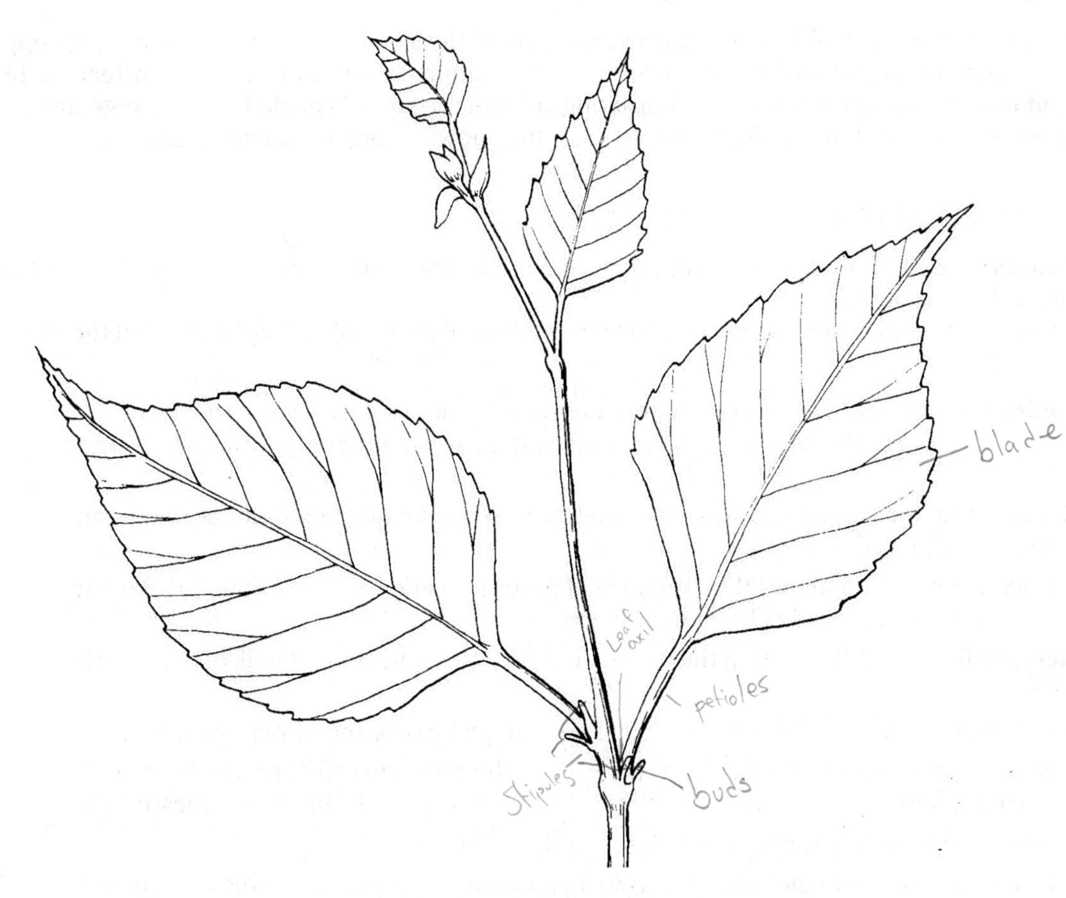

Lab Figure 2-6 Structures of dicot leaves.

Simple and Compound Leaves

A leaf that consists of a single leaf blade, as in Lab Figure 2-6, is a **simple leaf**. In a **compound leaf**, the blade is divided into two or more **leaflets**. All of the leaflets, together with their connecting parts and the petiole if present, make up a single leaf. The stalk at the base of the leaf is called a petiole, like a simple leaf. However, the stalks at the base of the leaflets are called **petiolules**. When the petiolules are missing, the leaflets are **sessile**. Stipules may be present on the petiolules in a few species.

There are two types of compound leaves. A **pinnately compound leaf** has all the leaflets attached to a central stalk called the **rachis**. The rachis is similar to the midrib of a pinnately veined leaf. A **palmately compound leaf** has all the leaflets attached to one point at the end of the petiole and does not contain a rachis.

Legumes have pinnately compound leaves. Soybeans and other beans, clovers, and sweetclovers have leaves with three leaflets that are called **trifoliolate**. Peanut has leaves with four leaflets, and some vetch have over a dozen leaflets in a leaf.

Figure 8-8 on page 102 shows both types of compound leaves. Be sure you understand the difference between simple leaves and compound leaves.

Leaf Shape and Leaf Margins

The differences in leaf shape and margin are valuable in identifying different species of plants. The **leaf margin** is the edge of the leaf blade. It can vary from smooth to serrated, to deeply lobed. The **leaf tip** can also vary in shape and can be used for identification.

Study the different shapes, margins, and tips shown in Lab Figure 2-7, and identify the examples provided by the instructor.

Leaf Arrangement

Another good tool for identifying plants is the arrangement of the leaves on the stem. A plant with **opposite leaf arrangement** has two full leaves at each node on opposite sides of the stem. **Alternate leaf arrangement** has one leaf at each node on alternating sides of the stem. **Whorled leaf arrangement** has three or more leaves at each node. Figure 8-9 on page 103 shows examples of leaf arrangement.

Structure of the Leaf Blade

The leaf blade is vitally important to the plant because that is where photosynthesis occurs. Read *Internal Structure of Leaves*, on page 103.

Lab Figure 2-8 shows a cross section of a leaf blade. Using Figure 8-10 on page 104, label the tissues in Lab Figure 2-8.

- ☐ **Epidermis:** a single layer of transparent cells on the upper and lower surfaces of the blade. The epidermis also has special pores called stomata that are formed by guard cells.
- ☐ **Guard cells:** form special pores or openings into the leaf called **stomata** (singular, stoma or stomate).
  - ☐ **Stoma:** a pore that allows the exchange of gases into and out of the leaf. It can open and close as the guard cells change shape.
- ☐ **Mesophyll:** the middle tissue in the leaf blade. Most plants have two kinds of mesophyll cells.
  - ☐ **Palisade cells:** cylindrical in shape and located just below the upper epidermis.
  - ☐ **Spongy cells:** rounded cells located between the palisade cells and the lower epidermis. Some plants, such as corn, contain only spongy cells in the leaf mesophyll.
  - ☐ **Xylem:** connects to the vascular tissue in the stem.
  - ☐ **Phloem:** connects to the vascular tissue in the stem. The vascular bundles, xylem and phloem, are visible as veins in the leaf blade.

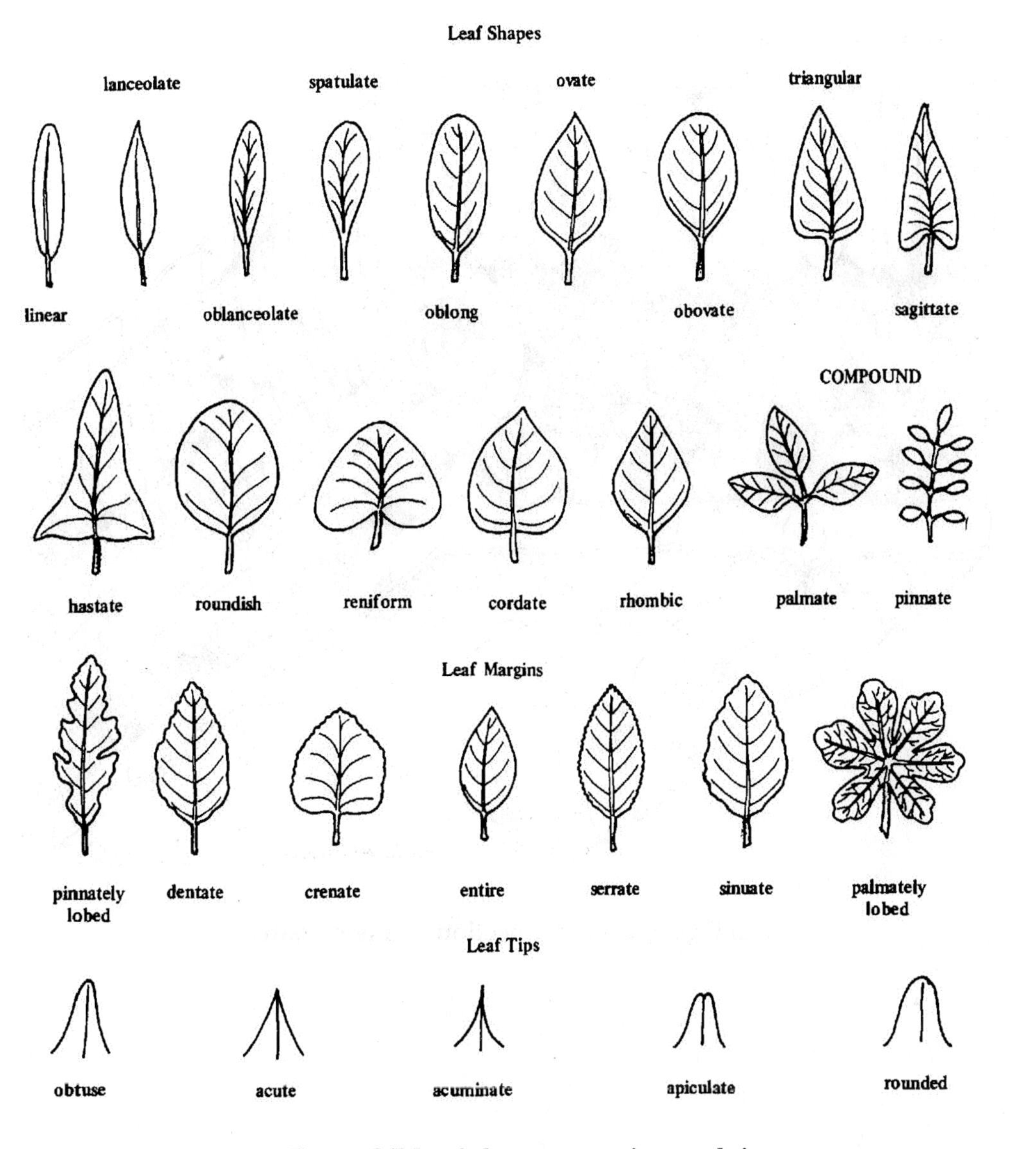

Lab Figure 2-7 Leaf shapes, margins, and tips.

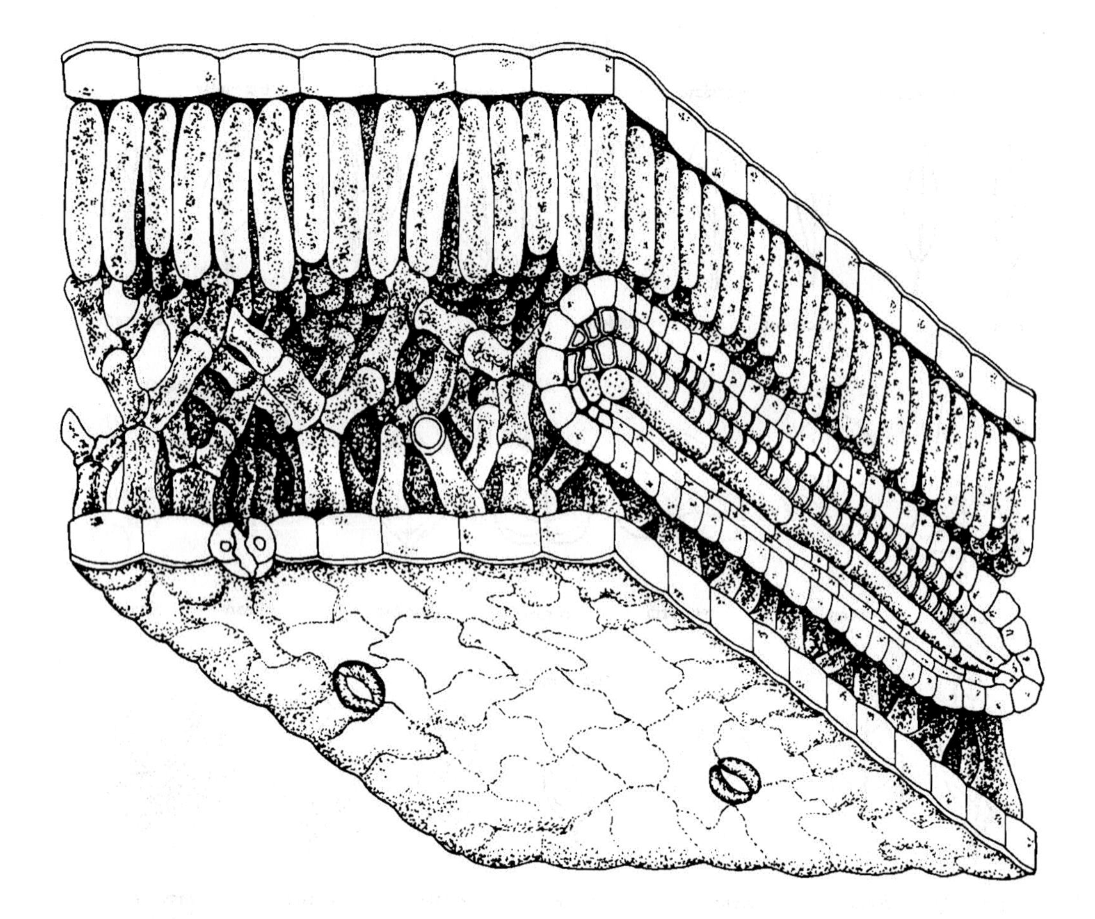

Lab Figure 2-8 Cross section of a leaf blade.

STUDY GUIDE

Complete the following table to help you learn the material in this exercise.

	Grasses	Legumes
Type of root system?	____________	____________
Arrangement of vascular bundles in roots?	____________	____________
Where does water absorption occur in roots?	____________	____________
Arrangement of vascular bundles in stems?	____________	____________
Cambium present?	____________	____________
Name for outer layer of cells?	____________	____________
Tissue that carries water and nutrients?	____________	____________
Tissue that carries sugars?	____________	____________
Position of xylem in relation to phloem?	____________	____________
Type of leaf arrangement?	____________	____________
Type of leaf venation?	____________	____________
Type of leaf margin?	____________	____________
Leaves simple or compound?	____________	____________
Three parts of leaves?	____________	____________
Interior tissue of leaves?	____________	____________

LABORATORY EXERCISE 3

Flowers and Fruits

INSTRUCTIONAL OBJECTIVES

Upon completion of this exercise you should be able to:

1. Define pollination, fertilization, and anthesis.
2. Identify the different parts of a flower. List their collective terms. Explain the difference between essential and accessory parts.
3. Explain the difference between the receptacle, peduncle, and pedicle. Understand that they are not part of a flower.
4. Explain the differences between regular and irregular, complete and incomplete, and perfect and imperfect flowers. Explain what staminate and pistillate flowers are and the difference between monoecious and dioecious plants.
5. Define inflorescence. Identify the different types described in the exercise.
6. Identify the different parts of a grass spikelet. Describe the difference between a spikelet and a floret.
7. Identify the different parts of a legume flower, including the different petals, and the free and fused stamens.
8. Explain how the keel affects pollination, and the role of insects in aiding the pollination of a legume flower.
9. Define a fruit. Explain the differences between dry and fleshy fruits, and dehiscent and indehiscent fruits.
10. Define ovule, funiculus, placenta, seed, and hilum. List the functions of each.
11. Define a legume fruit.

Chapter 10, Flowering and Reproduction, has information needed to complete this exercise. Read *Flower Structure*, pages 124-126, before proceeding with the exercise.

FLOWERS

As you learned in the first two laboratory exercises, there are many differences between monocots and dicots, and more specifically, grasses and legumes. This exercise continues the study of plant structures and their functions, concentrating on flowers, seeds, and fruits.

Dicot Flower Structure

Lab Figure 3-1 shows a typical dicot flower. Using Figure 10-1 on page 125 as a guide, label the structures in Lab Figure 3-1.

5 petals = 5 sepals
4 petals = 4 sepals

- □ **Sepals:** usually green, leaflike bracts or modified leaves that form the outermost whorl of floral parts. In the bud stage they protect the other parts from injury. The collective term for all the sepals is the **calyx**.
- □ **Petals:** usually colorful and fragrant to attract insects. Petals are located just within the calyx. Many produce nectar in specialized glands near the base. The collective term for all the petals is the **corolla**.
- □ **Stamens:** the male reproductive organs of the plant. The collective term for all the stamens is the **androecium**. Stamens in dicot flowers occur in multiples of four or five. Each stamen has two parts.
    - □ **Filament:** a stalk that supports the **anther**.
    - □ **Anther:** produces the pollen.
- □ **Pistil:** the female reproductive organ of a plant. If there is more than one pistil, they are collectively called the **gynoecium**. The pistil has three parts.
    - □ **Stigma:** located at the tip of the pistil, it is designed to attract and hold pollen.
    - □ **Style:** a stalk-like structure that connects the stigma to the ovary.
    - □ **Ovary:** contains the egg that develops within the ovule. The ovule at maturity becomes a seed. The ovary can contain from one to many ovules.
- □ **Receptacle:** the base of the flower located just below the calyx.
- □ **Pedicel:** a branch from the peduncle. It supports a flower when a group of flowers is present.
- □ **Peduncle:** a stem that supports one or more flowers. An example is the stem below the inflorescence and above the uppermost node in a plant.

The **essential flower parts** are the stamens and pistil, and the **accessory flower parts** are the sepals and petals. The collective term for the sepals and petals is the **perianth**.

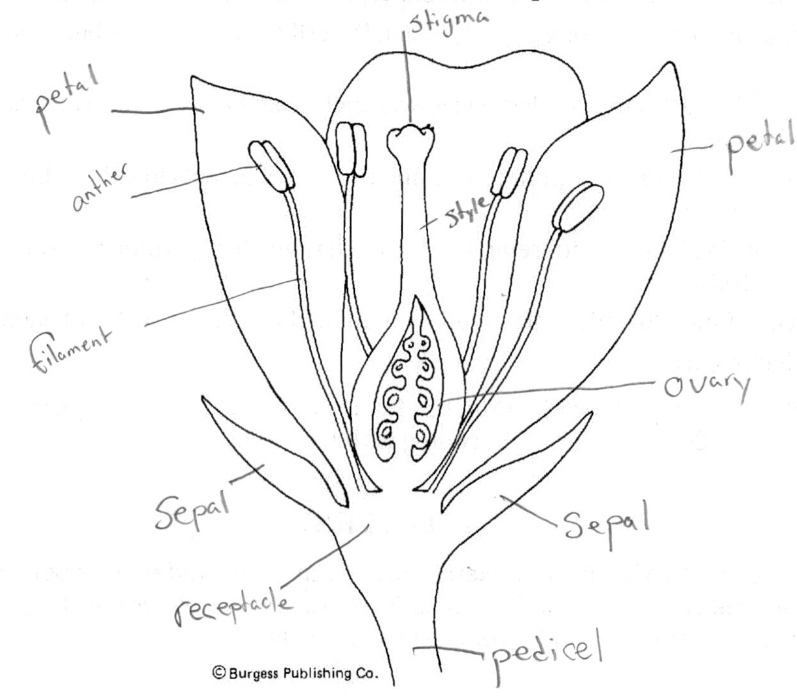

Lab Figure 3-1 Diagram of a typical dicot flower.

Grass Flower Structure

Lab Figure 3-2 shows a typical grass flower. Using Figure 10-2 on page 126 as a guide, label the structures in Lab Figure 3-2.

The calyx and corolla in grass flowers have been replaced with special modified leaves called **bracts.** The individual flowering unit of grasses is called a **spikelet** that is labeled as an inflorescence in Lab Figure 3-2. The spikelet contains the following parts:

- □ **Glumes:** two leaflike bracts at the base of the spikelet. In some species, the glumes completely enclose the rest of the spikelet. A spikelet will have one or more florets.
- □ **Rachilla:** stalklike structure that connects all the florets. It is the central axis of the spikelet.
- □ **Floret:** contains the stamens and/or the pistil and some bracts. In most grasses, there is more than one floret in each spikelet, with the uppermost floret undeveloped and sterile.
    - □ **Lemma:** first bract at the base of the floret. It may or may not have a bristlelike appendage on its tip or back called an **awn**. Lab Figure 3-2 does not show an awn.
    - □ **Palea:** bract located just above and opposite the lemma. It is usually smaller than the lemma and rarely has an awn. Together the lemma and palea enclose the stamens and pistil.
    - □ **Lodicules:** very small bracts at the base of the pistil. Lodicules serve no function other than for identification.
    - □ **Pistil:** the female reproductive organ of the plant. It has a single ovary with two feathery stigmas.
    - □ **Stamens:** the male reproductive organs of the plant. Most grasses have three stamens. In monocots, the stamens occur in multiples of three.

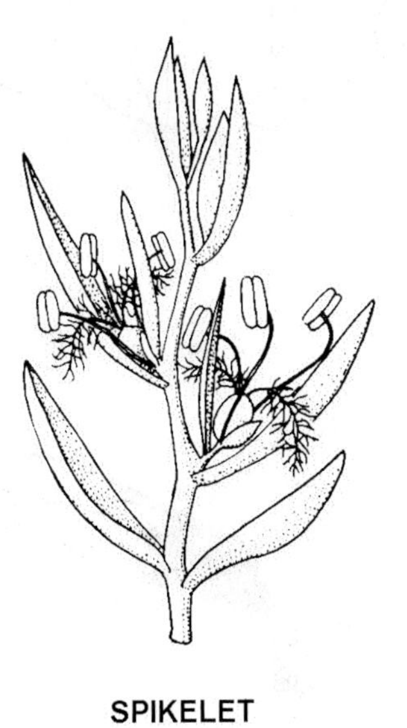

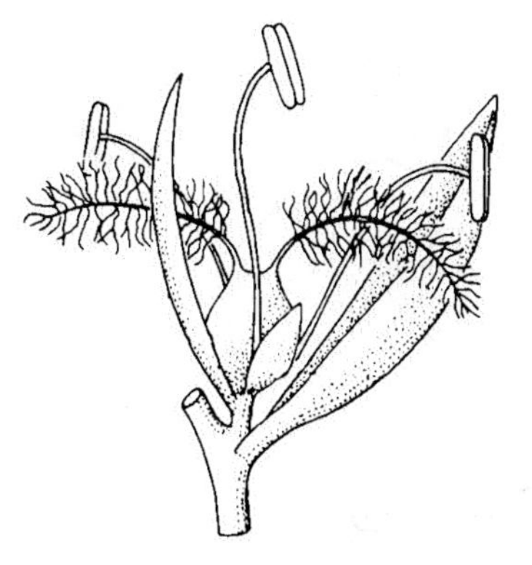

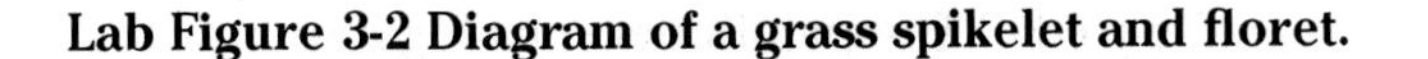

Lab Figure 3-2 Diagram of a grass spikelet and floret.

Legume Flower Structure

Most legumes have a special type of irregular flower called **papilionaceous**, with five petals that are not alike in shape and size. Lab Figure 3-3 shows several views of legumes flowers. However, there is not a corresponding drawing in Chapter 10. Using material supplied by the instructor, label the flower parts in Lab Figure 3-3.

- ☐ **Calyx:** five sepals that are all alike.
- ☐ **Standard:** the largest petal. It is located at the top of the corolla.
- ☐ **Wings:** two petals located on opposite sides of the corolla.
- ☐ **Keel:** two petals that have fused to form what looks like one petal in the shape of a boat keel. It is located at the bottom of the corolla and encloses the stamens and pistil.
- ☐ **Stamens:** there are ten stamens. Nine stamens have their filaments partially fused together. The tenth stamen is free from the rest.
- ☐ **Pistil:** one pistil with its stigma above the anthers.

In many legume species, the essential flower parts are enclosed within the keel that prevents pollination by wind. Since the stigma is above the anthers, pollen cannot fall to the stigma. In these cases, insects, particularly bees, are necessary for pollination. Insects trip the keel and expose the stamens and pistil. They also carry pollen between plants which aids in cross-pollination.

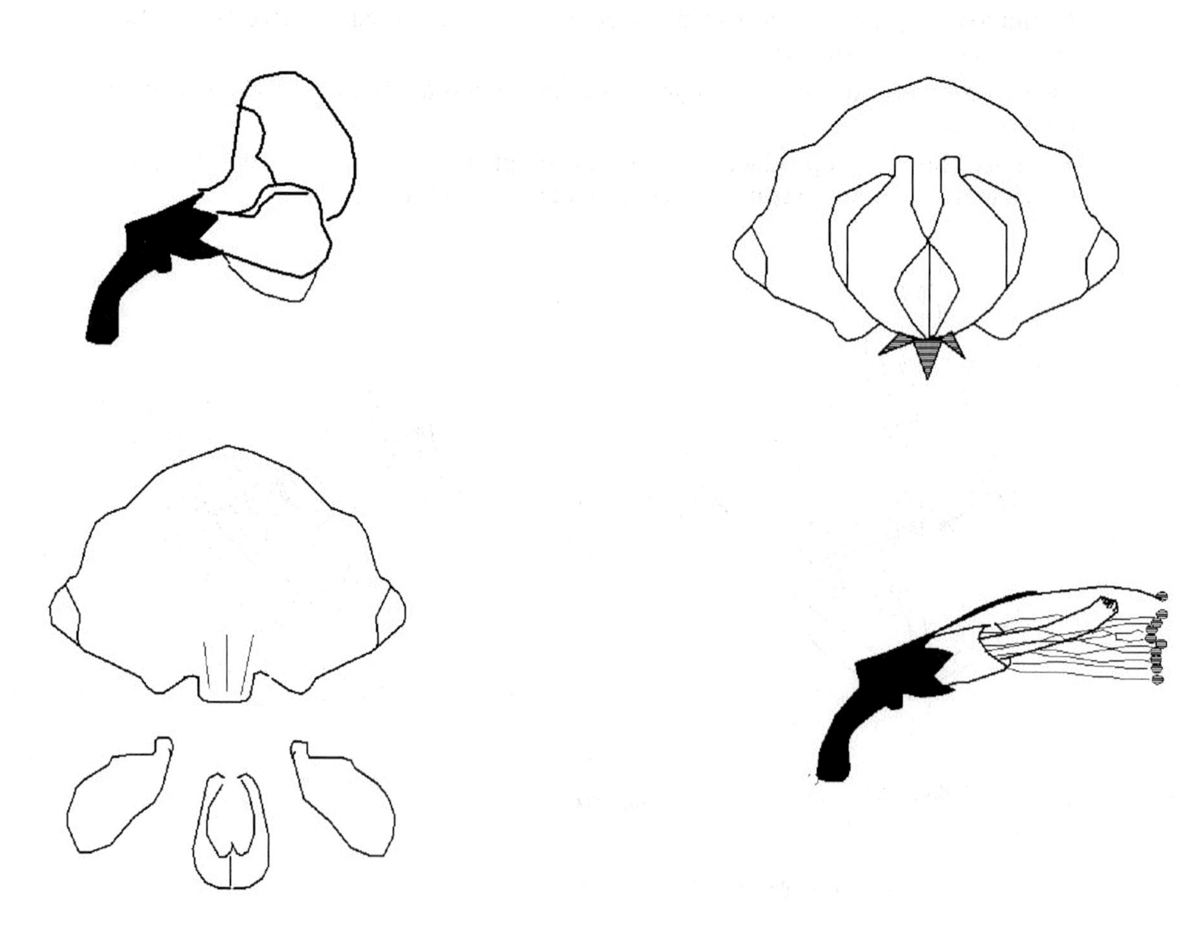

Lab Figure 3-3 The papilionaceous legume flower.

INFLORESCENCES

A group of flowers on the same peduncle is called an **inflorescence**. Read *Kinds of Flowers and Inflorescences*, pages 126-128. Using that information, and Lab Exercises 9-11, list the type of inflorescence found in each of the following crops.

Corn (ear) ______________________________

Corn (tassel) ______________________________

Grain sorghum ______________________________

Soybean ______________________________

Peanut ______________________________

Wheat ______________________________

Oats ______________________________

Barley ______________________________

Proso millet ______________________________

Sunflower ______________________________

Alfalfa ______________________________

Red clover ______________________________

Flax ______________________________

Safflower ______________________________

Sesame ______________________________

FRUITS

A **fruit** is a mature ovary with one or more seeds inside. Fruits are classified or grouped as those that are dry at maturity, and those that are fleshy at maturity.

Examples of dry fruits are the caryopsis, studied in Lab Exercise 1, which does not release seeds at maturity, and the legume, which splits open at maturity and releases its seed. Dry fruits that split open at maturity are called **dehiscent**. Those that do not split open are called **indehiscent**.

Examples of fleshy fruits are the **pome** (apple, pear), the **berry** (tomato), and the **drupe** (peach, plum). These are not important agronomically and are studied in this exercise.

The Legume Fruit

The legume fruit is dehiscent fruit and splits open along two lines called **sutures**. Fruits that split only along two lines are found exclusively within the Legume family. The common name for the legume fruit is the **pod**.

Lab Figure 3-4 shows several views of the legume pod. However, there is not a corresponding drawing in Chapter 10. Obtain a legume pod from the instructor. Using the material provided by the instructor, label the parts in Lab Figure 3-4.

- ☐ **Pod (ovary wall):** splits open at maturity.
- ☐ **Suture:** where the pod splits open.
- ☐ **Ovule:** develops into the seed by maturity.
- ☐ **Funiculus:** short stem that attaches the ovule to the ovary wall. The place where the seed is attached to the funiculus causes a scar called a **hilum**.
- ☐ **Placenta:** inside of the ovary wall. It transfers water and nutrients to the developing ovule.

Side view
Whole pod

Side view
Split open

Top View
Whole pod

Lab Figure 3-4 The legume fruit or pod.

STUDY GUIDE

Complete the following table to help you learn the material in this exercise.

	Grasses	Legumes
Petals and sepals present?	______	______
Glumes, lemma, palea present?	______	______
Name for basic flowering unit?	______	______
Number of stamens?	______	______
Fruit dehiscent?	______	______
Collective term for petals?	____________	
Collective term for sepals?	____________	
Female reproductive organ?	____________	
Parts of female reproductive organ?	____________	
Male reproductive organ?	____________	
Parts of male reproductive organ?	____________	
Short stalk at base of flower?	____________	
Essential flower parts?	____________	
Accessory flower parts?	____________	
Flower with both male and female parts?	____________	
Flower that is missing any part?	____________	
Short stalk at base of floret?	____________	
Stem that supports one or more flowers?	____________	
Structure that supports an awn?	____________	
Number of petals in a legume flower?	____________	
Common name for a legume fruit?	____________	

Part II

Plant Growth and Development

LABORATORY EXERCISE 4

Growth of Grasses and Legumes

INSTRUCTIONAL OBJECTIVES

Upon completion of this exercise, you should be able to:

1. Define mitosis.
2. Explain the differences between an apical meristem and an intercalary meristem, and the growth pattern associated with each.
3. Identify the four regions of growth in a young root.
4. Identify the four regions of growth in a young stem.
5. Define apical dominance.
6. Explain why a grass plant is better adapted to grazing and mowing than a dicot.
7. List the relative percentages of dry matter found in stalks, leaves, and grain in corn, sorghum, and soybean.
8. Draw a figure depicting dry matter accumulation during the life of an annual plant.
9. Define yield components. List the components described in the exercise.

PLANT GROWTH

Growth, or physical increase in size, occurs from cell division, called **mitosis**, which produces new cells in the plant and cell expansion which causes the newly produced cells to increase in size. In stems, the most increase in cell size occurs lengthwise resulting in increased stem length and subsequent plant height. This lengthwise cell expansion is called **cell elongation.** The same pattern of cell elongation also occurs in roots, which enables them to grow through the soil.

Cell expansion in leaves, flowers, and seeds may not involve cell elongation as such, but instead the cells may expand uniformly in all directions. The type of cell expansion that occurs in any given tissue will vary but that is not important in understanding crop growth and development. The most noticeable trait is in stem growth and its resulting effect on plant height.

The number of nodes in the stem of a crop is usually about the same no matter how tall the particular variety grows. Taller varieties or hybrids of any crop simply have longer internodes. Internode length is decided by the amount of cell elongation that occurs in the developing stem. The plant breeder can select for internode length and develop taller or shorter varieties as required.

ROOT GROWTH

Read *Root Structure and Growth,* on page 78, to better understand the principles of root growth. Lab Figure 4-1 shows a root tip. Using Figure 7-1 on page 79 as a guide, label the regions of growth in Lab Figure 4-1.

- □ **Meristematic region:** located at the tip of the root. The cells actively divide to produce more cells. The dividing cells are protected by the **root cap**.
- □ **Elongation region:** new cells produced by the meristem elongate lengthwise which causes the root to grow through the soil.
- □ **Differentiation region:** cells differentiate into the different tissues of the root, such as the vascular tissue. Growth continues from cell expansion but is much less than the growth found in the elongation region.
- □ **Maturation region:** tissues have formed and the cells reach their maximum size. No growth occurs in this region.

Most absorption of water and nutrients occurs in the region of differentiation and very young regions of maturation. The more mature tissue functions primarily in transporting water and nutrients to the rest of the plant. Therefore, a plant must continually produce new roots to continue absorption of water and nutrients. Older roots may be sloughed off by the plant.

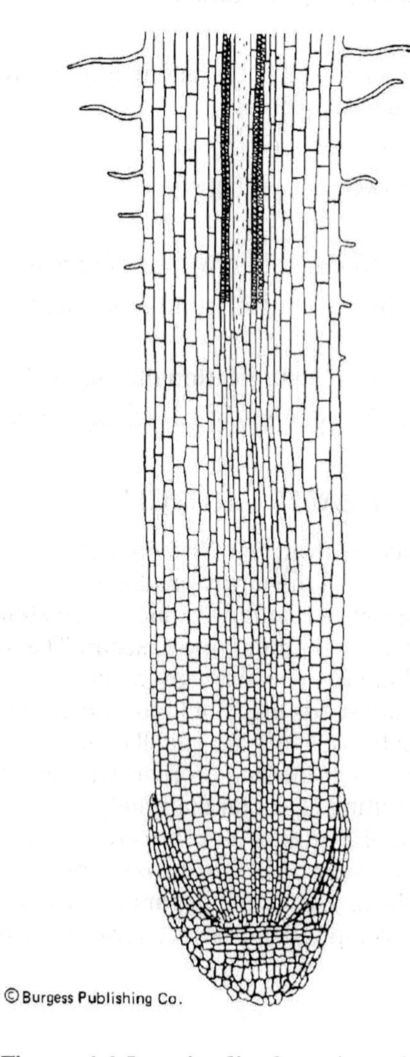

Lab Figure 4-1 Longitudinal section of a root tip.

STEM GROWTH

Read *Origin and Growth of Stems*, pages 96-97, which discusses the differences in growth between monocots and dicots. Note that dicots have apical meristems and grasses have intercalary meristems that result in entirely different growth patterns.

Dicot Stem Growth

Lab Figure 4-2 shows a dicot stem tip. Using Figure 8-5 on page 96 as a guide, label the regions of growth in Lab Figure 4-2.

- □ **Meristematic region:** located at the tip of the stem. The meristematic cells produce new cells *below* the original cells.
- □ **Elongation region:** located just below the meristematic region. New cells produced by the meristem elongate lengthwise which causes the stem to grow upward.
- □ **Differentiation region:** located just below the elongation region. Cells differentiate into the different tissues of the stem, such as the vascular tissue. Growth continues from cell expansion but is much less than the growth found in the elongation region.
- □ **Maturation region:** located just below the differentiation region. Tissues have formed and the cells reach their maximum size. No growth occurs in this region.
- □ **Apical meristem:** produces the new cells that develop into all the stem and leaf tissues.
- □ **Leaf primordium:** meristematic cells that develop into leaves.
- □ **Bud primordium:** dormant meristem that can produce a branch or flower.

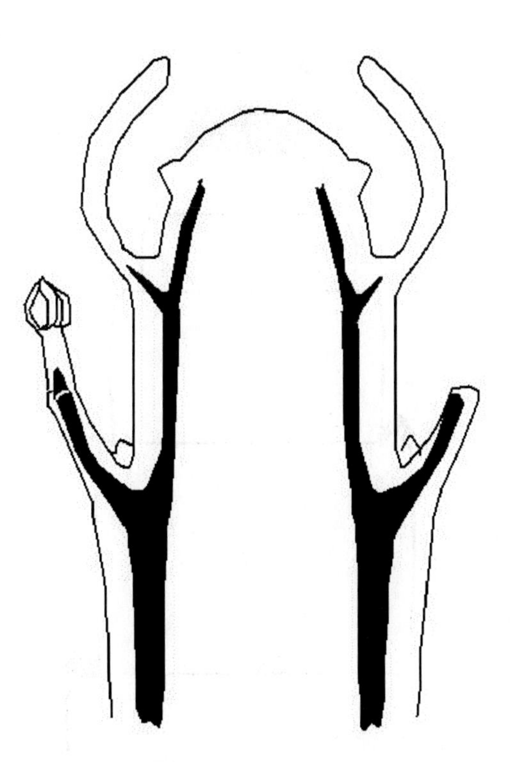

Lab Figure 4-2 Stem growth in dicots.

Grass Stem Growth

Lab Figure 4-3 shows a grass stem tip. Using Figure 8-5 on page 96 as a guide, label the regions of growth in Lab Figure 4-3.

- ☐ **Meristematic region:** located at the base of every internode. The meristematic cells produce new cells *above* the original cells.
- ☐ **Elongation region:** located just above the meristematic region. New cells produced by the meristem elongate lengthwise which causes the stem to grow upward.
- ☐ **Differentiation region:** located just above the elongation region. Cells differentiate into the different tissues of the stem, such as the vascular tissue. Growth continues from cell expansion but is much less than the growth found in the elongation region.
- ☐ **Maturation region:** located just above the differentiation region. Tissues have formed and the cells reach their maximum size. No growth occurs in this region.
- ☐ **Apical meristem:** produces the new cells that will continue to develop into all the stem and leaf tissues. In most grasses, it will eventually change to the floral primordium.
- ☐ **Intercalary meristem:** located at the top of each node. It produces cells that will develop into stem tissues.
- ☐ **Floral primordium:** meristematic cells that develop into the inflorescence.
- ☐ **Bud primordium:** dormant meristem that can produce a branch or flower.

The leaf primordium is not shown in Lab Figure 4-3.

Regrowth in Grasses and Dicots

Most grasses and forage legumes are well adapted to vigorous regrowth after mowing or grazing. However, grasses and legumes differ considerably in their method of regrowth. Regrowth is triggered by the apical dominance system discussed on page 97.

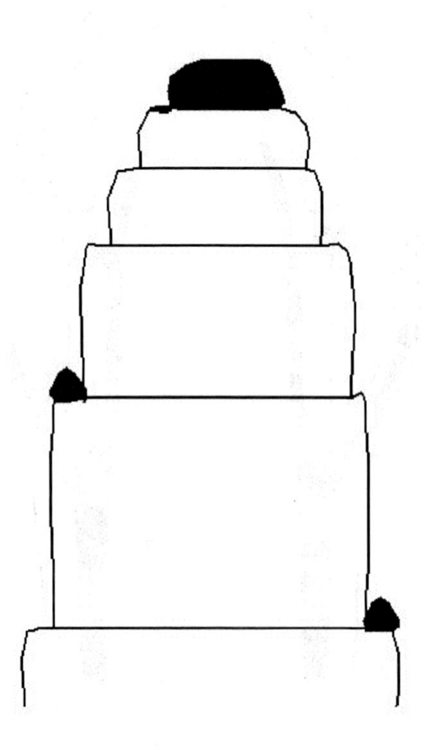

Lab Figure 4-3 Stem growth in grasses.

Most forage grasses initiate new tillers or stems from buds located on the crown. Figure 8-6 on page 100 shows a crown with tillers. Because there is an intercalary meristem at each node, cutting off the top of the stem will not remove all the meristems and growth is continuous. Removal of upper meristems can cause crown buds or axillary buds to become active, thus increasing the number of tillers. Meristematic tissues in grass leaves are found at the base of the sheath and blade, so removal of the tips of the leaf blades does not hinder growth.

When the top of a dicot is removed, the apical meristem is lost, leaving no other active aboveground meristems on the plant. New growth is initiated either from axillary buds at the nodes or from crown buds. Since these buds need to be activated, growth is not continuous. Meristematic tissue in dicot leaves is found along the margin of the leaf blade, so removal or injury of part of the leaf blade results in the cessation of growth or a deformed leaf. Therefore, regrowth is much faster in grasses than dicots, which makes grasses better adapted to mowing and grazing.

RATE OF GROWTH IN PLANTS

Growth can be defined as an increase in size or volume, but a more precise definition is an increase in the dry matter content of the plant. Dry matter is the plant material remaining after all the water has been removed.

As shown in Lab Figure 4-4, when a plant is young, it accumulates dry matter slowly. The growth rate increases and is very rapid during the middle of the plant's life. Growth then tapers off at maturity.

This sigmoid, or "S" curve, of growth is typical of all annual plants and perennials that are harvested for forage. The period of rapid dry matter accumulation is called the "grand period of growth" or the linear growth phase. During this time it is possible to see daily increases in size with many plants, such as corn. During this period, moisture and nutrient uptake are very critical for the plant.

Lab Figure 4-5 shows the distribution of dry matter in four grain crops through the growing season. Notice in all crops that the plant grows **vegetatively** early in the season (stems and leaves) and **reproductively** later in the season (grain or seeds). Most vegetative growth is completed by the time the plant flowers and begins reproductive growth.

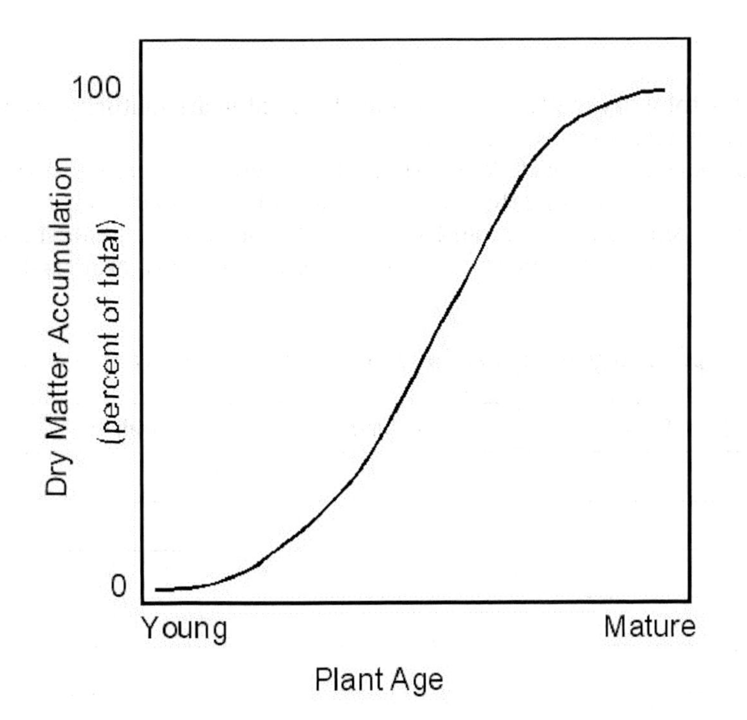

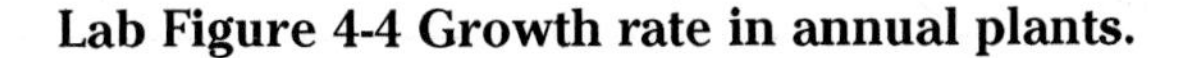
Lab Figure 4-4 Growth rate in annual plants.

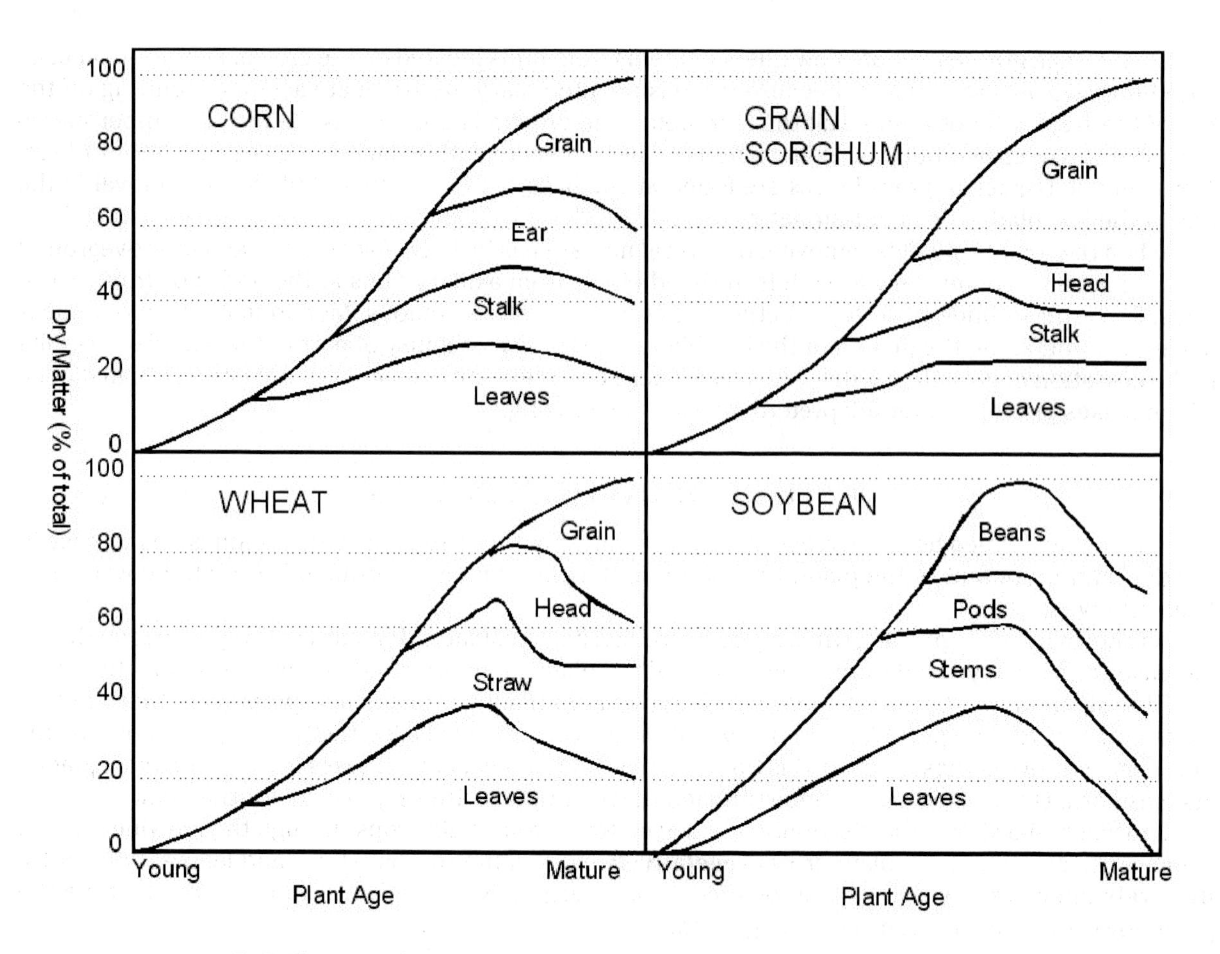

Lab Figure 4-5 Distribution of dry matter in four annual crops.

In the soybean, the total dry matter decreases as the plant nears maturity because the leaves drop off the plant leaving only stems with bean-filled pods.

At maturity, about 40% to 50% of the dry weight in most grain or cereal crops is found in the grain or seed, and the rest is divided among the stems, leaves, and other plant parts.

Study Lab Figure 4-5 and complete Table 4-1. Round all values to the nearest 5 percent. Use the difference between each line to determine the amount. For example, Corn grain is 40% (100 - 60).

Table 4-1 Proportion in percentage of dry matter in the aboveground parts of mature crop plants.

	Corn	Soybean	Sorghum	Wheat
Stalk or stem	________	________	________	________
Leaves	________	________	________	________
Grain or seed	________	________	________	________
Other plant parts	________	________	________	________
Lost from plant	________	________	________	________
Total	100	100	100	100

YIELD COMPONENTS OF CROPS

The yield of a grain crop plant can be partitioned into components that together contribute to the yield. Specific yield components differ for each crop and, in some cases, the use of the crop. For instance, if the crop will be cut for forage, its yield components will be different that if it was harvested for grain.

There are some similarities in yield components with grain crops. Yield components for all grain crops will include plants per acre; heads, ears, or pods per plant; kernels or seeds per head, ear, or pod; and weight per kernel or seed. Table 4-2 shows the specific yield components for four major crops.

Some crops may use other components to decide the value of the main components. For example, wheat may use tillers per plant and heads per tiller to calculate heads per plant. For forage crops, yield components are usually plants per acre and weight per plant. Weight per plant may be dry weight with hay crops and wet weight with silage crops.

Yield components are closely linked together and a change in one yield component will result in a subsequent change in another component. For example, an increase in plants per acre with corn may cause a decrease in kernels per ear if the plant population exceeds the maximum recommended for the growing conditions in the field. Decreasing plant population in wheat will increase the number of heads per plant by stimulating tillering. Decreasing the number of kernels may cause an increase in kernel weight.

The environment of the crop field decides how close the actual yield approaches the yield potential for the crop. Full yield potential is almost never achieved because environmental conditions usually limit one or more of the yield components during the growing season. Proper crop management assures that all yield components are at or near the maximum for the given conditions in the field.

Table 4-2 Yield components of four annual crops.

CORN	GRAIN SORGHUM	SOYBEANS	WHEAT
Plants per acre	Plants per acre	Plants per acre	Plants per acre
Ears per plant	Heads per plant	Pods per plant	Heads per plant
Kernels per ear	Kernels per head	Seeds per pod	Kernels per head
Weight per kernel	Weight per kernel	Weight per seed	Weight per kernel

STUDY GUIDE

Complete the following table to help you learn the material in this exercise.

Term for cell division?	________
Process that increases cell size?	________
Short extension of an epidermal root cell?	________
Name for protective tissue at tip of root?	________
Name for tissue located at tip of stem?	________
Name for stem meristem of grasses?	________
Name for stem meristem of legumes?	________
System that regulates activity of axillary buds?	________
Growth regulator involved in that regulation?	________
Plant part that holds most dry matter at maturity?	________

LABORATORY EXERCISE 5

Growth Stages of Corn and Soybean

INSTRUCTIONAL OBJECTIVES

Upon completion of this exercise, you should be able to:

1. Identify the growth stages of corn and soybean plants.
2. Describe the growth stages of corn and soybean.

GROWTH AND DEVELOPMENT OF CROPS

An understanding of the way a crop plant grows is essential to the expert management of that crop. Many critical decisions involve precise timing in relation to the crop's stage of growth. For example, certain herbicides and pesticides can only be applied at specific times in the crop's life cycle. The stage of growth at which damage occurs from pests or weather greatly influences the amount of yield loss. Uptake of water and nutrients varies as the crop progresses throughout its growth stages.

For any grain or seed crop, growth stages can be divided into two primary categories: vegetative growth and reproductive growth. During vegetative growth, the plant is producing stem and leaf tissue. Vegetative growth is designed to build the photosynthetic "factory" which will provide the energy the plant needs for maintenance of its vegetation and for development of the grain or seed. Reproductive growth involves the processes of flowering and seed development that will ultimately decide the yield of a grain crop. The energy needed for growth and development comes from **photosynthesis** in the leaves. In this exercise, you will study how corn and soybean plants grow and develop.

GROWTH STAGES OF CORN

A typical corn hybrid will produce about 20 to 21 leaves. All the leaves will be produced by the time the plant tassels. There will be some variation in the actual number of leaves depending on the hybrid and its maturity. Later maturing hybrids may produce a few more leaves than early maturing hybrids.

The number of kernels that develop on the ear, the rate of development of the kernels, and final kernel size vary between hybrids and can also be affected by environmental conditions. When comparing different maturing hybrids, the length of the reproductive growth period is not affected as much as the time for vegetative growth. Longer maturing hybrids will have longer vegetative periods than shorter maturing hybrids; and the differences in the length of the reproductive stages will not vary as much.

A hybrid that will normally mature in about 125 days after emergence will silk in about 65 days. However, the time spent in each growth stage is affected by environmental factors in the field.

Vegetative Growth Stages

Corn stems, like other crops of the grass family, grow in a "telescoping" fashion from the bottom up. Grasses, which also include wheat and sorghum, have stems with **meristems** or **growing points** at each node. Each growing point produces new cells that then elongate and develop into the various tissues of the stem. This "telescoping" type of growth is what makes grasses so adaptable to grazing and mowing.

Vegetative growth stages are designated using the letter "V". The first vegetative stage is called VE ("E" for "emergence"), and the last stage, VT ("T" for "tasseling"). The rest of the V stages are based on the number of fully developed leaves on the plant, and are designated as V1, V2, V3, etc. A fully developed leaf is one in which the collar is visible. The collar appears as a line between the leaf blade and the leaf sheath. (See Lab Figure 5-1.) The first leaf on the seedling is oval shaped and the rest will be longer and more pointed.

As the plant grows, the stem expands and the earliest leaves will be torn away and lost. About the V6 stage, it will be necessary to split the base of the stalk and look for internode elongation. The first node above the first elongated internode will usually be the **fifth node** and the fifth leaf will be attached to it. This internode will usually be almost 1/2 inch long. (See Lab Figure 5-1) The next node above it will be the sixth node and so forth. By tracing leaves to their point of attachment and counting the nodes, you can decide the specific V stage of growth. Using this method allows you to decide V stages regardless of any leaf loss.

Stage VE—Germination and Emergence

The first structure to emerge from the germinating seed is the **radicle,** or first root. As discussed in Lab Exercise 4, most of the physical increase in size of a growing plant comes from cell expansion due to water uptake. The germinating seed will absorb all water in its immediate vicinity, but that is a limited supply. It needs additional water to continue growing so it sends its embryonic root out to tap into additional water in the soil.

The next structure to emerge will be the **coleoptile.** This is a protective covering over the **plumule** that contains the first embryonic leaves. An embryonic stem called the **mesocotyl** elongates to push the coleoptile toward the soil surface. When the coleoptile receives sunlight, the mesocotyl stops elongating. The coleoptile will then split open and the first leaves will emerge. The VE stage is reached when the coleoptile reaches the surface.

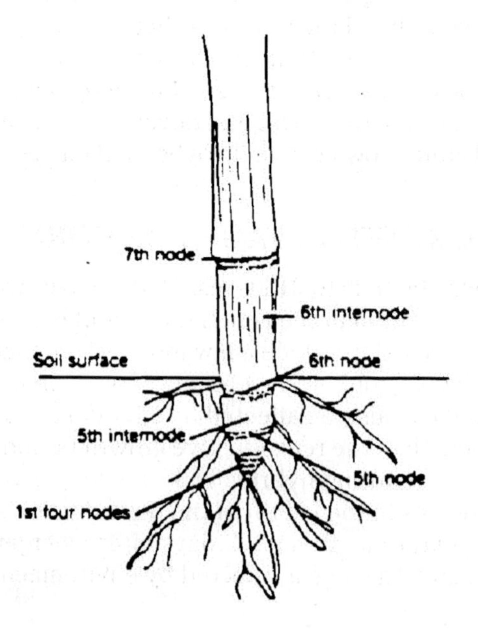

Lab Figure 5-1 Lower stalk split lengthwise.

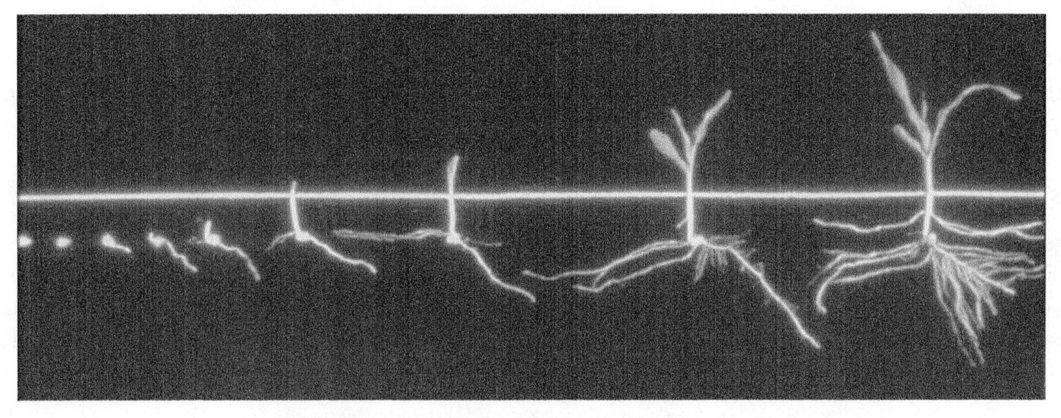

Lab Figure 5-2 Germination and emergence.

During this time the emerging seedling will be sending out additional roots from the seed called **seminal roots.** When VE is attained, new roots will begin emerging from the **coleoptilar node** at the top of the mesocotyl that is also the base of the coleoptile. These are called **secondary roots** or **nodal roots.** The secondary roots are the beginning of the main root system for the corn plant.

The growing point for the corn seedling is located at the coleoptilar node. It will usually be about 1 to 1 1/2 inches below the soil surface. At this time all stem tissue is located at the coleoptilar node and below. Therefore, everything that emerges from the soil is leaf tissue. This is important, as the seedling may be able to recover from any damage that might occur to emerged tissue at these early stages; for example, freeze damage or insect feeding. As long as the growing point is intact and healthy, the plant has a good chance of recovery.

Stages V1 to V5

Stage V1 is reached when the first leaf has fully emerged and its collar is visible. The first leaf is easy to recognize as it is more oval with a rounded tip. Subsequent leaves will emerge on alternate sides of the plant. Each subsequent stage is reached when the collar of each leaf becomes visible.

The growing point is still below the soil during these stages and everything above the soil is leaf tissue. By V5, the plant will be about 20 inches tall. What appears to be stem tissue is sometimes called a **pseudostem**; it is merely a tightly wound whorl of leaves. The plant is producing new nodes during this time and each new node produces a leaf that begins elongating upwards through the whorl. There is little internode elongation during this time.

Lab Figure 5-3 V3 plant.

By V5 all the nodes have been established and the tassel and ear shoots have been initiated although they are microscopic. Keep in mind that there are only five fully developed leaves above the soil, but the plant has essentially laid the complete foundation for its subsequent growth. The top of the stem is still below the soil surface.

Stages V6 to V11

At Stage V6, internode elongation begins and the growing point moves above the soil surface as shown in Lab Figure 5-5. Rate of growth rapidly increases in both the shoot and the roots. As the stem or stalk increases in diameter, smaller leaves near the base will be torn loose and may be completely lost. As discussed previously, it now becomes necessary to slit open the stalk and look for nodes to decide the exact stage. The first node above the first elongated internode is usually the fifth node as shown in Lab Figure 5-1.

Lab Figure 5-4 V6 plant.

Lab Figure 5-5 Dissected V6 plant.

Ear shoots and tillers will be visible if the plant is carefully dissected. **Tillers** or **suckers** will originate from the lowest nodes and may or may not become active. The amount of tillering in the plant is decided genetically, and will vary between hybrids. The presence or absence of active tillers has little or no effect on the yield.

By V9 the ear shoots and tassel will be readily visible if the plant is dissected. A potential ear shoot may begin to develop from every aboveground node except for the top six to eight; however, only the uppermost one or two will completely develop into a harvestable ear. The number of harvestable ears a plant will develop is decided mostly by plant population. Lower populations will encourage second ear development.

If growing conditions are good, the time between each leaf stage will continually decrease as the plant increases its rate of growth. There may be only two or three days between stages as the plant approaches V12. This rapid rate of growth will continue well into the reproductive stages. The water and nutrient requirements of the plant are also rapidly increasing during this time.

Lab Figure 5-6 V9 plant.

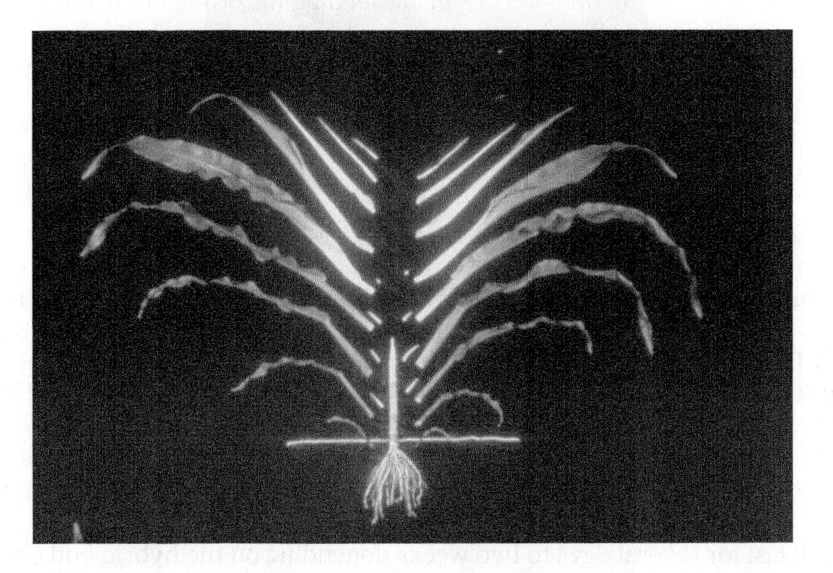

Lab Figure 5-7 Dissected V9 plant.

Stages V12 to V18

The number of potential kernels on each ear and the size of the ear are being decided during these stages. This is a critical time for establishing potential yield. If the corn plant experiences stress during this time, it will reduce ear size and kernels per ear that will result in lowered yield potential even if conditions improve later on.

During these stages, the corn plant will be progressing to a new leaf about every one to two days. It is now possible to estimate time to silking by subtracting the V stage from 21 (total number of leaves) and multiplying by two (days per leaf stage). For example, at V15 the corn plant is about 10 - 12 days from silking.

At V17, the tip of the upper ear shoot tips may be visible at the top of each leaf sheath. The tip of the tassel may also be visible.

By V18, the silks will be visible if the husks are removed from the developing ears. The silks will begin elongating from the basal flowers first and progress toward the ear tip. The plant is now about one week from silking.

There are genetic differences in the response of different corn hybrids to stress during this time. Hybrids that usually produce only one ear **(nonprolific hybrids)** will be more adversely affected by stress than **prolific hybrids** that usually produce more than one ear. Yields of prolific hybrids are more stable under variable stress but nonprolific hybrids will produce higher yields if no stress occurs. Therefore, hybrid selection should be based on the probability of stress occurring during this time. However, if the stress is severe, yields will be reduced drastically regardless of the type of hybrid planted.

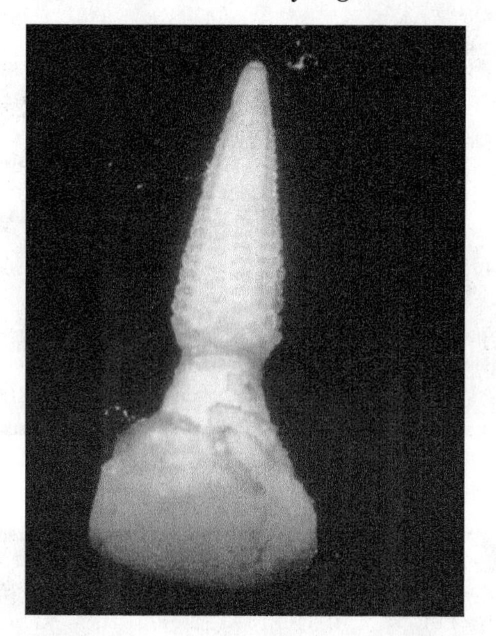

Lab Figure 5-8 Top ear of V12 plant.

Stage VT

The VT stage is reached when the last branch of the tassel is completely visible from the leaf whorl and the corn plant has almost reached its full height. At this stage, the silks have not yet appeared through the husks. The VT stage occurs about two to three days before silk emergence. However, the time between VT and R1 can vary considerably depending on hybrid and environmental conditions. **Pollination** will begin shortly after the tassel has fully emerged.

From VT to R1 is the most vulnerable time for hail damage to the corn plant because the tassel and all the leaves are exposed. Complete leaf removal by hail at this time will result in complete loss of the yield.

Pollination will last for several days to two weeks depending on the hybrid and environmental conditions. Single cross hybrids tend to have a shorter pollination time and stress may shorten pollination

Lab Figure 5-9 VT plant.

time. Pollen shedding will mostly occur in late morning or early evening. Every silk will need to receive a viable pollen grain for kernel development to proceed.

Reproductive Growth Stages

Reproductive growth stages are mostly concerned with kernel development. However, since the ear does not develop uniformly from base to tip, the middle of the ear should be used to decide the stage. Also, if the hybrid is prolific, the top ear should be used as it will be the most developed.

Stage R1—Silking

This stage begins when the first silks emerge from the tips of the husks. As each silk emerges, it hopefully will come in contact with a pollen grain. This process is called **pollination.** It will take about two to three days for all the silks to emerge and pollination to conclude. Silks grow about 1–1 1/2 inches per day but will quit elongating once they have been pollinated. Once a pollen grain has landed on a silk, it will "germinate" and send a **pollen tube** down through the silk to the ovary of the flower. **Sexual fertilization** then occurs. It takes about 24 hours from pollination to fertilization. Poor pollination and fertilization, and subsequent seed set, can result from insects, such as corn rootworm beetles, which feed on the silks.

The ear and shank have not yet reached full size at R1 although all the flowers are set. Further growth is due only to continued cell expansion. The ear flowers are quite small and are completely enclosed in glumes.

As discussed previously, stress during this time can be devastating to yield. Drought stress is especially harmful as it causes the silks and pollen grains to dry out. Stress at this time will usually cause poor filling of the ear tip.

Stage R2—Blister

Blister stage has been reached when the kernels resemble a blister. This usually occurs about 10–14 days after silking. The endosperm of the kernel is filled with a clear fluid and the embryo can be seen if the kernel is carefully dissected. By this time, the embryo has already formed its various parts such as the radicle and coleoptile; although much further development is yet to occur. The cob has reached its full size and the silks are turning darker in color.

Even though the endosperm is very watery, starch has begun to accumulate. The kernels will continue to develop rapidly and accumulate dry matter until maturity (R6). At R2, the kernels are at their highest moisture content (about 85% water) but that will steadily decline as the kernels fill with dry matter. Over 40% of the total dry matter above the ground accumulates in the grain of a corn plant.

Photosynthesis is continuing at a rapid rate and practically all of its production is going directly to the developing grain. The corn plant will also begin to mobilize nutrients from vegetative parts and translocate them to the grain. Uptake of nitrogen and phosphorus is continuing.

Stage R3—Milk

This stage occurs about 18–22 days after silking. It is reached when the kernels have turned yellow (on yellow corn) and the endosperm is filled with a milky white fluid. The embryo is now growing rapidly and is easily seen. The kernels are now about 80% water.

Stage R4—Dough

Dough stage is reached about 24–28 days after silking. The milky fluid has now thickened to a paste due to continued accumulation of starch in the endosperm. The embryo is nearing completion of its development and further grain filling will be practically all from starch accumulation. Water content will be about 70% and the kernels will have about one-half of their final dry weight.

Some kernels will begin denting as R4 nears completion. The cob will begin turning red or pink and most of the silks will be brown and dry.

Stage R5—Dent

At dent stage, about 35–42 days after silking, nearly all the kernels will be dented or denting. The denting is caused by drying of the endosperm material beginning at the top of the kernel. As the endosperm dries, the starch becomes hard and white, and forms a line near the top of the kernel that is visible when the kernel is viewed on the side opposite to the embryo. Drying will continue toward the base of the kernel and the starch layer can be seen as the starch line moves lower. The endosperm above the starch line is much harder and pressing one's thumbnail on the side of the kernel can help decide its location. Grain moisture is about 55% at the beginning of R5 and continues to steadily decrease as the starch line moves down.

Stage R6—Physiological Maturity

This stage occurs about 55 to 65 days after silking. **Physiological** maturity occurs when the kernels have reached maximum dry matter accumulation. The hard starch layer has progressed to the base of the kernel and a black or dark brown layer forms at the base of the embryo. This black layer forms when dry matter ceases moving into the kernel.

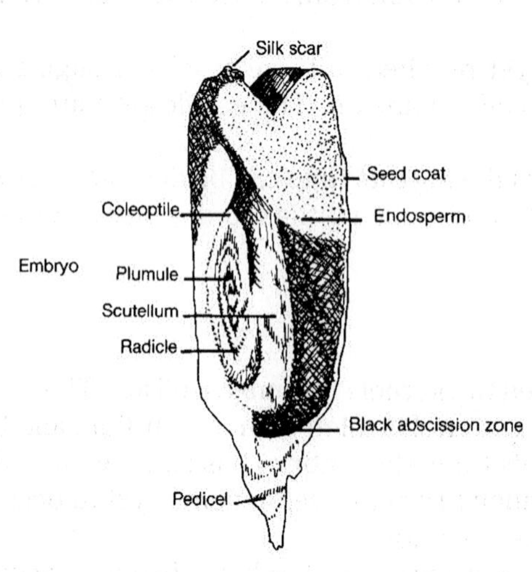

Lab Figure 5-10 Mature corn kernel.

Black layer forms first at the base of the ear and progresses toward the tip. Cut the kernel from base to top through the center of the embryo. The black layer will be easy to see at the base of the embryo. One word of caution about this procedure. Cut the kernels and look for the black layer shortly after removal of the ear or kernel from the plant. The black layer will form when dry matter stops moving into the kernel whether the kernel is mature or not. An ear or kernel removed from the plant will cease movement of dry matter into the kernel and the black layer will form. A wrong determination of the growth stage can occur if the kernels are not checked shortly after removal from the plant.

At R6, the plant has reached physiological maturity, but not harvest maturity. The grain moisture content will be about 30–35%, although this varies depending on the hybrid. The plant is not susceptible to stress at this time and a freeze may aid in speeding the drying of the grain by killing the plant.

The time needed to dry corn grain for harvest depends on the weather. Warmer temperature and lower humidity increase the rate of drying. There are also differences in hybrids, and breeders try to select for rapid dry down after physiological maturity. Corn for silage should be chopped as close to the beginning of R6 as possible for maximum yield.

GROWTH STAGES OF SOYBEAN

Soybean is a legume, which means that it has a much different type of growth than that discussed for grasses, such as corn, wheat, and sorghum. The main growing point of a soybean is located at the tip of the developing stem and is called an **apical meristem.** As discussed previously, grasses have intercalary meristems at each node and grow in a telescoping fashion. Since the apical meristem of a soybean is located at the tip of the stem, the plant is much more susceptible to damage from hail, freeze, insects, etc.

Another difference in the growth of soybean compared to grass crops such as corn, sorghum, or wheat is the method of flowering. Most soybean varieties are **indeterminate** while all grasses are **determinate.** Indeterminate plants flower over a longer period of time and continue to produce new stem and leaf tissue after flowering has begun. As already discussed with corn, the grass plant produces all its potential nodes and leaves *before* it initiates reproductive growth.

There are a few determinate varieties of soybean grown in the Midwest and the growth stages for them would be different. The growth stages described here are for indeterminate varieties.

Vegetative Growth Stages

As with corn, vegetative growth stages are designated with the letter "V". The first stage is VE ("E" for emergence) and the second is VC ("C" for cotyledon). The rest of the stages are based on the number of fully developed leaves on the plant and are designated as V1, V2, etc., similar to corn. The total number of V stages will vary depending on the variety and environmental conditions.

Lab Figure 5-11 An undeveloped soybean leaf.

The V stages following VC are based on the number of fully developed leaf nodes on the plants. A fully developed leaf node is one in which *the leaf above it* has unrolled or unfolded until the edges of the leaflets are no longer touching. Lab Figure 5-11 shows a leaf with the leaflets still touching. This can be confusing but it is important that you fully understand the system.

The first node is the one that holds the two **unifoliolate** leaves. These simple leaves are on opposite sides of the node. All subsequent leaves are **trifoliolate** which means they have three leaflets. There is only one trifoliolate leaf at each node and they appear on alternating sides of the stem.

When counting nodes, you begin with the node with the unifoliolate leaves, node 1, and progress up the stem. Stage V4 occurs when the leaflets of the first *five* nodes are unrolled. A node is not counted until the node *above* it has unrolled its leaves as previously described. Similarly, Stage VC is not counted until the unifoliolate leaves above the cotyledons have unrolled.

Stage VE—Germination and Emergence

Soybean has **epigeal emergence** as discussed in Lab Exercise 1. The cotyledons are pulled out of the soil by the elongating **hypocotyl**. The hypocotyl will elongate and arch toward the soil surface until it receives light. It will then straighten itself and pull the cotyledons out of the soil. Stage VE is reached when the arching hypocotyl breaks through the soil surface.

As the hypocotyl straightens, the cotyledons will fold down and begin photosynthesis. The cotyledons are embryonic seed leaves that were used by the seed for storing food needed to provide energy for germination and emergence.

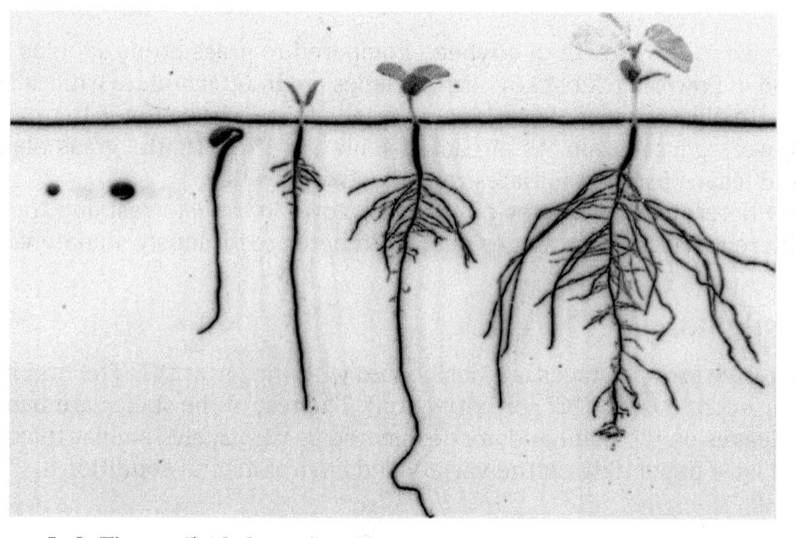

Lab Figure 5-12 Germination and emergence of soybean.

Stage VC—Cotyledon Stage

This stage is reached when the unifoliolate leaves have unrolled and their edges are no longer touching. This will occur a few days after VE.

The cotyledons are still supplying the seedling with food and are essential for continued growth and development. If one cotyledon is lost, the seedling will probably be able to continue growth; but if both are lost, yields could be reduced 8 to 9 percent. Cotyledons can be lost by insect feeding, disease, or physical damage during emergence. Damage is more likely if the soil is crusted. The seedling is also susceptible to damage by a late spring freeze since its growing point is above the soil.

Stages V1 to V2

Stage V1 is reached when the first trifoliolate leaf above the unifoliolate leaves has unrolled. As stated previously, the node with the unifoliolate leaves is counted as the first node for deciding vegetative growth stages.

By this time the leaves on the plant are conducting enough photosynthesis to provide the plant with the energy needed for growth and development. The cotyledons will begin to die and fall off the plant. The plant will reach V1 about 7 to 10 days after Stage VE. If growing conditions are satisfactory, the plant will reach new V stages about every five days until Stage V5.

Stage V2 is reached when the second trifoliolate leaf has unrolled. At this time, there will be three nodes with unrolled leaves; the unifoliolate node (Node 1), and two trifoliolate nodes (Nodes 2 and 3). Remember, when counting leaf stages, you do not count the uppermost node with unrolled leaflets.

At Stage V2, nodules will be forming on the roots. The nodules are round or oval-shaped growths on the roots and are easily seen if the plant is carefully dug up. The nodules are caused by infection of the roots by *Rhizobium* bacteria. Millions of these bacteria live in the nodules and provide nitrogen to the plant by fixing it from the atmosphere.

The root system will be sending out many branch roots by Stage V2. These early roots will be shallow within the soil unless the soil is very dry near the surface. Inter-row cultivation for weed control should be shallow to minimize pruning of roots.

Lab Figure 5-13 V2 plant.

Stages V3 to V6

Stage V3 is reached when the four nodes have unfolded leaflets. Plants will be about 7–9 inches tall at V3 and 12–14 inches tall by Stage V6.

During these stages, the plants will begin producing branches at some nodes and flowers at others. Branches and flowers originate from **axillary buds** that are located within the upper angle between the stem and the **leaf petiole.** The petiole is the stalk-like structure that connects the leaflets to the stem. The angle formed by the stem and petiole is called the **leaf axil.**

Axillary buds are similar to the main growing point at the top of the stem that is producing new nodes, internodes, and leaves. If an axillary bud produces a branch, it will be identical to the apical bud found at the tip of the main stem of the plant; producing nodes, internodes, leaves, and eventually flowers. The amount of branching on a soybean plant is decided by the variety, the plant population, and the row spacing. Lower populations and wider rows increase branching in varieties that tend to branch. The first branch will usually sprout from the first trifoliolate node (Node 2).

Lab Figure 5-14 V6 plant.

Axillary buds also allow the plant to recover from damage to the apical bud from hail, insect feeding, etc. The apical bud inhibits the activity of many of the axillary buds. If damaged, one or more of the axillary buds will resume growth; thus minimizing the overall damage. A 50% loss of leaves at Stage V6 will only result in about 3% loss in yield.

At about V5 in most varieties, many of the axillary buds at upper nodes, which are not branching, will appear bushy. They are beginning to produce the flowers that will open at Stage R1. Flowers on soybean develop in clusters called **racemes.** A raceme is a short stem-like structure that produces two or more flowers along its length. The numbers of flowers, and subsequent pods, that will develop at each node are dependent on variety and environmental conditions. Stress will reduce flower production.

By Stage V5, the total number of nodes that will be produced by the plant is decided. Some of the uppermost nodes, however, will not fully develop (have unrolled leaflets) in indeterminate varieties.

By Stage V6 (seven nodes have unfolded leaflets), probably both the cotyledons and unifoliolate leaves have fallen off the plant. However, the nodes are still visible for deciding vegetative leaf stage.

Reproductive Growth Stages

There are eight reproductive growth stages, each designated by the letter "R". Stages R1 and R2 describe flowering; R3 and R4, pod development; R5 and R6, seed development; and R7 and R8, maturation. Vegetative growth will continue after flowering begins, especially in indeterminate varieties.

Each R stage is decided by the beginning of each developmental stage and will continue through at least part of subsequent stages. For example, flowering will continue through part of pod development and seed development.

Stages R1 and R2—Bloom

Stage R1 begins when a flower is open on the plant. Flowering will usually begin when the plant reaches Stage V7 to V10. The first flowers will usually open on the third to sixth nodes of the main stem and flowering progresses upward and downward from there. Flowers on each raceme will open from the base to the tip.

Stage R2, **full bloom,** occurs when there is a flower open at one of the two uppermost nodes on the main stem, which has a fully developed leaf. Remember a fully developed leaf means that the leaf above it has unfolded leaflets. By now, the plant will usually be 17 to 22 inches tall with 8 to 12 fully developed nodes. The plant will likely double its height by maturity and quadruple its dry weight. Stage R2 marks the beginning of a period of rapid growth and nutrient uptake that will continue through R6.

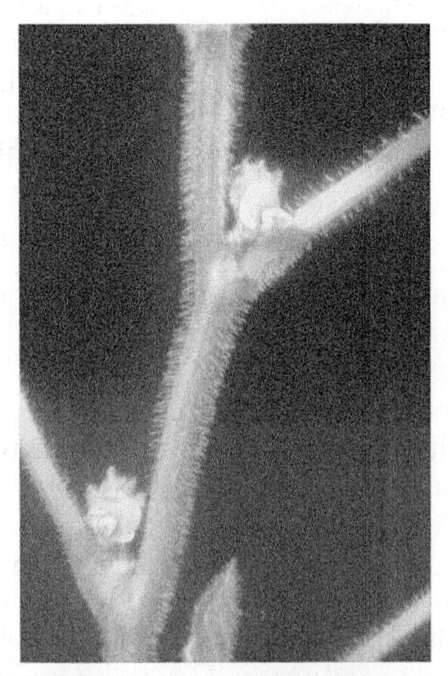

Lab Figure 5-15 Beginning of flowering on main stem.

Lab Figure 5-16 R1 plant.

Since the apical growing points on the main stem and branches are still producing new leaves, and there are a number of axillary buds that can still initiate growth, loss of leaves during this time will result in little yield loss. A 50% leaf loss at Stage R2 will result in only about a 6% yield loss.

Stage R3—Beginning Pod

Stage R3 is reached when a pod 3/16 inch long has developed at one of the four uppermost nodes with a fully developed leaf. At this time there is likely to be developing pods, withering flowers, and open flowers on the plant. The most developed pods will be on the lower nodes where flowering first began.

At this stage, the plant is entering a critical time that decides its final yield potential. As discussed in Lab Exercise 4, soybean yield can be divided into four basic components: plant population, pods per plant, seeds per pod, and seed weight. Assuming that population is adequate, the last three components are being set beginning with Stage R3.

The soybean plant has the potential for much greater production than it finally achieves. Over half of the flowers produced by the plant abort and never contribute to yield. About one-half of the lost flowers abort before they begin pod development, and the other half after pod development begins. This over production of flowers is not totally wasted as it allows the plant to compensate for changes in environmental conditions. Stress from R1 through R3 usually will not greatly affect the yield because the plant can still produce many more flowers, pods, and seeds until Stage R5.

Stage R4—Full Pod

This stage is reached when a pod reaches 3/4 inch long at one of the four uppermost nodes with a fully developed leaf. At this time, the plant will likely have 13 to 20 fully developed nodes and be 28 to 39 inches tall.

Flowering may be completed near the beginning of R4 or beginning of R5. The number of nodes that develop and subsequent plant height are affected by seeding date. Later seeded plants will begin and complete flowering sooner because the photoperiod (daylength) is shorter at an earlier date in the plant's life.

The plant is now entering a period of rapid, steady increase in dry weight in the pods. Seed development is beginning and many pods on the lower nodes have reached full size. This period of rapid growth increases the sensitivity of the plant to stress.

Lab Figure 5-17 Raceme on R4 plant.

Stage R5—Beginning Seed

Stage R5 is reached when the seed is 1/8 inch long in a pod at one of the four uppermost fully developed nodes on the main stem. The plant will have 15 to 23 fully developed nodes and be 30 to 45 inches tall. Rapid seed growth occurs during this stage and the plant is moving dry matter and nutrients from roots, stems, and leaves to the developing seeds.

At the beginning of R5, reproductive development will likely range from newly opened flowers to pods with seeds 1/3 inch long. Midway between R5 and R6 the plant will reach its maximum height, node number, and leaf area; nitrogen fixation rate will peak; and seeds will begin a period of rapid, steady increase in dry weight and nutrient accumulation.

Final yield of the crop will depend on the rate of dry weight accumulation in the seeds and the amount of time that the seeds can accumulate dry matter. There is little difference between varieties in rate of dry matter accumulation. However, higher yielding varieties will usually accumulate dry matter for a longer period of time.

Stage R6—Full Seed

Stage R6 is reached when a pod at one of the four uppermost fully developed nodes contains a green seed that fills the pod cavity. Although this stage is signified by a bean whose width equals its pod cavity, there will be a variety of bean sizes within the pods.

The rate of growth of the beans and the whole plant continues rapidly, but begins to slow as the plant approaches R7. Dry weight and nutrient accumulation in the plant reaches maximum in late R6 and reaches maximum in the seed about R7. Root growth is completed during R6.

Rapid leaf yellowing and death will begin during this stage and continue to about R8. Loss of leaves will begin on the lower nodes and progress upwards. Three to six trifoliolate leaves may have been lost before rapid leaf loss begins.

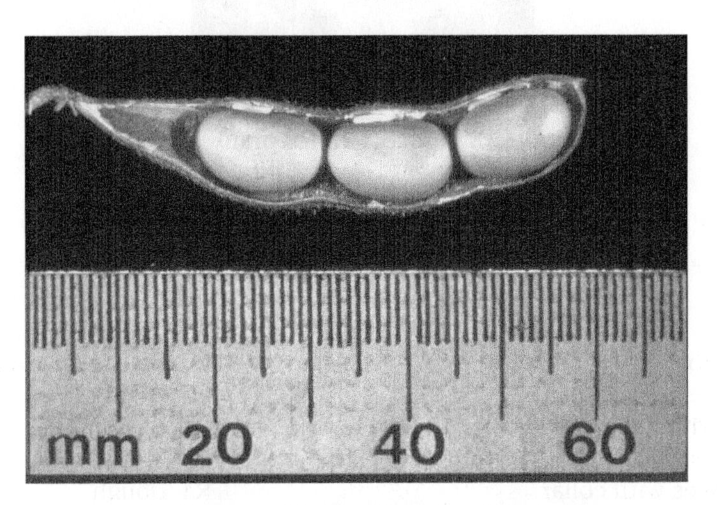

Lab Figure 5-18 Filled pod on R6 plant.

Stage R7—Beginning Maturity

This stage begins when one normal pod on the main stem has reached its mature pod color; which is usually brown or tan depending on the variety. Note that this stage differs from the previous R stages. The mature pod can be located at *any* node, not just the four uppermost nodes.

At R7 the plant is rapidly approaching physiological maturity. Pods that have matured will not accumulate additional dry matter. The seeds will be at about 60% moisture and are fully developed. Stress that occurs from now until R8 will have little effect on yield.

Stage R8—Full Maturity

Full maturity is reached when 95% of the pods have reached their mature pod color. All leaves will likely be lost from the plant. About 5 to 10 days of drying weather will allow soybean to dry to less than 15% moisture for harvesting.

Harvest of soybeans should be timed when the beans are as close as possible to 13% moisture. Earlier harvest will result in the need for drying for safe storage. Delayed harvest increases loss from bean shattering before harvest and during harvest, increased bean seed splitting, and loss of dry weight for sale.

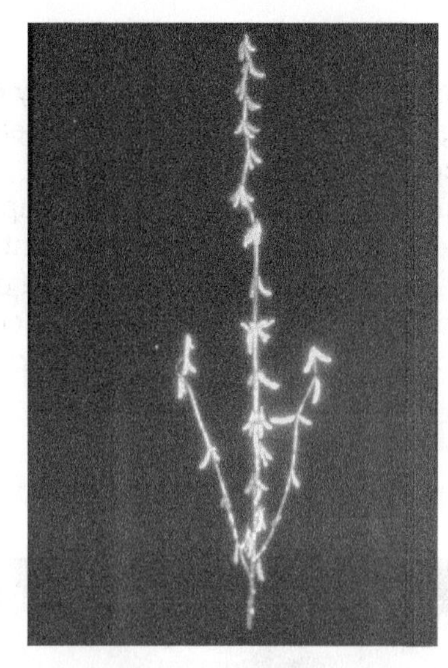

Lab Figure 5-19 R8 plant.

SUMMARY

Stages of Development of Corn

Vegetative Substages	Reproductive Substages
VE Emergence	R1 Silking
V1 One leaf with collar	R2 Blister
V2 Two leaves with collar	R3 Milk
V3 Three leaves with collar	R4 Dough
*	R5 Dent
V(n) nth leaf	R6 Physiological maturity
VT	Tasseling

Stages of Development of Soybean

Vegetative Substages	Reproductive Substages
VE Emergence	R1 Beginning bloom
VC Unifoliolate leaves unroll	R2 Full bloom
V1 One fully developed node	R3 Beginning pod
V2 Two fully developed nodes	R4 Full pod
V3 Three fully developed nodes	R5 Beginning seed
*	R6 Full seed
V(n) n fully developed nodes	R7 Beginning maturity
	R8 Full maturity

Corn photographs are taken from *How a Corn Plant Develops*, Special Report No. 48, Iowa State University. Used with permission.
Soybean photographs are taken from *How a Soybean Plant Develops*, Special Report No. 53, Iowa State University. Used with permission.

STUDY GUIDE

Briefly describe the following growth stages in corn:

VE ________________________________

V3 ________________________________

V6 ________________________________

V12 ________________________________

VT ________________________________

R1 ________________________________

R2 ________________________________

R3 ________________________________

R4 ________________________________

R5 ________________________________

R6 ________________________________

Briefly describe the following growth stages in soybean:

VE ________________________________

VC ________________________________

V1 ________________________________

V6 ________________________________

R1 ________________________________

R2 ________________________________

R3 ________________________________

R4 ________________________________

R5 ________________________________

R6 ________________________________

R7 ________________________________

R8 ________________________________

LABORATORY EXERCISE 6

Effects of Light, Temperature, and Plant Nutrition on Crop Growth

INSTRUCTIONAL OBJECTIVES

Upon completion of this exercise, you should be able to:

1. Define phototropism.
2. Define photoperiod and photoperiodism
3. Explain how a long-day, short-day, and day-neutral plant will respond to different photoperiods.
4. Explain the importance of light on germination of small seeds.
5. Calculate growing degree days using the two methods described in the exercise.
6. Define macronutrient and micronutrient. Name at least four examples of each.
7. Define mobile and immobile nutrient. Name at least three examples of each.
8. Explain how the mobility of a nutrient affects the first appearance of deficiency symptoms.
9. List the three factors that can affect the severity of a nutrient deficiency. Explain how each factor operates.
10. Define induced deficiency.
11. Define chlorosis and necrosis. Describe the appearance of each on a plant.
12. List the general deficiency symptoms of nitrogen, phosphorus, potassium, iron, and zinc.
13. List the relative percentages of N, P, and K in grain, stem, and leaves of corn, sorghum, wheat, and soybean.
14. Explain how nutrient removal from the soil changes when a crop is harvested for grain or forage.

Chapter 5, Soils, Chapter 10, Flowering and Reproduction, and Chapter 12, Climate, Weather, and Crops, have information that you will use to complete this exercise. Begin by reading *Light and Photoperiodism* and *Temperature* on pages 128–130.

LIGHT

Light affects plant distribution, growth, and development. Every plant is adapted to a specific light requirement that must be met in order to survive. Light requirement includes **intensity** or the amount of light energy received during a period of time and most crop plants require high light intensities for maximum production. Light **quality** is a second requirement that plants have. Quality is mostly the wavelength, or color, of light. Plants use mostly visible light for photosynthesis and other plant processes. Light **duration** is the third light requirement of plants. Duration is usually expressed as the **photoperiod**, or the hours of daylight per day.

Phototropism

Phototropism is the influence of light on plant movement or orientation. If exposed to light from only one direction, most plants will grow toward the light source. This type of phototropism is caused by increased cell elongation on the side of the stem away from the light. The increased cell elongation is caused by higher concentrations of auxin on the dark side of the stem.

Stems exhibit positive phototropism; they grow towards the light. Roots, however, exhibit negative phototropism; they grow away from the light. Lab Figure 6-1 illustrates phototropism from a sideways light source.

Some plants exhibit very active phototropism during normal growth. For instance, soybean leaflets orient themselves at right angles to the sun and "follow" the sun from sunrise to sunset. Sunflower heads will face the sun until the plants near maturity.

Etiolation

Excessive lengthening of the internodes, called **etiolation**, is caused by insufficient light intensities on the stem. It results in tall, spindly plants. The excessive elongation is caused by the same increase in auxin described in phototropism. Auxin will accumulate anytime stem tissue receives insufficient light intensity. The most common cause of etiolation is excessive plant population of crop plants and weeds. Other causes are shading and extended cloudy weather.

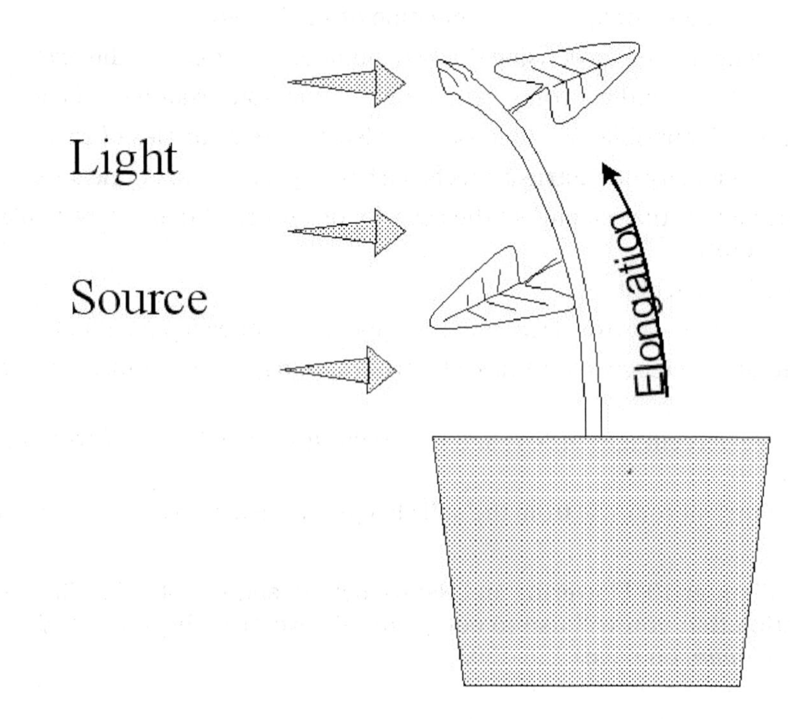

Lab Figure 6-1 Phototropic response to a directional light source.

Photoperiodism

As discussed in Chapter 10, photoperiodism is the response of plants to changes in daylength. Photoperiodism allows plants to respond to seasonal changes that may result in more favorable or unfavorable growing conditions.

Using Figure 10-4 on page 129 as a guide, and the reading assignment in Chapter 10, fill out the following table that illustrates the response of flowering to changing photoperiod. In each case, show whether the plant will flower or remain vegetative.

Table 6-1 Flowering response to photoperiod.

Environment	Plant Type		
	Long-day	Short-day	Day-neutral
Long day—short night	____________	____________	____________
Short day—long night	____________	____________	____________
Short day—night broken by a short period of light	____________	____________	____________
24 hours continuous light	____________	____________	____________

Light and Seed Germination

Some seeds, especially weed seeds, are photoperiod sensitive which assures that they will not germinate during the wrong season. However, many other seeds are light sensitive but are not photoperiod sensitive. These seeds require light of a very short duration to germinate. Phytochrome also functions in this system, but germination is triggered when P_r is converted to P_{fr} by light. This light requirement for germination occurs mostly in small seeds such as pigweed and lettuce, and assures the seed that it is close enough to reach the soil surface with its limited food reserves.

TEMPERATURE

Read *Temperature* on pages 161–163. Temperature affects nearly all of the physiological processes in the plant including growth, flowering, and even photoperiodic response. This exercise, however, concentrates on the effect of temperature on the rate of growth in crop plants.

A plant's response to changing temperatures during the growing season can be measured using a **heat unit** system called **growing degree days**. With most crops, daily changes in temperature give a better measure of growth than time (days after planting). Growing degree days (GDD) are calculated on a daily basis and accumulate throughout the growing season from emergence to maturity.

The upper and lower temperature limits for optimum growth vary with the crop. For corn and soybeans, the limits are about 86° and 50° F. For small grains, the limits are about 90° and 40°F. Other crops may have different limits. Each degree of temperature above the minimum for growth contributes toward development of the plant and is called a **growing degree day unit**.

There are two common ways to calculate GDD. The first method puts limits on the maximum values used in the equation.

Method 1

$$GDD = \frac{\text{Min. Temperature (50)} + \text{Max. Temperature (86)}}{2} - 50$$

Maximum temperatures above 86° F are entered as 86 since the additional heat does not contribute to crop development. Minimum temperatures below 50°F are entered as 50 since temperatures below this level do not reverse development.

The second method does not put any limits into the equation regardless of the daily maximum or minimum.

Method 2

$$GDD = \frac{Min.\ Temperature + Max.\ Temperature}{2} - 50$$

There is little difference between the two methods in accumulated GDD for a growing season. Method 1 accumulates GDD faster than Method 2 in the spring and fall when the minimum temperature is likely to drop below 50° F. Method 1 is slower than Method 2 in the summer when the maximum temperature is likely to be above 86° F. Method 2 allows the use of mean temperatures during a day, week, month, or growing season. However, it is important that only one method be used throughout the growing season.

The National Weather Service (NWS) uses Method 1 as do many corn seed companies. The NWS begins accumulating GDD on March 1. For an individual crop, GDD should be accumulated from emergence. Local GDD information can be obtained from the NWS or the local or state extension services.

Calculate the growing degree day units for the example below. The total accumulated GDD are shown to check your work.

Table 6-2 Calculation of growing degree day units.

Day	Maximum	Minimum	Method 1 GDD	Method 2 GDD
1	94	73	29.5	33.5
2	90	70	28.0	30
3	85	68	26.5	26.5
4	87	65	25.5	26
5	75	62	18.5	18.5
6	71	48	10.5	9.5
7	79	52	15.5	15.5
		Total GDD:	**154.0**	**159.5**

PLANT NUTRITION

Plant nutrition is one of the most critical factors in crop production. Most crops require the addition of at least one nutrient in the form of fertilizer to reach a profitable yield. *Soil Fertility*, Pages 51–55, gives background information on soil nutrients and fertilizers. In this exercise, we will concentrate on plant nutrition and the deficiency symptoms associated with each nutrient.

Classification of Nutrients

Essential nutrients are the elements that are necessary for growth and reproduction in plants. As discussed in Chapter 5, nutrients can be divided into two broad categories, macronutrients and micronutrients. Remember, although micronutrients are used in only small amounts, they are just as essential as macronutrients. A micronutrient deficiency can cause just as much injury to a plant's health as a deficiency of macronutrients.

Nutrients can also be classified according to their relative mobility within the plant. **Mobile nutrients** are readily moved within the plant. If a plant is exhibiting a deficiency of a mobile nutrient, the nutrient will move from older tissue to points of growing tissue to enable the plant to continue growth. This results in the deficiency symptoms first appearing in the older tissue at the bottom of the plant. Mobile nutrients include nitrogen, phosphorus, potassium, magnesium, and chlorine. **Immobile nutrients** cannot be moved from one tissue to another. The deficiency symptoms will first appear in very young tissue and may severely damage meristematic growing points. Immobile nutrients include sulfur, calcium, iron, manganese, zinc, copper, molybdenum, and boron.

Factors Affecting the Severity of a Nutrient Deficiency

Several factors can affect the severity of a particular nutrient deficiency. **Soil factors** such as soil pH and concentration of other elements can affect the availability of certain nutrients to plants. Lab Figure 6-2 shows the availability of nutrients at different pH values. Maximum availability is shown by the widest part of the bar. Nutrients are most available at the pH range of many soils. Plants grown on acid soils can exhibit deficiencies of phosphorus, calcium, potassium, sulfur, and others. Plants grown on alkaline soils can be deficient in iron, zinc, copper, and others.

Excessive concentrations of some elements can make other elements unavailable to plants. For instance, high concentrations of calcium or magnesium can tie up iron and make it unavailable to plants even though it is in abundant supply in the soil. This is called an **induced deficiency** because it is indirectly caused by another factor either internal or external to the plant. Many plant nutrients are affected by an imbalance of other nutrients. Another example of induced deficiency is phosphorus fixation in acid soils.

A second factor that can affect the severity of a nutrient deficiency is **species difference**. Sensitivity to a deficiency of a particular nutrient, especially micronutrients, can vary greatly among different species of plants. For example, sorghum is very sensitive to an induced deficiency of iron while wheat is not affected under the same conditions.

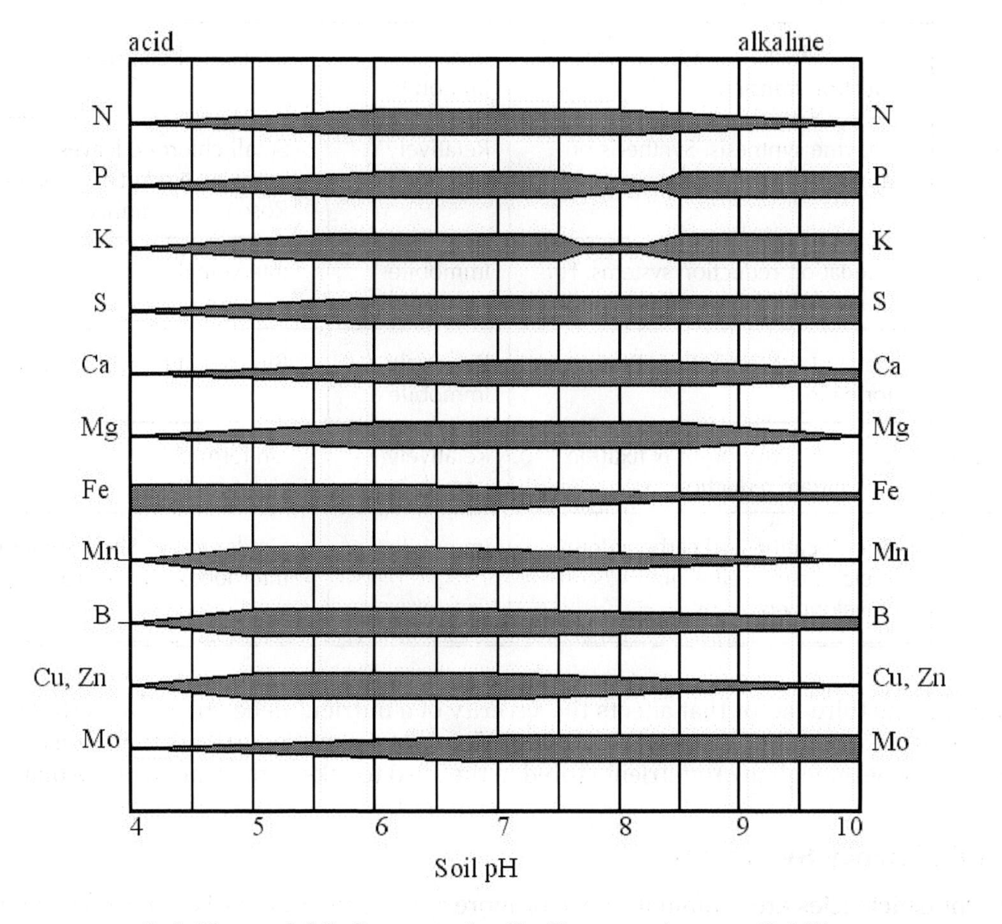

Lab Figure 6-2 Influence of soil pH on nutrient availability.

Table 6-3 Functions and deficiency symptoms of nutrients.

Nutrient	Role in Plant	Mobility	Deficiency Symptoms
Nitrogen (N)	Component of proteins, nucleotides, enzymes, alkaloids. 40–50% of protoplasm.	Mobile	Uniform chlorosis of leaves followed by necrosis. Stunting if severe.
Phosphorus (P)	Component of nucleotides, phospholipids, enzymes, ATP. Synthesis of carbohydrates and proteins.	Mobile	Purplish color with purple veins in leaves. Stunted growth. Poor root growth.
Potassium (K)	Functions as enzyme cofactor to regulate photosynthesis, carbohydrate translocation, protein synthesis, etc.	Very Mobile	Marginal necrosis in leaves. Leaves may curl.
Calcium (Ca)	Component of cell wall.	Immobile	Flattened tops. Growing tips of both roots and stems may die. Leaves distorted.
Magnesium (Mg)	Component of chlorophyll. Enzyme activator.	Mobile	Interveinal chlorosis, red-purple color on leaves.
Sulfur (S)	Component of all proteins. Plant hormones. Plant flavors such as mustard.	Immobile	Uniform chlorosis of whole plant. Thin stems.
Iron (Fe)	Chlorophyll synthesis. Enzymes for electron transfer.	Very Immobile	Interveinal chlorosis.
Zinc (Zn)	Enzyme synthesis. Synthesis of auxin.	Relatively Immobile	Small chlorotic leaves with rough margins. Necrotic spots on leaves. Rosette appearance.
Manganese (Mn)	Oxidation-reduction systems. Formation of O2 in photosynthesis.	Immobile	Interveinal chlorosis.
Copper (Cu)	Catalyst for respiration. Protein formation.	Relatively Immobile	Bluish-green, necrotic leaves.
Molybdenum (Mo)	Enzyme activator for N fixation and nitrate reduction	Relatively Immobile	Uniform chlorosis.
Boron (B)	Translocation and carbohydrate metabolism. Cell wall development.	Immobile	Similar to Ca. Shortening of upper internodes. Brittle growing points.

Seed size is the third factor that affects the severity of a nutrient deficiency. Plants that germinate from large seeds are generally less sensitive to the deficiencies of micronutrients than those from small seeds. There may be enough micronutrient stored in the seed to take care of much of the plant's needs.

Common Deficiency Symptoms

Most nutrient deficiencies are exhibited as one or more symptoms. **Chlorosis** is a yellow color caused by a lack of chlorophyll. There are three types of chlorosis that occur in the leaves. With **interveinal**

chlorosis, the veins remain green while the rest of the leaf tissue is chlorotic. With **intraveinal chlorosis**, the veins are chlorotic while the rest of the leaf is green. With **uniform chlorosis**, all the tissue is chlorotic.

The second common nutrient deficiency is **necrosis**. Necrosis causes death of leaf tissue, usually characterized by brown color. **Stunting** is the third common nutrient deficiency. It is caused by a general slowdown in the plant's metabolism or cell division. Stunting can also be caused by shortened internodes due to less cell elongation. The fourth common nutrient deficiency is **distortion**. It is usually characterized by leaves that become spongy, thickened, or curled.

Functions and Deficiency Symptoms of Nutrients

Table 6-3 lists the functions and most common deficiency symptoms for several nutrients. Note that chlorosis is the most common deficiency symptom. Pay particular attention to the mobility of each nutrient as that will determine where the symptoms will occur.

NUTRIENT UPTAKE AND DISTRIBUTION IN CROP PLANTS

A crop plant takes up soil nutrients throughout the growing season and incorporates them into the stems, leaves, and seeds. Nutrients may be either removed from the field in the harvested seed or forage, or returned to the soil in the crop residue. The relative proportions of the nutrients that are harvested depend on the plant species and whether the plant was harvested for grain, forage, or other purpose. For instance, corn harvested for silage will remove more nutrients than corn harvested for grain.

Lab Figures 6-3, 6-4, and 6-5 show the distribution of nitrogen, phosphorus and potassium in annual crop plants. There is significant translocation of nitrogen and phosphorus from the stems and leaves to the developing grain or seed. There is less translocation of potassium. Therefore, most of the nitrogen and phosphorus taken up by the plant is in the grain, but most of the potassium is in the stems and leaves. Sometimes, nutrients are lost from the aboveground parts of the plant through translocation to the roots or by leaching from the leaves.

Study Lab Figures 6-3, 6-4, and 6-5 and complete the following table. Round all values to the nearest 5 percent. Use the difference between each line to determine the amount to complete the table. For example, nitrogen in corn grain is 60% (100 - 40).

Table 6-4 Percentage of nutrients on a dry weight basis in the aboveground parts of mature crop plants.

	N	P	K
Corn			
Grain	______	______	______
Stem	______	______	______
Leaves	______	______	______
Other parts	______	______	______
Soybean			
Beans	______	______	______
Stem	______	______	______
Leaves	______	______	______
Other parts	______	______	______
Lost from shoot	______	______	______
Sorghum			
Head	______	______	______
Stem	______	______	______
Leaves	______	______	______
Wheat			
Grain	______	______	______
Straw	______	______	______
Leaves	______	______	______
Other parts	______	______	______
Lost from shoot	______	______	______

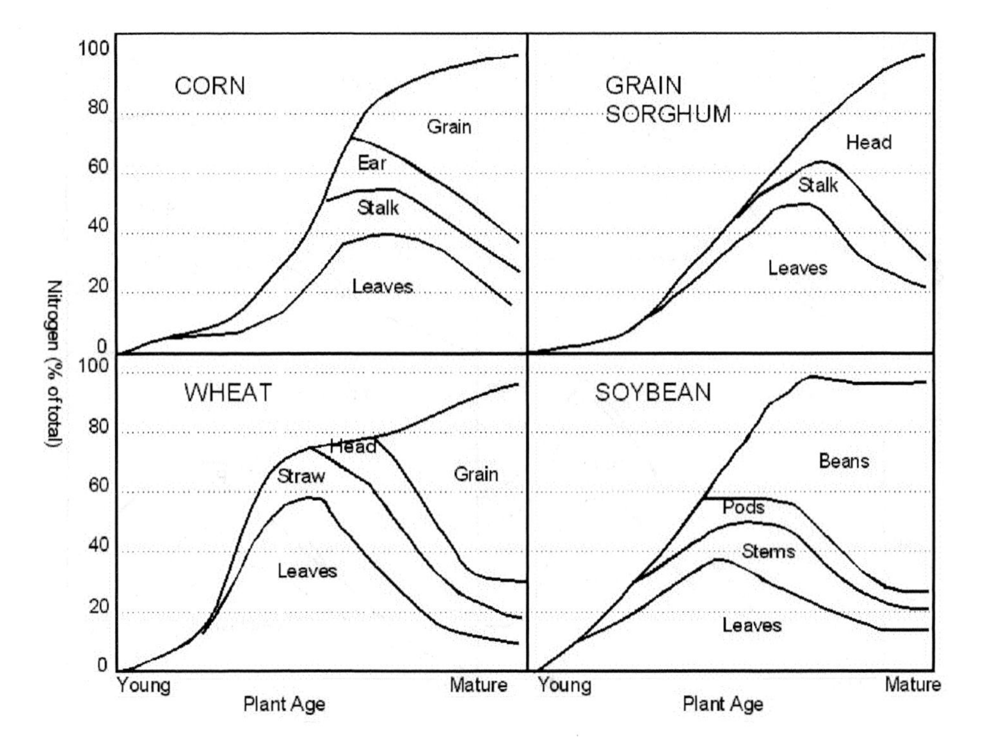

Lab Figure 6-3 Nitrogen distribution in four crops.

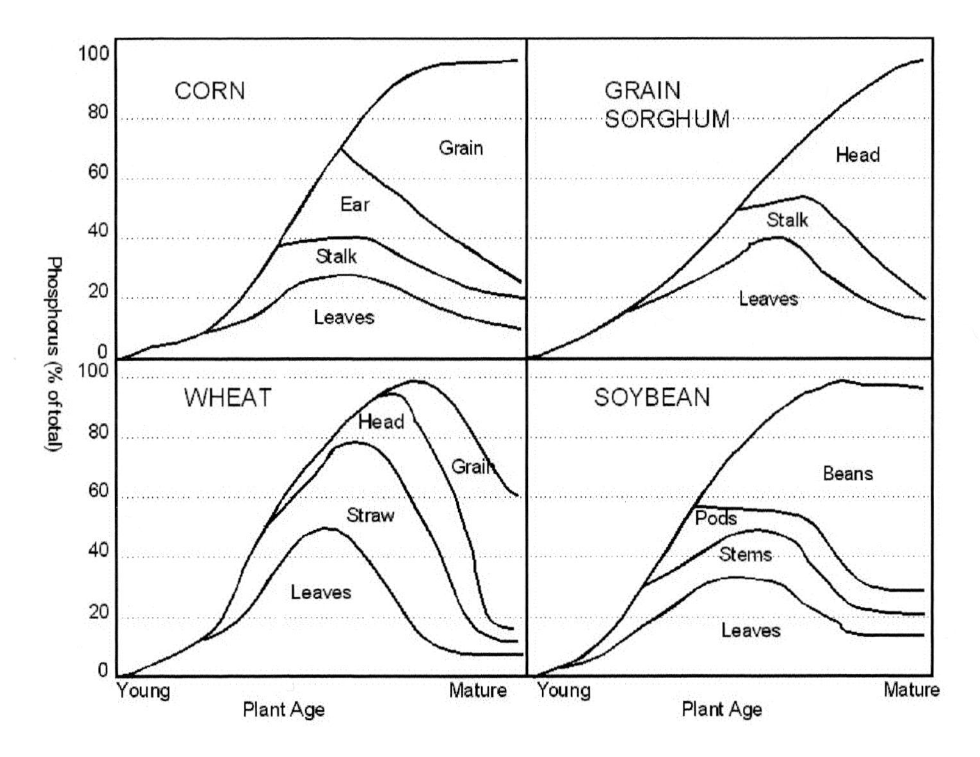

Lab Figure 6-4 Phosphorus distribution in four crops.

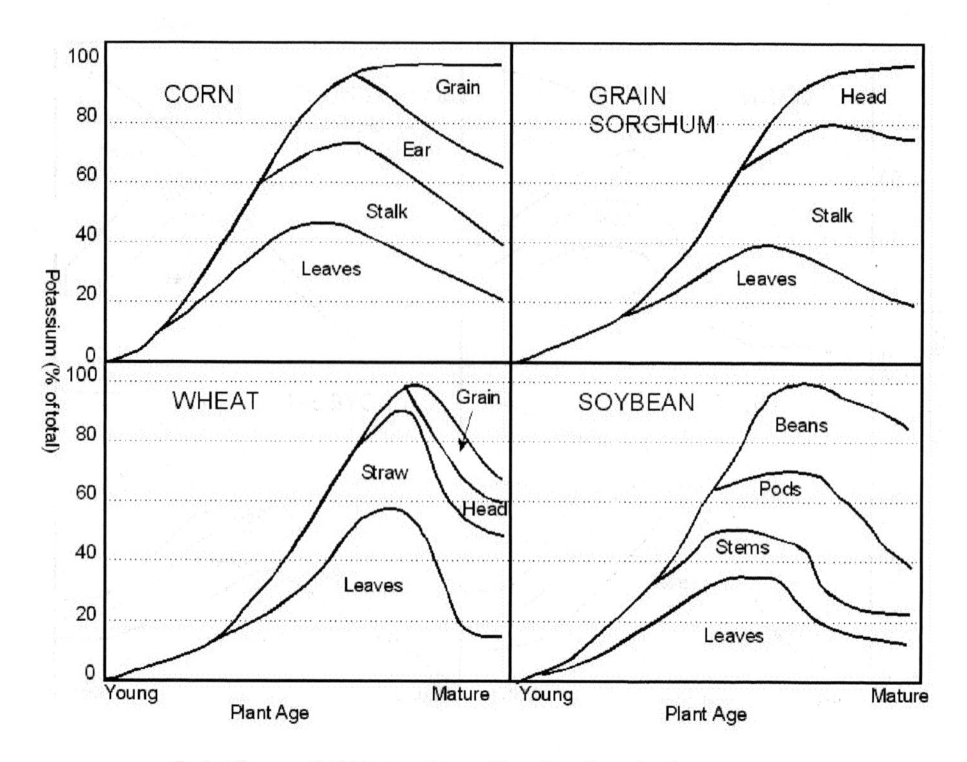

Lab Figure 6-5 Potassium distribution in four crops.

STUDY GUIDE

Complete the following table to help you learn the material in this exercise.

Plant response to directional light?	________________
Photoperiod response of spring flowering plants?	________________
Pigment involved in photoperiodism?	________________
Effect of acid soils on phosphorus?	________________
Most common nutrient deficiency symptom?	________________
Nutrients used in relatively small quantities?	________________
Plant part that holds most nitrogen at maturity?	________________

Briefly describe deficiency symptoms and mobility of the following nutrients.

	Symptom?	Mobile?
Nitrogen	______________________________	____
Phosphorus	______________________________	____
Potassium	______________________________	____
Sulfur	______________________________	____
Iron	______________________________	____
Zinc	______________________________	____

Part III

Agronomic Calculations

LABORATORY EXERCISE 7

Fertilizers, Seeding Rates, and Residue Management

INSTRUCTIONAL OBJECTIVES

Upon completion of this exercise, you should be able to:

1. List the information needed to properly calculate the amount of fertilizer necessary to correctly supply the crop with nutrients.
2. Correctly solve the practice problems provided which calculate the amount of fertilizer to apply to a crop field.
3. List the information needed to properly calculate the seeding rate for a crop.
4. Correctly solve the practice problems provided which calculate the seeding rate for a crop.
5. List the steps to estimate the percent residue cover over a crop field using the line-transect method.
6. Correctly estimate the percent residue cover in the demonstration provided.

INTRODUCTION

There are many types of arithmetic calculations that need to be done in the management of a crop. In this laboratory exercise, and in Exercise 8, Pesticide Application, you will learn how to do many of these calculations.

FERTILIZER APPLICATION CALCULATIONS

Fertilizers contain mostly nitrogen, phosphorus, and potassium, but may also contain other nutrients as well, such as micronutrients. **Fertilizer analysis** is usually expressed as percent nitrogen (N), phosphate (P_2O_5), and potash (K_2O). For example, a fertilizer with an analysis of 18-46-0 contains 18% N, 46% P_2O_5, and 0% K_2O. Notice that only 64% of the fertilizer in this example contains available plant nutrients; the remaining 36% may be carrier or filler to make the fertilizer easier to handle, or may be other elements such as hydrogen or oxygen that have no nutrient value. In addition to these nutrients, a fertilizer may contain other nutrients, such as sulfur or zinc.

Fertilizers are added to field crops for several reasons. Most commonly, the fertilizer is applied to correct an existing or potential nutrient deficiency. Sometimes a fertilizer is applied to help increase a plant's disease resistance. For instance, adding sulfur to potatoes increases resistance to certain diseases. Adequate soil fertility prevents the uptake of toxic minerals or compounds. For example, adequate amounts of other nutrients can inhibit uptake of strontium in soils high in this element.

When calculating the amount of fertilizer to apply to any field, you need to know the following information:

1. The amount of actual nutrients needed, usually on a per acre basis. This is determined by taking a representative sample of the soil, sending it to a testing laboratory for nutrient analysis, and obtaining fertilizer recommendations based on the analysis.
2. The nutrient analysis of the fertilizer that will be applied.
3. The size of the field or area that will receive the application of fertilizer.
4. The rate at which the applicator, such as a sprayer or spreader, will apply the fertilizer. This is usually adjusted to apply the proper amount.

It is always advisable to measure the amount of fertilizer that the applicator is applying to a given area to assure the accurate rate of application. If the rate applied by the applicator is too low, the crop will likely experience a nutrient deficiency during the growing season. If applied at too high a rate, the crop may be harmed by the excess nutrients, the environment may be harmed, or, at the very least, money will be wasted.

To measure the actual amount of dry fertilizer applied, the applicator is driven over a measured distance while the fertilizer is collected using one or more containers. To calculate the amount applied, use the following formula where distance and width are measured in feet:

$$Pounds / acre = \frac{pounds\ applied}{(Distance\ X\ Width)\ X\ 43{,}560ft^2\ /Acre} \tag{1}$$

To measure the actual amount of liquid fertilizer applied, use the same technique described above and the following formula:

$$Gallons/acre = \frac{gallons\ applied}{(Distance\ X\ Width)\ X\ 43{,}560ft^2\ /Acre} \tag{2}$$

To calculate the amount of liquid fertilizer to apply in a field, use the following formula:

$$Gallons\ fertilizer/acre = \frac{pounds\ nutrient/acre \div pounds/gallon}{\%\ nutrient\ \div 100} \tag{3}$$

For a dry fertilizer, use the following formula:

$$Pounds\ fertilizer/acre = \frac{pounds\ nutrient/acre}{\%\ nutrient\ \div 100} \tag{4}$$

The next step is to adjust the applicator to apply the correct amount as calculated in Equations 3 or 4. The actual method to adjust the applicator depends on the model, and is supplied by the manufacturer.

When the fertilizer contains more than one nutrient, the amount to apply is based on only one nutrient. It is necessary, however, to calculate the amount of the other nutrients that are being applied so that they are figured into the overall fertilizer program for the field. The following example illustrates this:

You need to apply 150 pounds/acre of nitrogen, and 20 pounds/acre of phosphate. You have purchased a 18-46-0 fertilizer that you will use to apply the phosphate, and a 33-0-0 to supply the remaining nitrogen. The amount of phosphate to apply using the 18-46-0 is calculated as follows.

$$Pounds\ 18-46-0\ /\ acre = \frac{20\ pounds\ phosphate\ /\ acre}{46\ \div\ 100} = 43.5 \quad (5)$$

The next step is to calculate how much N is also being applied. To do this, use the following formula:

$$Pounds\ nutrient/acre = pounds\ fertilizer/acre\ X\ (\%\ nutrient \div 100) \quad (6)$$

or, for this example:

$$Pounds\ N/acre = 43.5\ pounds/acre\ X\ (18\%\ N \div 100) = 7.8 \quad (7)$$

To calculate the amount of 33-0-0 to apply to provide the remaining N requirements, use the following formula:

$$Pounds\ fertilizer/Acre = \frac{pounds\ required - acre\ pounds\ applied}{\%\ nutrient\ \div\ 100} \quad (8)$$

or, for this example:

$$Pounds\ 33-0-0/Acre = \frac{150\ lbs/acre - 7.8\ lbs\ applied}{33\%\ N \div 100} = 431 \quad (9)$$

For a liquid fertilizer, you would divide the answer in Equation 9 by pounds/gallon.

Some fertilizers, such as phosphate or zinc, can be most efficiently applied in a band at seeding as discussed in Chapter 5. When band applying a fertilizer, application rates are based on individual rows. To calibrate the fertilizer application equipment on the planter, simply substitute row width, in feet, for width in Equation 1, and use the same procedure described previously for calibrating the applicator.

Work the practice fertilizer problems at the end of this exercise.

SEEDING RATE CALCULATIONS

Seeding rate of any crop is dependent on the type of crop and expected yield goal as discussed in Chapter 6. When calculating the seeding rate, you need to know the following information:

1. Percent germination. This information is found on the tag attached to the seed bag.
2. Expected field emergence. The average is 85% for most crops, but if percent germination is less than 90%, it will need to be adjusted as shown below.
3. Target plant population. This information is supplied by the seed grower, or can be obtained from your local Extension Service.
4. The row spacing you will use.

As discussed earlier with fertilizer application, it is advisable to measure the amount of seed that your planter is applying. Too low a seeding rate results in lower yields and increased weed problems. If the seeding rate is too high, the crop plants may experience stress from increased competition and higher water use.

To measure the amount of seed that the planter is applying it is necessary to drive the planter over a measured distance while collecting the seed, similar to the method used to measure fertilizer application rate described previously. To calculate the amount applied, use the following steps:

First calculate the length of the row that equals one acre:

$$Feet/row/Acre = \frac{43,560\ ft^2\ /acre}{row\ spacing, feet} \quad (10)$$

Note that row spacing is in feet, not inches. To convert to feet, divide inches by 12.
Next, calculate the adjusted seeding rate:

$$Adjusted\ seeding\ rate\ = \frac{recommended\ plant\ population}{\%\ expected\ field\ emergence \div 100} \quad (11)$$

Use 85% expected field emergence, for grain crops. If percent germination is less than 90, multiply 0.85 by percent germination/100.
Next, calculate the number of seeds that should be applied per foot of row:

$$Recommended\ seeds/foot\ of\ row = \frac{adjusted\ seeding\ rate}{feet/row/acre} \quad (12)$$

Then, after driving the planter over a measured distance, calculate the seeds/foot that was applied:

$$Applied\ seeds/foot\ of\ row = \frac{seeds\ applied}{distance\ traveled,\ feet} \quad (13)$$

and compare to the recommended rate (Equation 12). Adjust the planter as needed using the manufacturer's instructions. Every row should be measured and adjusted as needed.
Sometimes, it is easier to understand seeding rate expressed as inches between seeds instead of seeds/foot. To calculate inches between seeds, divide 12 by seeds/foot.
The following illustration provides an example. Soybeans will be seeded in 30 inch rows at a final population of 150,000 plants/acre.
Calculate the feet/row/acre:

$$Feet/row/acre = \frac{43,560\ ft^2\ /acre}{2.5\ feet} = 17,424 \quad (14)$$

Next, calculate the adjusted seeding rate: We will use 85% for the expected field emergence.

$$Adjusting\ seeding\ rate = \frac{150,000}{85 \div 100} = 176,470 \quad (15)$$

Now, calculate the seeds/foot that the planter should be applying:

$$Recommended\ seeds/foot\ of\ row = \frac{176,470}{17,424} = 10.1 \quad (16)$$

Inches between seeds would be:

$$Inches/seed = \frac{12}{10.1} = 1.2 \quad (17)$$

or a little under 1 1/4 inch.
When determining actual plant population after the crop has emerged, a common method is to measure the length of the row that equals 1/1000 acre. That is determined by dividing feet/row/acre by

1000. For example, 1/1000 acre of 30 inch rows would be 17,424/1000 or 17.4 feet. After counting the plants in that distance, multiply by 1000. For example, 150 plants would be equivalent to 150,000 plants per acre. This procedure should be repeated several times in different locations in the field and an average taken of all the counts.

Work the practice seeding problems at the end of this exercise.

DETERMINING RESIDUE COVER

Adequate cover of the soil surface by plant residues is important in protecting the soil from erosion as discussed in Chapter 5. A simple method for estimating crop residue cover is called the **line-transect method.**

The line-transect method involves stretching a 100 ft. or a 50 ft. measuring tape across the field surface, and then count the number of times a piece of crop residue is located beneath each foot mark on the tape. The step-by-step procedure is:

1. Find a representative area for the whole field. Avoid end rows, or areas that have been affected by flooding, drought, weed or insect infestation, or other factors that might have resulted in lower yields.
2. Stretch a 50 ft. or 100 ft. measuring tape diagonally across the crop rows. Anchor one end so the tape is fully extended across several passes of the tillage implements used.
3. Check for crop residue at each foot mark of the tape. Do not move the tape while counting. Look at the same side of the tape at each foot mark. Look straight down at the tape and foot mark. Count only those foot marks that have residue directly under them. Do not count if questionable.
4. Calculate percent cover. The numbers of foot marks on a 100 ft. tape that are directly over residue determine the percent cover. For example, if 23 foot marks are directly over residue, the residue cover is 23%. If a 50 ft. tape is used, multiply by two the number of foot marks you count.
5. Repeat this procedure at least three times over the field and obtain an average for the locations.

Using the demonstration material available, decide the percent residue cover in each case.

PRACTICE PROBLEM SET 1

1. Lester Lotsacorn sampled his soil, sent the sample to a testing laboratory, and the analysis shows that he needs to apply 150 lb/A N and 25 lb/A banded phosphate to the corn crop he will be seeding. He can purchase 45-0-0 and 18-46-0 fertilizer. How much of each fertilizer should he apply per acre?

2. How much of each fertilizer should he purchase for his 200-acre field?

3. Lester will band apply the phosphate during seeding his crop, and needs to calibrate the fertilizer applicator on his planter. How much phosphate fertilizer should his applicator apply in each row over a 200 ft. distance if the row spacing is 36 inches?

1 Acre = 43,560 ft²

$\frac{\text{lb } 18\text{-}46\text{-}0}{\text{Field}} = \frac{54 \text{ lb } 18\text{-}46\text{-}0}{43{,}560 \text{ ft}^2} = \frac{600 \text{ ft}^2}{\text{Area}} = \frac{32{,}400}{43{,}560} = \frac{.74 \text{ lbs } 18\text{-}46\text{-}0}{\text{Area}}$

$\frac{200 \text{ ft} \times 36 \text{ in}}{1 \text{ ft}/12 \text{ in}} = \frac{7200}{12} = 600 \text{ ft}^2$

4. Lester is calibrating his nitrogen fertilizer applicator that spreads dry fertilizer over a 50 ft width. After traveling 200 ft. while collecting the fertilizer, the applicator had applied 80 pounds. How many pounds/acre is his applicator applying?

$\frac{\text{lb } 45\text{-}0\text{-}0}{\text{Acre}} \quad \frac{80 \text{ lbs } 45\text{-}0\text{-}0}{10{,}000 \text{ ft}^2} \times \frac{43{,}560 \text{ ft}^2}{\text{Acre}} = \frac{80 \times 43{,}560}{10{,}000} = \frac{348 \text{ lbs } 45\text{-}0\text{-}0}{\text{Acre}}$

50 x 200 = 10,000 ft²

5. If the applicator was properly adjusted, how many pounds of the 45-0-0 should he be applying over the 200 ft distance?

$\frac{\text{lb } 45\text{-}0\text{-}0}{\text{Area}} \quad \frac{312 \text{ lbs } 45\text{-}0\text{-}0}{\text{Acre}} \times \frac{\text{Acre}}{43{,}560 \text{ ft}^2} \times \frac{10{,}000 \text{ ft}^2}{\text{Area}} =$

$\frac{312}{1} \times \frac{1}{43{,}560} \times \frac{10{,}000}{1} = \frac{3120000}{43{,}560} = \frac{71.6 \text{ lb } 45\text{-}0\text{-}0}{\text{Area}}$

6. Now Lester needs to calibrate his planter to seed his corn crop. He wants a final plant population of 22,000 plants/A and the germination rate on the bag is listed as 90%. How many seeds/ft should his planter be dropping?

$\frac{\text{Seeds}}{\text{Row Ft}} = \frac{22{,}000 \text{ plants}}{\text{ACR}} \times \frac{\text{ACR}}{14{,}520 \text{ RW FT}} \times \frac{100 \text{ Seeds}}{85 \text{ Plants}} = 1.78$

$\frac{43{,}560 \text{ ft}^2}{3 \text{ ft}}$ = 14,520 RW Ft

$\frac{22000 \times 1 \times 100}{1 \times 14{,}520 \times 85} = \frac{1.78 \text{ seeds}}{\text{Row FT}}$ (1.8)

7. How many inches between seeds should he have?

$\frac{\text{Inches}}{\text{Seed}} = \frac{12}{1.78} = \frac{12}{1.8} = 1.8 \text{ inches/seeds}$

8. After completing his preplant tillage, Lester checks the percent residue cover in his field. Using a 100 ft. measuring tape, he counts 34 pieces of residue under the foot marks. What is the percent residue cover?

9. After emergence, Lester wants to check the plant population in his field. What measurement should he use to measure the length of row equal to 1/1000 acre?

PRACTICE PROBLEM SET 2

1. Sally Stubblefield has soil sample results showing that she needs to apply 90 lb/A of nitrogen and 50 lb/A of phosphate for her wheat crop. She will use 18-46-0 starter fertilizer at planting and topdress with 34-0-0 in the spring. How much of each fertilizer should she apply?

$\frac{\text{lb 18-46-0}}{\text{Acre}} = \frac{50 \text{ lb } P_2O_5}{1 \text{ Acre}} = \frac{100 \text{ lb 18-46-0}}{46 \text{ lb } P_2O_5} = \frac{5000}{46} = \frac{109 \text{ lb 18-46-0}}{\text{Acre}}$

$109 \times .18 = 19.62$

$\frac{\text{lb 34-0-0}}{\text{Acre}} = \frac{90 - 19.62 \text{ lb N}}{\text{Acre}} \times \frac{100 \text{ lb 34-0-0}}{34 \text{ N lb}} = \frac{7038}{34} = \frac{207 \text{ lb 34-0-0}}{\text{Acre}}$

2. Sally will apply the phosphate with her grain drill at planting and needs to calibrate it. How much fertilizer should she apply in each 250 ft. of row if the row spacing is 8 in.

$\frac{250 \text{ ft} \times 8 \text{ in}}{1 \text{ ft}/12} = \frac{2000}{12} = 166.666 \text{ ft}^2$

$\frac{\text{lb 18-46-0}}{\text{Area}} = \frac{166.666 \text{ ft}^2}{\text{Area}} \times \frac{50 \text{ lb } P_2O_5}{43,560 \text{ ft}^2} \times \frac{100 \text{ lb 18-46-0}}{46 \; P_2O_5} = \frac{833330}{2003760} = \frac{.42 \text{ lb 18-46-0}}{\text{Area}}$

3. Then Sally calibrates the seeding rate of the drill by collecting seed from 3 rows for 100 ft. and finds the 3 rows have sown 0.71 lb. of seed. What is this seeding rate?

4. Sally wanted to plant 90 lb/A. How much seed should she have collected from the three rows?

5. If Sally's seed has 14,000 seeds/lb, and the germination rate is 90%, what should be the plant population if she plants 90 lb/A?

6. Sally takes her yard stick to the field to check the population. She finds an average of 42 plants/yard of row. What is the plant population?

7. Sally needs to calibrate her 40 ft fertilizer spreader the following spring. If she drives 150 ft and the spreader drops 38 lb of 34-0-0, how much fertilizer/A is she applying?

8. How much fertilizer should have been spread in the 150 ft length if it was set properly?

Answers to Practice Problem Set 1:

1. 54 lb 18-46-0, 312 lb 45-0-0
2. 62,400 lb 45-0-0 and 10,800 lb 18-46-0
3. 0.74 lb
4. 348 lb
5. 71.6 lb
6. 1.8 seeds/ft
7. 6.67 inches
8. 34%
9. 14.5 ft

Answers to Practice Problem Set 2

1. 109 lb 18-46-0, 207 lb 34-0-0
2. 0.42 lbs
3. 155 lb/A
4. 0.41 lbs
5. 1,071,000 plants/A
6. 914,760 plant/A
7. 276 lb
8. 28.5 lb

LABORATORY EXERCISE 8

Pesticide Application

INSTRUCTIONAL OBJECTIVES

1. Define pesticide. List two important things that must be done before a pesticide is applied.
2. Describe the different kinds of pesticide formulations.
3. Explain the difference between active ingredient and inert ingredient.
4. List what a pesticide label must contain by law. Explain why this information is important.
5. Define selective and nonselective pesticides.
6. Define translocated and non-translocated pesticides.
7. List the times a pesticide can be applied in relation to the crop.
8. List the four factors that influence the rate of application of a pesticide.
9. List the basic steps in calibrating a sprayer.
10. Solve the practice problems provided.

PESTICIDES

A pesticide is a chemical that is used to kill or inhibit plant and/or animal life. There are four general categories of pesticides: *rodenticides, fungicides, insecticides,* and *herbicides*. Each pesticide is designed to do a specific job and care must be exercised when using it to make certain that it is used safely and does not become a danger to other living things.

Pesticides are inherently dangerous and become more so when carelessly applied. The small amount of time needed to make certain that pesticides are used and applied correctly is insignificant when compared to the safety of the operator, the environment, and the crops involved.

When applying pesticides, strict attention must be given to two important items: (1) the information on the pesticide label, and (2) the calibration of the application equipment.

Pesticide Formulations

Most pesticides are purchased in one of these types of formulation:

Dry: dusts, granules, etc., which are applied in the dry form.

Liquids: forms a liquid solution with water or other solvent.

Wettable powders: a dry powder that remains suspended in a carrier solution if kept agitated.

Emulsifiable concentrate: a liquid that will not mix with water or other carrier but will remain suspended as droplets within the carrier if kept agitated.
Soluble powders: a dry powder or granule that will dissolve and form a true solution when mixed with water or other solvent.

Active Ingredients

The portion of a pesticide formulation that is chemically active is called the **active ingredient(s)**. It is the chemical in the formulation that has pesticidal properties (makes it a pesticide). Pesticides are rarely sold as pure compounds; instead, they are mixed with other compounds, called **inert ingredients**, which have no pesticidal properties.

Inert ingredients are added to make pesticides easier to handle, store, transport, or mix with a carrier. In dry formulations, inert ingredients are added to reduce the toxicity of the material, making it easier and safer to apply.

Pesticide Labels

By law, a pesticide label must contain the following information:

Name and Address of manufacturer.
Name of product: the commercial name, such as Atrazine 4L.
What the product is: a selective herbicide, contact insecticide, etc.
Ingredients: chemical name of the product, percent active ingredients and percent inert ingredients.
EPA Registration number: proof that the product was approved by the U.S. Environmental Protection Agency
Establishment number: identifies the specific facility that produced the product.
Classification statement: based on hazards, intended use, and environmental. Pesticides are classified either as "general use" or "restricted use".
Directions for use: specific pests controlled, crops on which it can be applied, timing and rate of application, etc.
Signal words and symbol: three key words that indicate the level of toxicity to humans and animals:
- **Danger:** highly toxic. A taste to a teaspoonful can kill the average person.
- **Warning:** moderately toxic. A teaspoonful to a tablespoonful can kill.
- **Caution:** low toxicity. A tablespoonful to more than a pint is needed to kill the average person.

Precautionary statements: health and safety guidelines while handling the product. Includes ways the pesticide may enter the body, and protective clothing and equipment statement.
Statement of practical treatment: first aid guidelines including a "Note to Physicians".
Environmental hazard statement: common sense reminders to avoid environmental contamination. May include specific warning statements such as "The product is highly toxic to bees".
Reentry statement: if necessary, specific time before one can enter treated field without protective clothing.
Storage and disposal statement: outlines methods for proper storage of chemical and proper disposal of empty containers.

Methods and timing of Application

Read *Chemical Weed Control* in Chapter 15 of the text. This section outlines differences in selectivity, site of action, time of application, and length of pesticidal activity for herbicides that can be applied to all pesticides. There are also some variations of methods that can be used to apply pesticides:

Broadcast application—the pesticide is applied uniformly to the entire surface of the soil or plant canopy.
Banded application—the pesticide is applied in bands or strips that cover only a portion of the soil or canopy surface. Bands can be between rows or over the rows.
Spot application—the pesticide is applied only to very small areas, such as a single plant.

Lab Figure 8-1 Broadcast application (left) and band application (right).

Directed application—the pesticide is aimed at just a portion of the plant, such as the base of the stem or the leaves.

Examples of broadcast and band applications are shown in Lab Figure 8-1.

APPLICATOR CALIBRATION

Pesticides are important tools in crop production. However, they are only safe and effective when properly applied in strict accordance with label directions. An extremely important step in proper application of pesticides is accurate calibration of the equipment that will apply it. In this exercise, we will concentrate on spray equipment. However, the basic factors discussed with sprayers can also be applied to other types of equipment such as dry applicators.

The rate of application of a pesticide is influenced by four factors:

Speed of travel: determines the area covered by the applicator in a given time.
Pressure: determines the rate at which the spray solution will pass through the nozzles.
Nozzle size and type: determines the amount of material that will be sprayed in a given time and the pattern of coverage over the surface of the soil.
Concentration of pesticide: determines the amount of active ingredient in a given volume of spray mixture.

When applying solid materials with dry applicators, nozzle size and pressure are replaced by size of discharge chutes and speed of applicator.

Steps to Calibrate a Sprayer

Sprayer nozzles are the least expensive, yet most often overlooked part of a sprayer. It is important to make sure that the nozzles are properly installed and calibrated.

Select the proper nozzle size and type and install on the sprayer. All nozzles should be the same. Nozzle height and spacing for uniform coverage of spray are specific for each nozzle type. Next, select a pressure that will give the desired spray pattern and gallons per acre. Information on nozzle height, spacing, and pressure is available from all quality dealers.

Next, determine the speed at which the sprayer will travel. Using the tractor gear and throttle setting that will be used when spraying, record the time required to travel 100 feet. Repeat, traveling in the opposite direction. If the time differs by more than one or two seconds, repeat the measurement, using a longer distance of travel. These measurements should be taken in the field that will be treated, or in a field with similar soil conditions. Speed on a hard or paved surface will be different. DO NOT rely on the tractor's speedometer for determining speed, as most speedometers are not accurate enough. Use the following formula to calculate applicator speed in feet/minute:

$$Feet/minute = \frac{60\ sec/min\ X\ feet\ traveled}{seconds} \quad (1)$$

Add the carrier, such as water, to the spray tank but do not add the pesticide. Attach pint jars to several nozzles. Using the proper pressure, record the seconds needed to fill the pint jars. If the time needed to fill the jars varies more that a few seconds replace the nozzles that require different times. Determine the output of the nozzles using the following formula:

$$Gallons/minute/nozzle = \frac{60\ seconds/minute}{8\ pints/gallon\ X\ seconds/pint} \quad (2)$$

Now calculate the gallons/minute output of the entire sprayer by multiplying Equation 2 by the number of nozzles on the sprayer:

$$Gallons/minute/sprayer = gallons/minute/nozzle\ X\ nozzles \quad (3)$$

The next step is to calculate the width that the sprayer will cover by multiplying the number of nozzles by the nozzle spacing and converting to feet:

$$Sprayer\ width = \frac{nozzles\ X\ nozzle\ spacing,\ inches}{12\ inches/ft} \quad (4)$$

You are now ready to calculate the gallons/acre that the sprayer will apply:

$$Gallons/acre = \frac{43,560\ ft^2/acre\ X\ gal/min/sprayer}{ft/min\ X\ sprayer\ width,\ ft} \quad (5)$$

Using gallons/acre as calculated in Equation 5, calculate the amount of pesticide to add to each gallon of spray solution. Use the pesticide label to determine the rate at which the pesticide will be applied to the crop. Units per acre can be gallons, quarts, pints, pounds, etc. Use the following formula substituting the appropriate unit:

$$Units/gallon = \frac{recommended\ rate/A}{gallons/A} \quad (6)$$

The last step calculates the amount of pesticide to add to the sprayer tank:

$$Units/tank = units/gallon\ X\ tank\ capacity,\ gallons \quad (7)$$

Example Problem

The following example illustrates the use of the calculations just discussed. Atrazine 4L herbicide will be applied to a field that will be seeded to corn at the rate of 6 pints/acre. It takes 13 seconds to travel 100 feet with the sprayer. The nozzle output is 12 seconds/pint, and the 15 nozzles are spaced 20 inches apart. How much herbicide should be added to a 250 gallon tank?

Using Equation 1:

$$Feet/minute = \frac{60\ sec/min\ X\ 100\ feet}{13\ seconds} = 461.54 \quad (8)$$

Using Equation 2:

$$Gallons/minute/nozzle = \frac{60\ seconds/minute}{8\ pints/gallon\ X\ 12\ seconds/pint} = 0.625 \quad (9)$$

Using Equation 3:

$$Gallons/minute/sprayer = 0.625\ gallons/minute/nozzle\ X\ 15\ nozzles = 9.375 \quad (10)$$

Using Equation 4:

$$Sprayer\ width = \frac{15\ nozzles\ X\ 20\ inches}{12\ inches/ft} = 25\ feet \quad (11)$$

Using Equation 5:

$$Gallons/acre = \frac{43,560\ ft2/acre\ X\ 9.375\ gal/min/sprayer}{461.54\ ft/min\ X\ 25\ ft} \quad (12)$$

Using Equation 6: (Note that the units are pints.)

$$Pints/gallon = \frac{6\ pints/A}{35.4\ gallons/A} = 0.17 \quad (13)$$

Using Equation 7:

$$Pints/tank = 0.17\ pints/gallon\ X\ 250\ gallons = 42.5 \quad (14)$$

If you have questions regarding the steps just discussed, or if you don't understand the mathematical logic used in the calculation, ask your instructor for help.

Other Calculations

When calculating the amount of pesticide to apply to a small area, different calculations may be used. Rates for pesticides specially formulated for lawns and gardens are usually given on a square foot or 100 square feet basis. To determine how much to apply, measure the area to be treated and calculate the amount needed for that area. For example, if the label rate is 1 pound/100 ft^2, and you have 300 ft^2, apply 3 pounds.

When applying a pesticide whose label uses acres for the rate, divide the area to be treated by 43,560 ft^2/acre to calculate the fraction of an acre that will be treated. One of the practice problems illustrates this.

When comparing costs of two formulations of the same pesticide, calculate the cost of each formulation on an active ingredient basis. Divide the cost per unit by the percent active ingredient (a.i.) listed on the label, **expressed as a decimal**. If the units are different; for example, pints (liquid) and pounds (dry), convert the liquid to a weight. The seller of the pesticide should be able to supply information on the weight of liquid pesticides. Be sure that units are the same. For example, if liquid weight is given as pounds/gallon, you will need to divide it by 8 to convert it to pints. One of the practice problems illustrates this.

Band application of pesticides, especially herbicides, is a common way to save money and minimize residual carryover. When calculating the amount of pesticide to add to a sprayer that will apply bands, calculate the normal amount that would be applied, and then divide the result by the fraction of soil that will be sprayed. For example, if 12 inch bands will be sprayed over 36 inch rows, then the fraction sprayed is 1/3, so divide the result by 3. Two of the practice problems illustrate this.

Work the practice problems that follow.

PRACTICE PROBLEM SET 1

1. A tractor pulling a sprayer travels 100 feet in 25 seconds. The average nozzle output is measured as 1 pt/15 sec, and the ten nozzles are 20 inches apart. How much 80% wettable powder should be added to a 200-gal tank if the recommended application rate is 3 lb/acre?

2. A sprayer has 10 nozzles spaced 36 inches apart along the boom. The average time needed to fill a pint jar is 8 sec. The sprayer travels 440 feet/minute. How many gallons/acre are being applied?

3. Your front lawn measures 60 feet by 25 feet. How many ounces of herbicide would be needed if the label rate is expressed as 3.75 pounds/acre? (1 pound=16 ounces)

4. How many gallons will be needed to spray the above lawn if the recommended rate of spray solution is 30 gal of solution/acre?

5. How many acres will you be able to spray with one tankful if you apply 30 gallons/acre with a sprayer that covers a width of 60 ft, while applying 3 lb/acre of pesticide, and have a tank capacity of 600 gal?

6. You have a chance to buy two different formulations of the same pesticide. Brand A is a liquid that contains 40% a.i., costs $18.95/gal, and weighs 9 lb/gal. Brand B is a wettable powder that contains 80% a.i. and costs $3.49/lb. What is the cost per lb of a.i. for each?

7. If you would normally broadcast apply 4 lb/acre of a pesticide, how much would you apply if the pesticide was applied in 12 inch bands every 36 inches?

8. A sprayer is set up to apply a 10-inch band of spray every 30 inches. It will apply 20 gallons/acre. If herbicide application is normally 3 lb/acre broadcast, how much should be added to its 500-gal tank?

PRACTICE PROBLEM SET 2

1. During calibration a sprayer travels 180 feet in 16 seconds. What is its speed in feet/minute?

2. The nozzles on the sprayer discharge an average of 20 ounces of liquid in 28 seconds. What is the output in gallons/minute/nozzle? (1 pint=16 ounces)

3. The sprayer has 20 nozzles spaced 30 inches apart. How many gallons/A is it applying?

4. If the pesticide label says to apply 6.5 ounces/A how many ounces/gallon should you apply with this sprayer?

5. If the sprayer has a 2000 gallon tank, how many ounces of the pesticide should you add? How many pounds of the pesticide should you add if it weighs 16 ounces per pound?

Answers to Practice Problem Set 1

1. 11 lb
2. 30.9 gal/A
3. 2 oz
4. 1 gal
5. 20 acres
6. Brand A: $5.26/lb a.i., Brand B: $4.36/lb a.i.
7. 1.33 lb/acre
8. 25 lb.

Answers to Practice Problem Set 2

1. 675 ft/min
2. 0.335 gal/min/nozzle
3. 8.65 gal/A
4. 0.75 oz/gal
5. 1500 oz/tank, 93.75 lb/tank

Part IV

Plant Identification

LABORATORY EXERCISE 9

Grain Crops

INSTRUCTIONAL OBJECTIVES

Upon completion of this exercise, you should be able to:

1. Identify both the kernels and the inflorescences of the grain crops described in the exercise.
2. Define a cereal grain.
3. Identify the type of inflorescence found on each plant.
4. Identify the three head types of wheat.
5. Identify the market class of a sample of wheat kernels. List the uses for each market class, and where it is grown in the United States.
6. State the scientific names for each crop species.
7. Discuss the light, temperature, and moisture requirements of corn, sorghum, and wheat.
8. List four Agronomic Groups for corn and grain sorghum.
9. Identify the kernel types of dent, flint, pop, and sweet corn. Identify the white and yellow endosperm kernel types of grain sorghum.
10. List two market classes each for corn and grain sorghum.
11. Identify wheat, barley, and oats using auricle and ligule characteristics.
12. Identify club wheat, common wheat, oats, rye, two-row and six-row barley, rice, proso millet, foxtail millet, and grain sorghum using inflorescence characteristics.

INTRODUCTION

As discussed in Chapter 2, a **grain crop** is a grass grown for its dry, edible seed. In this exercise, you will study the vegetative, floral, and seed characteristics of several common grain crops. A review of the characteristics of grass leaves studied in Lab Exercise 2, and inflorescences studied in Lab Exercise 3 will help you in this exercise.

SMALL GRAINS

Small grains are relatively short plants with small seeds as compared to other grain crops such as corn and sorghum. **Fall-seeded small grains** are winter wheat, winter barley, winter oats, and rye. **Spring-seeded small grains** are spring wheat, spring barley, spring oats, and rice. Although not always classified as a small grain, proso millet is also included in this exercise. Table 9-1 lists information for identification of small grains and millet. Table 9-2 lists agronomic information on these crops.

Identification characteristics for small grains include presence or absence of auricles and ligules, head or inflorescence type, presence or absence of awns, number of florets, and kernel characteristics. Kernel characteristics include shape of the suture and its cheeks, presence or absence of a beard or hull, kernel color, and overall kernel shape. Lab Figure 9-1 shows some examples of kernel characteristics of wheat.

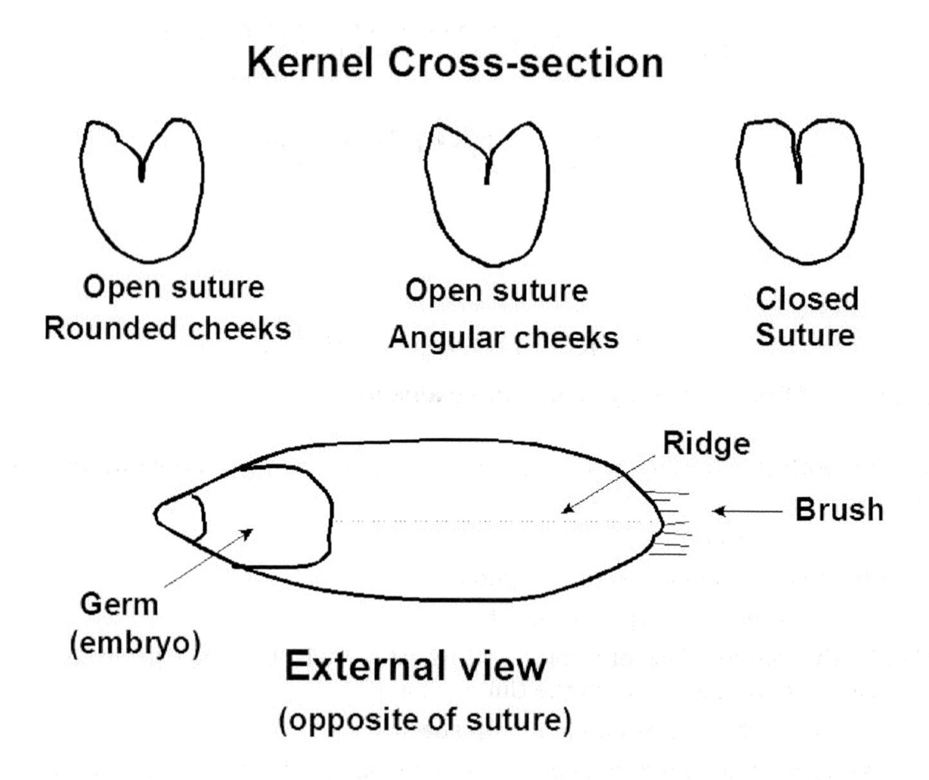

Lab Figure 9-1 Identification characteristics of wheat kernels.

Kernel Market Classes of Wheat

Table 9-1 describes three general types of wheat: common, durum, and club. Wheat kernels are marketed using the following classes, which determine baking quality and use.

Hard red winter: long, slender, vitreous kernel, widest near germ end; rounded cheeks; smooth ridge on back; no definite ring at base of brush. Used in bread. Grown in Central and Southern High Plains and Midwest. Produced on common wheat.

Hard red spring: about same size as hard red winter wheat; angular cheeks; ridge on back to one with dimple to other side; pointed brush with definite line demarcation. Used in bread. Grown in Northern High Plains. Produced on common wheat.

Soft red winter: barrel-shaped kernel with widest point near middle; sometimes wrinkled back; open suture; usually soft and chalky; large germ. Used in biscuits and pastry. Grown in Midwest and eastern United States. Produced on common wheat.

White: there are two types depending on type of wheat. Used in pastry, crackers, and breakfast food. Grown in western and northwestern United States.

Common white: kernel shape same as soft red winter wheat, but kernel is white and starchy.

Club: described under club wheat in Table 9-1.

Durum: described under wheat plant type. Used in spaghetti and macaroni. Grown in Northern High Plains.

Table 9-1. Identification information for small grains. (2 pages)

Crop	Scientific Name	Collar characteristics		Inflorescence characteristics			
		Auricle	Ligule	Type	Spikelets	Florets	Kernels
Common Wheat	*Triticum aestivum*	Hairy, medium	Medium, rounded	Spike	Awned or awnless. If awned, shorter and more spreading than durum	Usually three per spikelet	Depends on market class
Durum wheat	*Triticum durum*	Hairy, medium	Medium, rounded	Compact spike	Long, coarse awns parallel to rachis. Glumes sharply keeled.	Several per spikelet	Long. Pinched pointed germ. Hard, translucent, amber colored. Open suture, angular cheeks. High ridge down back.
Club Wheat	*Triticum aestivum* subspecies *compactum*	Long	Long, pointed	Short, compact, club-shaped. spike	Usually awnless. Set at wide angle to rachis. Very short rachis joints.	Several per spikelet	White, starchy. Shorter and smaller than common white wheat. Irregularly shaped, hump toward germ end. Pointed brush, angular cheeks, open suture.
Six-row Barley	*Hordeum vulgare*	Long	Long, pointed	Spike	Six rows on rachis. Rough or smooth awns.	Three per spikelet	Lemma and palea adhere to kernel. Two-thirds have curved sutures, one-third bottlenecked at germ end.
Two-row Barley	*Hordeum vulgare*	Long	Long, pointed	Spike	Two rows on rachis. Smooth awns.	Three per spikelet, one fertile	Lemma and palea adhere to kernel. Straight sutures. Plumper than 6-row.
Rye	*Secale cereale*	Short	Short	Long, drooping spike	Awned or awnless.	Barbs on lemmas. Kernels partly exposed from glumes.	Large plump to small slender. Slate to light brown to dark brown color. Wrinkled appearance.
Oats	*Avena sativa*	None	Long, square-shaped	Open, upright panicle	Awned or awnless. Lemma and palea enclose the kernel.	Two or more per spikelet	Lemma and palea attached to the kernel, ridged. Grayish-yellow color.

Crop	Scientific Name	Collar characteristics		Inflorescence characteristics			
		Auricle	Ligule	Type	Spikelets	Florets	Kernels
Rice	*Oryza sativa*			Loose, open panicle	Awned or awnless. Glumes small. Lemma and palea enclose the kernel.	One per spikelet	Lemma and palea enclose the kernel. Yellow to yellowish-brown color. Longitudinal ridges.
Foxtail millet	*Setaria italica*			Tight, drooping panicle	One to three bristles at base. Lemma and palea enclose the kernel.		Dull, not shiny. Rounded. Color varies.
Pearl millet	*Pennisetum glaucum, or P. typhoideum*			Dense, spike-like panicle	Cluster of bristles at base. Unequal glumes. Lemma and palea enclose the kernel.	Two per spikelet. Lower staminate, upper fertile.	Cone shaped. Pearl gray to green in color.
Proso millet	*Panicum miliaceum*	None	Short, thick	Open, spread-ing panicle	Awnless.		Glossy. Rounded. Color varies, usually white to yellow. Larger than foxtail millet.

Table 9-2. Agronomic information for small grains.

Crop	Area of origin	Kernels per pound	Seeding rate (bu/acre)	Expected Yield (bu/acre)	Growth stages** (Approximate Nebraska date)*				
					Seeding	Jointing	Boot	Flowering	Maturity
Winter wheat	Middle East	12 to 20 thousand	½ to 1	25-90	September	Mid-April	Mid-May	Late May to Early June	Late June to Early July
Spring wheat	Middle East	12 to 20 thousand	1 to 1½	20-70	Late March to early April	Mid-May	Early June	Mid to late June	Mid to late July
Spring barley	Middle East	13,000	1 to 1½	30-90	April	Mid-May	Late May to Early June	Early to mid-June	Mid-July
Winter barley	Middle East	13,000	1 to 1½	30-90	September	Mid-May	Late May to Early June	Early to mid-June	Mid-July
Winter rye	Southwest Asia	18,000	1 to 1½	20-80	Late September	Late April to Early May	Late May to Early June	Late June to Early July	Late July to Early August
Spring oats	Europe	14,000	2	35-110	Early April	Mid-May	Late May to Early June	Early to mid-June	Mid-July
Proso millet	China or SE Asia	80,000	10-20 lbs/A	20-60	June	Early July	Late July	August	September
Rice	India or SE Asia	15,000	90-140 lbs/A		Primarily grown in Louisiana, Texas, California, Arkansas. Long growing season				

*The dates of these stages vary greatly from SE to NW Nebraska.

** Jointing: Stem elongation when nodes can be felt in lower stem. Occurs shortly after floral initiation.
Boot: Head swells inside the flag (uppermost) leaf sheath.
Flowering: Pollen shedding
Maturity: Plant has finished developing its grain.

Other information for small grains:

Culture: Wheat is grown in rotation with fallow in semiarid regions.
Seeding depth: ¾ to 1½ inches. Winter wheat, up to 4 inches
Germination temperatures: 32° minimum, except millet and rice.

Light: C_3, long-day
Row spacing: 7 to 14 inches, or broadcast
Water requirement: 10-15 inches (wheat)

CORN AND SORGHUM

Corn is a native of Central America or Southern Mexico. Sorghum is a native of Africa. Practically all of the corn and sorghum grown today are hybrids. Both grain and forage types of sorghum are grown, but we will study only grain sorghum.

Corn is marketed in the following classes:

Sweet corn: kernels translucent and wrinkled in appearance at maturity; 2,000 kernels/pound.

Popcorn: small kernels; some pointed, some rounded; hard, flinty endosperm around outside of kernel with soft, starchy endosperm in the center; usually small and slender ear; kernel color varies from white to dark strawberry red; 3,000 kernels/pound.

Dent corn: the common type of field corn; kernels dented; soft, starchy endosperm in center of kernel with hard, flinty endosperm on the sides; kernel color varies, usually yellow or white; 1,200 kernels/pound.

Other classes: flour corn, wax corn, flint corn, pod corn.

Lab Figure 9-2 shows the endosperm characteristics of popcorn, dent corn, flint corn, and flour corn.

Grain sorghums are marketed in the following classes:

White endosperm: pericarp color may be white or red; endosperm color is cream to tan; 13,000 to 24,000 kernels/pound.

Yellow endosperm: transparent pericarp; yellow endosperm visible; 13,000 to 24,000 kernels/pound.

Bird-resistant: dark pericarp; purple vitreous endosperm; 13,000 to 24,000 kernels/pound.

Table 9-3 lists some additional information for corn and sorghum.

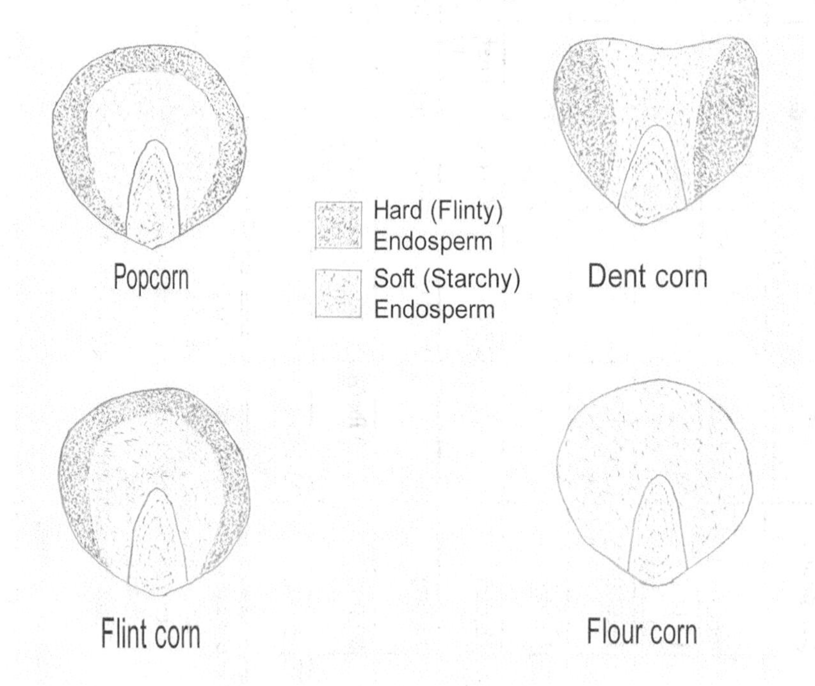

Lab Figure 9-2 Endosperm characteristics of corn market classes.

Table 9-3. Corn and sorghum information. (2 pages)

Characteristic	Corn	Sorghum
Scientific name	*Zea mays*	*Sorghum bicolor*
Origin:	Selected from mutation of the grass teosinte about 7,000 years ago in Central America. By 1492 there were over 200 varieties planted throughout the Americas. Many modern dent hybrids are crosses of flint and dent varieties.	Selected from wild type in S.E. Asia or E. Africa 5,000 to 7,000 years ago. Many modern grain hybrids are crosses of milo and kafir varieties.
General description:	Large-seeded summer annual. Stem is a stalk, 8-20 alternate leaves, height 5-10 feet, fibrous root system over 5 feet deep.	Small-seeded summer annual similar to corn. Leaf sheaths are waxy, tillers are common, and dwarfing genes have been incorporated to make hybrids 3-5 feet tall for combine harvesting. Ancestors were 3-12 feet tall.
Varietal type:	Hybrid since 1930's	Hybrid since 1950's
Inflorescences:	Ear is a pistillate spike. Tassel is a staminate panicle.	Head is a panicle inflorescence. Can be either open or closed. Perfect flowers.
Spikelets:	Paired, both are fertile; 2 florets/spikelet with one fertile floret for both ear and tassel.	Single spikelets; 2 florets/spikelet; with one fertile floret.
Temperature requirements:	Warm-season crop. Minimum germination temperature: 50° F. Optimum for growth: 70-75° F.	Warm-season crop. Minimum germination temperature: 70° F. Optimum for growth: 80-85° F.
Water requirement:	28-30 inches	22-26 inches
Light requirement:	Short-day C_4 plant	Short-day C_4 plant
Critical stress period:	Pollination to blister stage. Drought stress during flowering may result in a time gap between start of pollination and start of silk emergence resulting in poor pollination.	Boot stage (just before flowering). Sorghum is more drought resistant than corn due to a waxy cuticle, a finer, more divided root system, and no delay between pollination and stigma receptivity.
Major pests:	Rootworms, corn borer, cutworms, spider mites.	Sorghum midge, greenbug aphids.
Major diseases:	Goss's wilt, Stewart's wilt, seedling blights, northern leaf blight, corn lethal necrosis.	Charcoal stalk rot, seedling blights.
Seeding depth:	1 ½ to 2 inches.	1 to 1 ½ inches.
Seeding date:	Early spring, based on 50° soil temperature.	Late spring, based on 70° soil temperature.
Population: Dryland: **Irrigated:**	14,000 to 18,000 depending on rainfall. 20,000 to 28,000 depending on hybrid.	35,000 to 75,000, depending on rainfall. 80,000 to 120,000 depending on hybrid.

Characteristic	Corn	Sorghum
Row spacing:	30-38 inches.	10-40 inches.
Yields: Dryland: **Irrigated:**	50-150 bushels/acre depending on rainfall. 120-150 bushels/acre depending on hybrid	40-100 bushels/acre depending on rainfall. 120-200 bushels/acre depending on hybrid.
Agronomic groups:	Dent, flint, pop, sweet, flour, waxy, pod, opaque (high lysine)	Grain sorghum, forage sorghum, sorgo, broomcorn
Grain market classes:	White or yellow: dent, flint, waxy	White, yellow, brown, mixed

STUDY GUIDE

Complete the following table to help you learn the material in this exercise.

		Name	Uses
Wheat market classes:	1.	____________	____________
	2.	____________	____________
	3.	____________	____________
	4.	____________	____________
	5.	____________	____________
Wheat head types:	1.	____________	
	2.	____________	
	3.	____________	
Sorghum market classes:	1.	____________	
	2.	____________	
	3.	____________	
Corn market classes:	1.	____________	
	2.	____________	
	3.	____________	
	4.	____________	

Identification characteristics:

	Ligule	Auricle
Wheat	____________	____________
Barley	____________	____________
Oats	____________	____________

	Scientific Name	Inflorescence Type
Common wheat		
Durum wheat		
Club wheat		
Six-row barley		
Two-row barley		
Rye		
Oats		
Rice		
Foxtail millet		
Pearl millet		
Corn		
Sorghum		

LABORATORY EXERCISE 10

Forage Crops and Range Plants

INSTRUCTIONAL OBJECTIVES

Upon completion of this exercise, you should be able to:

1. Define forage crops and range plants. List at least four examples of each.
2. Explain the differences between a warm-season grass and a cool-season grass. List at least four examples of each.
3. Define forage legumes. List at least four examples of each.
4. Identify both the plants and seeds of the plants studied in the exercise.
5. Explain the difference between a bunchgrass and a sod-forming grass. Identify each grass studied as a bunchgrass or sod-former.
6. Identify the grasses studied by growth habit.
7. Identify the plants studied by life span.
8. Identify the plants studied as native or introduced.
9. State the scientific name of each species studied.

INTRODUCTION

As discussed in Chapter 2, a **forage crop or range plant** is grown for its vegetation which is fed primarily to ruminant livestock. In this exercise, you will study the vegetative, floral, and seed characteristics of several forage crops and range plants.

Identification characteristics include presence or absence of stipules, ligules, auricles, or plant hairs. Plants with hairs are **pubescent**, plants without hairs are **glabrous**. A review of the characteristics of leaves studied in Exercise 2, and inflorescences studied in Exercise 3 will help you in this exercise.

Bunchgrasses and Sod-Forming Grasses

A **sod-forming** grass spreads by rhizomes and/or stolons, and forms a dense mat or sod. It also usually produces new plants by seed. A **bunchgrass** spreads only by seed and if rhizomes or stolons are present, they are very short and sometimes nonfunctional. A bunchgrass does not form a dense sod. It grows in clumps or bunches formed by the numerous tillers.

Warm-Season and Cool-Season Grasses

A **warm-season grass** grows mostly during the summer months and remains dormant the rest of the year. Most of our warm-season grasses are perennials. A **cool-season grass** grows mostly in the spring and autumn, and usually remains dormant during the summer and winter.

FORAGE AND RANGE PLANTS IDENTIFICATION KEYS

The following keys can be used to aid in identification of the perennial plants listed in Tables 10-1, 10-2, and 10-3. The keys are based on pairs of identifying statements with the same numbers. If the first statement is *true* then you continue to the next numbered statement directly below it. If the first statement is *false* then you continue to the second statement with the same number. You continue comparing statement pairs until you have identified the plant in question.

Grass Identification Key

1. Inflorescence is a panicle.
    2. Collars with visible ligules.
        3. Ligules membraneous.
            4. Flattened stems. Bunchy panicle. ORCHARDGRASS.
            4. Hairy nodes. Slender, golden-brown panicle. INDIANGRASS.
        3. Hairy ligules in "V" at base of leaf blade. Open, spreading panicle. SWITCHGRASS.
    2. Collars with no ligules.
        5. "M" or "W" mark about 2/3 way up leaf blade. Open panicle, no awns. SMOOTH BROMEGRASS.
        5. No marks on leaf blades.
            6. Erect culms with corms at the base. Cylindrical spike-like panicle. TIMOTHY.
            6. No corms at base.
                7. Stems or leaves coarse.
                    8. Slightly coarse leaf blades. Smooth stems. Long, narrow panicle. TALL FESCUE.
                    8. Coarse stems, smooth leaves. Semi-dense spike-like panicle. REED CANARYGRASS.
                7. Stems and leaves smooth. Leaf tips boat-shaped. Open panicle. KENTUCKY BLUEGRASS.
1. Inflorescence is a spike or raceme.
    9. Inflorescence is a spike.
        10. Plants spread by stolons or rhizomes.
            11. Plants spread by stolons. Very short stems. Plants dioecious. BUFFALOGRASS.
            11. Plants spread by rhizomes. Leaves blue-green. Visible auricles. WESTERN WHEATGRASS.
        10. Plants are bunchgrasses with no or very short rhizomes.
            12. Leaf blades with hairy margins. Smooth stems. Short spikes on one side of peduncle. SIDEOATS GRAMA.
            12. Leaf blades smooth, glabrous.
                13. Spike looks like one half of a feather. BLUE GRAMA.
                13. Spike is evenly shaped.
                    14. Spike flattened. Overlapped spikelets. CRESTED WHEATGRASS.
                    14. Leaves glossy, mostly at base of stem. Mostly awnless spike. PERENNIAL RYEGRASS.
    9. Inflorescence is a raceme.
        15. Racemes in groups of 3 at top of peduncle. BIG BLUESTEM.
        15. Many single racemes along peduncle. LITTLE BLUESTEM.

Legume Identification Key

1. Leaves compound with more than 3 leaflets.
    2. Leaflets even-numbered with tendrils at tip of leaf. HAIRY VETCH.
    2. Leaflets odd-numbered.

1 Five leaflets. BIRDSFOOT TREFOIL.

    3. Leaflets 9 to 25. No tendrils. CROWNVETCH.

1. Leaves compound with 3 leaflets.
   4. Leaflets with pubescence.
      5. Pubescence on entire leaflets. Leaflets with inverted "V". Prominent stipules with red veins. RED CLOVER
      5. Pubescence on leaflet margins. Rest of leaflet glabrous. Large ovate stipules. KOREAN LESPEDEZA
   4. Leaflets glabrous.
   1. Stems not upright, rooting at nodes. White blotches on leaves. WHITE & LADINO CLOVER.
      6. Stem upright.
         7. Leaflets with serrations along entire margin.
            8. Serrations very fine. Capitulum inflorescence. ALSIKE CLOVER.
            8. Serrations more prominent. Raceme inflorescence with white or yellow flowers. SWEET-CLOVER.
         7. Leaflets with serrations only on margin 1/3 from tip. Raceme inflorescence with purple flowers. ALFALFA.

Table 10-1 lists information on warm-season grasses, Table 10-2 lists information on cool-season grasses, and Table 10-3 lists information on forage legumes.

Table 10–1. Identification characteristics of warm season forage grasses.

Name	Scientific Name	Height (ft)	Growth Habit	Origin	Life Span	Plant and Inflorescence	Seed
Big bluestem	*Andropogon gerardii*	3 - 6	Sod: rhizomes	Local native.	Perennial	Plant often purplish. Three racemes. Spikelets paired, one sessile and one pedicellate. Sessile spikelet, fertile, bearing prominent awn. Upper spikelet staminate.	Racemes broken with rachis extending about halfway along one side of lower spikelet. Pedicel along other side about same distance. The staminate spikelet usually lost in threshing; 150,000 seeds/pound.
Little bluestem	*Schizachyrium scoparium (Andropogon scorparius)*	2 - 5	Bunch	Local native.	Perennial	Numerous slender, curved, spike-like racemes, one to a branch. Rachis slender and hairy. Spikelets paired, sessile, fertile, awned.	Similar to big bluestem, but pedicel somewhat longer, more slender, and curved away from the spikelet. 260,000 seeds/pound.
Switchgrass	*Panicum virgatum*	3 - 6	Sod: rhizomes	Local native.	Perennial	Hairy ligule; hairs in 'V' at base of blade. Very open panicle.	Smooth and shiny, narrowly ovate, grayish. 370,000 seeds/pound.
Side-oats grama	*Bouteloua curtipendula*	2 - 3	Bunch, with short rhizomes	Local native.	Perennial	Culms erect, glabrous. Leaf blade hairy on margin. 35-80 short spikes with 3 to 7 spikelets. All spikes on one side of branch. One fertile floret per spikelet.	Usually short spikes stripped from the main axis. 200,000 seeds/pound.
Blue grama	*Bouteloua gracilis*	1 - 2½	Bunch	Local native.	Perennial	Culms short, erect, smooth, leafy at the base; leaf blades glabrous. 1-3 spikes. Each spike is like ½ of a feather.	Threshed as spikelets from rachis. 900,000 seeds/pound.
Buffalograss	*Buchloe dactyloides*	< 1	Sod: stolons	Local native.	Perennial	♂: 2-3 one-sided spikes, 4-8" tall ♀: bur-like cluster in lower leaves, 2-3" tall.	Bur-like spike with 4 to 5 spikelets. 50,000 seeds/pound.
Indiangrass	*Sorghastrum nutans*	3 - 8	Sod: rhizomes	Local native.	Perennial	Hairy nodes; prominent ligule. Slender, golden-brown panicle. Spikelets paired, sessile, fertile, awned.	Enclosed in hairy, bronze-yellow hulls. Long twisted awn attached. 170,000 seeds/pound.
Sudangrass	*Sorghum bicolor*	4 - 8		Native of Africa.	Annual	Similar to sorghum but narrower leaves. Open panicle.	Enclosed by glumes. Color mixed. 55,000 seeds/pound.

Table 10–2. Identification characteristics of cool season forage grasses.

Name	Scientific Name	Height (ft)	Growth Habit	Origin	Life Span	Plant and Inflorescence	Seed
Smooth bromegrass	*Bromus inermis*	2 - 3	Sod: rhizomes	Native of Europe.	Perennial	'M' mark on leaves ⅔ way up blade. Open erect panicle. No awns. Can be a major weed.	Large, flattened. Slanting scar and hairs on rachilla joint. Caryopsis shows through papery palea. Rachilla long and slender. 137,000 seeds/pound.
Kentucky bluegrass	*Poa pratensis*	1 - 2	Sod: rhizomes	Native of Europe.	Perennial	Leaf tips boat-shaped. Open panicle. Double mid-rib.	Small, boat-shaped. Rachilla joint persistent with slanting scar. 2,200,000 seeds/pound.
Orchardgrass	*Dactylis glomerata*	1½-2½	Bunch: short rhizomes	Native of Europe.	Perennial	Flattened sheaths and stems. Long membranous ligule. Bunch-type panicle. Spikelets on one side of panicle branch.	Medium-sized, awns present. Lemma tip with a quarter twist. Long slender rachilla, two kernels sometimes attached by rachilla. 590,000 seeds/pound.
Crested wheatgrass	*Agropyron cristatum (Agropyron desertorum)*	1 - 1½	Bunch. Rhizomes rare.	Native of Russia.	Perennial	Flat leaves. Flattened spike. Spikelets out at wide angle from rachis, highly overlapped.	Long, slender. Lemma unrolled, short, tapers to short awn point, palea barbed on edges. Wedge-shaped rachilla. Seeds often do not thresh free. 190,000 seeds/pound.
Western wheatgrass	*Pascopyrum smithii (Agropyron smithii)*	1 - 1½	Sod: rhizomes	Local native.	Perennial	Blue-green leaves rolled up when dry. Spike. Auricles at collar.	Lemma smooth, rounded back, lacking an awn. Rachilla short wedge-shaped, similar to crested wheatgrass. Usually thresh apart and do not remain in groups as does crested wheatgrass. 110,000 seeds/pound.
Tall fescue	*Festuca arundinacea*	2 - 3	Bunch	Native of Europe.	Perennial	Slightly coarse leaves. Long, narrow panicle.	Awnless. Dark purple color. Short, peg-shaped rachilla. Caryopsis about twice the size of lemma and palea. 227,000 seeds/pound.
Timothy	*Phleum pratense*	2 - 3	Bunch	Native of Europe	Perennial	Erect culms with a corm at the base. Dense, cylindrical spike-like panicle. One floret per spikelet.	Shiny, tan, football shaped. 1,230,000 seeds/pound.
Reed canarygrass	*Phalaris arundinacea*	2 - 8	Sod: rhizomes	Native of N. temperate lat.	Perennial	Coarse stems. Leaves broad, smooth, light green. Semi-dense spike-like panicle. Hydrophyte.	Waxy appearance. ⅛ inch long. Similar to switchgrass. Olive-gray color. 550,000 seeds/pound.
Perennial ryegrass	*Lolium perenne*	1 - 2	Bunch	Native of Asia & N. Africa	Short-lived perennial	Many long, narrow leaves near base, nearly naked seed stalk. Leaves glossy. Mostly awnless spike. Spikelets edgewise to rachis.	Awnless, containing lemma and palea. About ¼ inch long. 330,000 seeds/pound.

Table 10–3. Identification characteristics of forage legumes.

Name	Scientific Name	Height (ft)	Growth Habit	Origin	Life Span	Plant and Inflorescence	Seed
Alfalfa	*Medicago sativa*	1 - 3		Native of Persia or Asia Minor	Perennial	Well developed crown. Broad stipules. Serrations on distant ⅓ of leaflets. Raceme with purple flowers.	½ mitten-shaped, ½ kidney-shaped. May have squarish ends due to closeness in pod. Greenish-yellow to light brown in color, turning dull, dark reddish-brown with age. 220,000 seeds/pound.
Sweetclover White: Yellow:	*Melilotus alba officinalis*	3 - 8	Can be a weed.	Native of Asia.	Annual or biennial	Entire leaflet margin serrated. Elongated raceme with white or yellow flowers.	Light brown uniform color, mitten-shaped, and kidney-shaped. 250,000 seeds/pound.
White & ladino clover	*Trifolium repens*	< 1	Stolons	Native of Europe and Asia.	Perennial	Petioles and leaves glabrous. White blotches on leaves. Capitulum with white flowers.	Small, ½ canary-yellow, ½ brown. Heart-shaped. 700,000 seeds/pound.
Alsike clover	*Trifolium hybridum*	1 - 1½		Native of Sweden.	Short-lived perennial	Leaves & stems glabrous; stems erect. Indistinct vein on stipules. Capitulum with pink/white flowers.	Small. Dark green, yellow, black, sometimes mottled. Heart-shaped. 680,000 seeds/pound.
Red clover	*Trifolium pratense*	1 - 2		Native of Eurasia.	Short-lived perennial	Stems and leaves pubescent. Leaves palmately compound. Leaflets with inverted "V". Distinct red veins on stipules. Capitulum with red flowers.	Larger than white or alsike. Lemon-yellow to violet "two-tone" color. Mitten- to heart-shaped. 260,000 seeds/pound.
Korean lespedeza	*Kummerowia stipulacea (Lespedeza stipulacea)*	1		Native of E. Asia.	Annual	Branched stems. Large ovate stipules. Raceme in leaf axil with purple flowers.	Shiny black, oval. 240,000 seeds/pound.
Birdsfoot trefoil	*Lotus corniculatus*	1 - 2½		Native of Europe	Perennial	Similar to alfalfa in growth habit. Leaves opposite with 5 leaflets in a sessile pinnate leaf. Umbel with yellow to orange flowers.	Oval to spherical. Light to dark brown. 375,000 seeds/pound.
Crownvetch	*Coronilla varia*	1 - 1½		Native of Europe or Asia.	Perennial	No tendrils on pinnate leaves. Glabrous leaves and stems. Umbel with white, rose, or violet flowers.	Rod-shaped. Yellow ocher to mahogany color. 140,000 seeds/pound.
Hairy vetch	*Vicia villosa*	1 - 1½		Native of Europe.	Annual, biennial, perennial	Pinnate leaflets have tendrils. Stems and leaves pubescent. Raceme with lavender flowers.	Spherical. Dull black. Variable in size, smaller than soybeans. 21,000 seeds/pound.

STUDY GUIDE

Complete the following table to help you learn the material in this exercise:

Warm season grass	**Scientific name**	**Growth Habit**	**Origin**	**Life Span**
Big bluestem	____________	____________	____________	____________
Little bluestem	____________	____________	____________	____________
Switchgrass	____________	____________	____________	____________
Indiangrass	____________	____________	____________	____________
Side-oats grama	____________	____________	____________	____________
Blue grama	____________	____________	____________	____________
Buffalograss	____________	____________	____________	____________
Sudangrass	____________	____________	____________	____________
Cool season grass				
Smooth bromegrass	____________	____________	____________	____________
Kentucky bluegrass	____________	____________	____________	____________
Orchardgrass	____________	____________	____________	____________
Crested wheatgrass	____________	____________	____________	____________
Tall fescue	____________	____________	____________	____________
Forage legumes				
Alfalfa	____________	____________	____________	____________
Sweetclover	____________	____________	____________	____________
White clover	____________	____________	____________	____________
Red clover	____________	____________	____________	____________
Korean lespedeza	____________	____________	____________	____________
Crown vetch	____________	____________	____________	____________
Hairy vetch	____________	____________	____________	____________

LABORATORY EXERCISE 11

Oil and Protein Crops Plus Specialty Crops

INSTRUCTIONAL OBJECTIVES

Upon completion of this exercise, you should be able to:

1. Identify both the plants and seeds of the plants studied in this exercise.
2. List the life cycles, origin, and uses of the crops studied.
3. State the scientific name of the plants studied.
4. Name the oil crop which is produced in larger quantities than the others studied in the exercise.
5. Explain where, in relation to the ground, the peanut seed develops.
6. List some uses of plant protein and oil.
7. Define a multigerm seed, and state which crop produces multigerm seed.
8. List four criteria for successful production of a new crop in a region.

INTRODUCTION

Crops that are high in oil and protein are increasingly important in fighting malnutrition in the world and providing balanced, healthy diets. Many plant oils are highly unsaturated, which helps to reduce blood cholesterol levels. Some common uses of plant oils are cooking, margarine, salad dressings, food products and animal feeds, paints, varnishes, lacquers, soaps, detergents, and plastics. They are also used as lubricants in industry. Most oil and protein crops are legumes and soybean is the major oil crop in the U.S. Specialty crops have specific uses such as sugar, starch, drugs, bio-fuel, etc.

DEVELOPMENT OF NEW CROPS

The introduction and development of a new crop in a region is complex and time consuming. But like soybean crop development in the U.S. that took decades to accomplish, the effort can be well rewarded. To minimize the risks associated with new crop development, a decision making process should be employed from the beginning. From the production system, to the market system, to the consumption system, the new crop being tested must meet four key requirements to ensure success. The new crop must be environmentally possible, physically possible, economically feasible, and institutionally permissible. A new crop is **environmentally possible** if it is adapted to the climate and soils of the region. At the production level, the most frequent **physical** constraint of new crop development is finding farm machinery capable of planting, cultivating, or harvesting the crop. **Economic feasibility** means that markets must exist or be developed that will make the crop profitable for the producer in the long run. How a new crop would fit into current government programs or pass certain regulations would determine whether or not the production of the crop is **institutionally permissible**.

Table 11-1 lists characteristics of oil, protein, and specialty crops. Although not listed in Table 11-1, corn is also grown for its oil content.

Table 11–1. Identification characteristics of oil, protein, and specialty crops. (2 pages)

Name	Scientific Name	Use	Height (ft)	Origin	Life Span	Plant and Inflorescence	Seed
Soybean	*Glycine max*	Oil and protein	2 - 3	Native of E. Asia.	Annual	Erect, branching pubescent stems and leaves with large pinnate leaves. White or purple flowers in axillary racemes.	Spherical. Various colors. 1,000 to 8,000 seeds/pound.
Field bean (Dry, edible bean)	*Phaseolus vulgaris*	Protein	2 - 5	Native of S. America	Annual	Twining to erect, branching stems with large pinnate leaves. Nearly glabrous. Various colors in axillary racemes.	Kidney-shaped to oval with squarish end. Various colors. 1,000 to 2,000 seeds/pound.
Peanut	*Arachis hypogaea*	Oil and protein	2 - 3	Native of S. America	Annual	Twining to erect stems. Leaves pinnately compound with 4 leaflets. Broad stipules. Bright yellow flowers begin aboveground but after fertilization send a "peg" below surface where the pod develops. Raceme.	Large. Reddish-brown in color. 1,000 seeds/pound.
Cotton	*Gossypium hirsutum*	Oil, protein, and fiber.	2 - 7	Native of C. and S. America	Perennial killed by frost. Grown as annual.	Stems erect, branching. Simple heart-shaped leaves with three to seven lobes. Fruit is a leathery capsule or boll with 3 to 5 cells.	Linted: fibers remain attached to the seed. Must delint before planting. Delinted: most fibers removed. Ovate to tear-shaped. Dark brown to black. 4,000 seeds/pound.
Castor	*Ricinus cummunis*	Oil, lubricant	3 - 12	Native of Africa	Perennial killed by frost. Grown as annual.	Leaves simple, six to eleven palmate lobes, toothed. Monoecious with pistillate flowers at top of spike, staminate flowers below.	Large, ½ to ⅝ inch in length. Shiny, smooth, and gray with tan to brown mottling. Two lobe-like appendages on the hilum end of the seed. 1,000 seeds/pound.
Flax	*Linum usitatis-simum*	Oil and fiber	1½ - 3	Native of Mediter-ranean	Annual	Much branched near the top, with small, simple linear leaves with entire margin. Yellow to blue flowers in panicle at top of plant. Fruit a brownish, globular, 5-celled capsule.	Flattened, egg-shaped. Shiny brown or yellow, waxy. 82,000 seeds/pound.
Tobacco	*Nicotiana tobacum*	Drug	4 - 6	Native of C. and S. America	Annual	Simple leaves with sticky hairs. Flowers in terminal raceme. Fruit a capsule.	Extremely small, reddish-brown. 5,000,000 seeds/pound.
Sesame	*Sesamum indicum*	Oil and seed	3 - 5	Native of south Eurasia	Annual	Tubular, two-lipped flower about ¾ inch long with a pink or yellow corolla. Spike inflorescence.	Resembles flax seed in size and shape. Color creamy white to black. 100,000 seeds/pound.

Name	Scientific Name	Use	Height (ft)	Origin	Life Span	Plant and Inflorescence	Seed
Safflower	*Carthamus tinctorius*	Oil	1 - 3	Native of India and Near East	Annual	Branched at top with white or yellowish smooth pithy stems and branches. Flowers globular and thistle-like, 1/2 to 1½ inches in diameter with white, yellow-orange, or red florets in head inflorescence.	Resemble small sunflower seeds. White or cream-colored. 8,000 to 13,000 seeds/pound.
Sunflower	*Helianthus annuus*	Oil and seed	5 - 10	Native of America	Annual	Rough hairy stem 1 to 3 inches in diameter. Flower is head or disk 3 to 24 inches in diameter with yellow ray florets.	Elongated rhomboid achene. Oil varieties mostly dark brown to black. Seed varieties often striped. 3,000 to 9,000 seeds/pound.
Sugarbeet	*Beta vulgaris*	Sugar	1	Native of Europe	Biennial	Large fleshy taproot. Simple leaves in a rosette. Harvested for sugar at end of first growing season.	Looks like grapenuts. Rough irregular. One embryo per seed (monogerm), or more than one embryo per seed (multigerm). 22,000 seeds/pound.
Common buckwheat	*Fagopyrum esculentum*	Food and feed	2 - 3	Native of China	Annual	Stems reddish, strongly grooved. Leaves simple, triangular to cordate. Flowers with petal-like sepal, no petals, white to pink, in axillary or terminal racemes or cymes.	Triangular to pyramid-shaped. Dark brown, 20,000 seeds/pound.
Potato	*Solanum tuberosum*	Starch	1½ - 2	Native of C. and S. America	Annual	Erect, branched stems, slightly pubescent. Leaves pubescent, pinnately compound. Five-petaled flowers in a terminal compound cyme. Stolons enlarge to form tubers.	Tuber: large to small, globular to oblong. Starchy interior. Contains many nodes (eyes), which are used to produce new plants.
Rapeseed / canola	*Brassica napus*	Rapeseed: lubricant, biofuel. Canola: oil	2 - 3	Native of Asia	Annual	Erect, branched stems arising from a rosette. Leaves lobed, 4 to 12 inches long, on short petioles. Four-petaled flowers of varied color on elongated racemes. Fruit 2-celled, dehiscent, 15 to 40 seeds.	Small, round, yellow to black. 115,000 seeds/pound.
Crambe	*Crambe abyssinica*	Oil, lubricant	2 - 4	Native of Mediter-ranean Region	Annual	Stems erect. Leaves pinnately-lobed, 4 inches long and 3 inches wide. Petioles 8 inches long, hairy Flowers small, white, 4-petaled. Inflorescence a panicle-like raceme. Fruit a pod	Round, ⅛ inch diameter, enclosed in a hull. Light brown color. 27,000 seeds/pound.

STUDY GUIDE

Complete the following table to help you learn the material in this exercise:

	Scientific name	Use(s)	Origin	Life Span
Soybean				
Field bean				
Peanut				
Cotton				
Castor				
Flax				
Sesame				
Safflower				
Sunflower				
Rapeseed/canola				
Mungbean				
Sugarbeet				
Buckwheat				
Tobacco				
Potato				

LABORATORY EXERCISE 12

Weeds

INSTRUCTIONAL OBJECTIVES

Upon completion of this exercise, you should be able to:

1. Define prohibited noxious weed, restricted noxious weed, and common weed.
2. List the prohibited and restricted noxious weeds and weed seeds in your state.
3. Identify the weed species, by plant and seeds, presented in the exercise.
4. List the life cycle of each weed.
5. List the method by which each weed spreads or propagates itself (seed, etc.).
6. State the scientific name of the weeds studied.

INTRODUCTION

A weed is defined as a plant out of place, or growing where it is not wanted. Many plants listed as weeds by the Weed Science Society of America are also grown as crops in certain areas. These include johnsongrass, sweetclover, and others.

NOXIOUS WEED LAWS

Most states have noxious weed laws which prohibit and/or restrict the presence of certain weeds on property and the presence of certain weed seeds in crop seed offered for sale. Read *Common or Noxious*, page 187, for information on this topic.

List below the noxious weeds of your state.

WEED IDENTIFICATION KEYS

The following keys can be used to aid in identification of the weeds listed in Table 12-1. The keys are based on pairs of identifying statements with the same numbers. If the first statement is *true* then you continue to the next numbered statement directly below it. If the first statement is *false* then you continue to the second statement with the same number. You continue comparing statement pairs until you have identified the plant in question.

GRASS AND SEDGE WEEDS

1. Stems triangular in cross-section. Leaves stiff. Plants with rhizomes. Yellow-green flowering spike. YELLOW NUTSEDGE.
1. Stems round or flattened in cross-section.
    2. Fruit a spiny bur. SANDBUR.
    2. Fruit not a spiny bur.
        3. Stems or branches terminate with a single flowering spike.
            4. Spike dense and cylindrical with bristles. Seeds round or elliptical, not sharply pointed, no terminal bristle.
                5. Bristles tawny-yellow. A few long slender hairs on upper side of blade at base. Seeds broadly oval, ⅛ inch or more long. YELLOW FOXTAIL.
                5. Bristles pale green or pale yellow. Leaves without hairs on upper surface or entirely covered with fine, short hairs. Seeds narrowly oval, ⅛ or less long.
                    6. Leaves with fine hairs on upper surface. Plants up to 6 feet tall. GIANT FOXTAIL.
                    6. Leaves without hairs on upper surface. Plants less than 3 feet tall. GREEN FOXTAIL.
            4. Spike flattened. Seeds oblong and pointed at tip. QUACKGRASS
        3. Stems or branches do not terminate in a single flowering spike. Inflorescence of various type, or compound.
            7. Inflorescence of 2 to 5 spikes in finger-like fashion. No awns present. Stems and leaf sheaths hairy. Seeds narrowly oval with brownish inner hull. LARGE CRABGRASS.
            7. Inflorescence not of 2 to 5 spikes.
                8. Spikelets without awns.
                    9. Panicle inflorescence. Plant erect, densely hairy. WITCHGRASS.
                        9. Inflorescence slender, almost spike-like, consisting of an aggregation of short spikelets. Stems usually more than ¼ inch in diameter. BARNYARD GRASS.
                8. Spikelets with short or long awn.
                    10. Some or all awns spirally twisted at base.
                        11. Spikelets about ¾ inch long excluding awns in an open panicle inflorescence. Seeds oblong, dull gray to black. Annual plant. WILD OATS.
                        11. Spikelets about ¼ inch long excluding awns, 1-seeded, in an open panicle inflorescence. Seeds plump. Perennial plant with rhizomes. JOHNSONGRASS.
                    10. Awns not spirally twisted at base.
                        12. Open panicle inflorescence.
                            13. Stems erect and hairy. Awns long and conspicuous. Spikelets with awns.
                                14. Florets spreading, only slightly overlapping. DOWNY BROME.
                                14. Florets much overlapped, clinging closely. HAIRY CHESS.
                            13. Stems erect and hairy. Awns long and conspicuous. Spikelets without awns. JAPANESE BROME.
                        12. Panicle dense and irregular. Stems usually more than ¼ inch in diameter. BARNYARD GRASS.

BROADLEAF WEEDS

1. Flowers are orange, yellow, or cream colored.
    2. Leaves and stems covered with stiff spines. BUFFALOBUR.
    2. Leaves and stem not covered with spines. Fruit may be spiny.
        3. Plants prostrate and spreading in every direction. Leaves pinnately compound. PUNCTUREVINE.

3. Plants erect or ascending.
4. Leaves lobed, but not compound, opposite.
5. Leaves almost fern-like with many deep lobes
6. Plants annual, leaves much divided and smooth above. COMMON RAGWEED.
6. Plants perennial, leaves less divided and rough above. WESTERN RAGWEED.
5. Leaves palmately lobed with 3 to 5 conspicuous lobes. Plants over 10 feet tall. GIANT RAGWEED.
4. Leaves entire, or not deeply lobed.
7. Four petals per flower in a cross-like pattern. Plants bristly-hairy. WILD MUSTARD.
7. Four petals per flower in a cross-like pattern.
8. Plants with milky sap. Leaves simple, linear with smooth margins. LEAFY SPURGE.
8. Plants without milky sap.
9. Flowers without petals. Leaves grass-like. Stems triangular. YELLOW NUTSEDGE
9. Flowers with petals.
10. Flowers in heads, each head resembling a single flower. Yellow ray flowers. Leaves rough. SUNFLOWER.
10. Flowers separate, not in heads. Plants velvety or wooly.
11. Fruit a berry enclosed in an angular, papery husk. Plant perennial. GROUND CHERRY.
11. Leaves heart-shaped. Plant annual. VELVETLEAF.
1. Flowers are not orange, yellow, or cream colored.
12. Flowers mostly white.
13. Plants are parasitic. Stems are threadlike and yellow with only minute scale-like leaves. FIELD DODDER.
13. Plants not parasitic.
14. Plants with milky sap. Leaves opposite. HEMP DOGBANE.
14. Plants without milky sap.
15. Plants prickly or spiny. Leaves with spines on midrib. HORSE NETTLE.
15. Plants not prickly or spiny.
16. Four petals per flower in a cross-like pattern.
17. Plants hairy
18. Seed pod triangular. Plant covered with tiny, branched hairs. SHEPHERD(S PURSE.
18. Two-celled heart-shaped seed pods. Plants perennial. HOARY CRESS.
17. Plants not hairy. Heart-shaped seed pods with 10 to 12 seeds. FIELD PENNYCRESS.
16. Petals not four in number.
19. Plants vine-like, trailing, twining, or climbing. Leaves heart or arrow-shaped.
20. Flowers small and clustered in upper leaf axils. Membranous sheaths at base of leaves. Seeds triangular. WILD BUCKWHEAT.
20. Flowers large and funnel shaped.
21. Leaves heart-shaped. Flowers about 2" across. TALL MORNING GLORY.
21. Leaves arrowhead-shaped.
22. Leaves with narrow basal lobes. Flowers about 1" diameter. FIELD BINDWEED.
22. Leaves with squared off basal lobes. Flowers 1½ to 2" diameter. HEDGE BINDWEED.
19. Plants erect, ascending, sometimes prostrate, not vine-like.
23. Leaves not toothed or lobed. Plants prostrate. Leaves blue-green. KNOTWEED.
23. Leaves wavy-toothed or lobed. Fruit a small berry, black when mature. BLACK NIGHTSHADE.
12. Flowers not mostly white.
24. Flowers mostly green.
25. Leaves lobed, but not compound, opposite.
26. Leaves almost fern-like with many deep lobes

        27. Plants annual. Leaves much divided and smooth above. COMMON RAGWEED.
        27. Plants perennial.
            28. Leaves less divided and rough above. WESTERN RAGWEED.
            28. Leaves wooly-white beneath. Fruit spiny. SKELETON LEAF BURSAGE.
    26. Leaves palmately lobed with 3 to 5 conspicuous lobes. Plants over 10 feet tall. GIANT RAGWEED.

25. Leaves entire, or not deeply lobed, alternate or opposite.
    29. Plants a twining vine. Membranous sheaths at base of leaves. Leaves alternate Seeds triangular. WILD BUCKWHEAT.
    29. Plants not a twining vine.
        30. Stems encircled by a white or brownish membranous sheath.
            31. Plants with a basal rosette of leaves plus stem leaves.
                32. Leaves alternate and arrowhead-shaped. Plant is perennial. RED SORREL.
                    33. Leaves alternate and oblong with wavy margins. CURLY DOCK.
            31. Plants without a basal rosette of leaves. Plants prostrate. Leaves blue-green. KNOTWEED.
        30. Stems not encircled by a white or brownish membranous sheath.
            32. Leaves entire, not toothed or lobed.
                33. Stems with milky juice.
                    34. Leaves alternate, simple, and linear with smooth margins. Seeds in three-lobed capsule. LEAFY SPURGE.
                    34. Leaves opposite, erect, narrow, smooth. Seed pod long and slender. HEMP DOGBANE.
                33. Stems without milky juice.
                    35. Leaves sessile (not stalked), narrow.
                        36. Upper leaves narrow or cylindrical and spine-tipped. RUSSIAN THISTLE.
                        36. Leaves narrow, pubescent. KOCHIA.
                    35. Leaves petioled (stalked). Plants little branched with main stem predominant. Flowers in terminal spikes.
                        37. Spikes dense, somewhat bristly to the touch. Stem may have reddish markings above a reddish root. Perfect flowers. ROUGH PIGWEED.
                        37. Spikes less dense, not bristly to the touch. Leaves narrow. Plants dioecious. Male plants somewhat prostrate. WATER HEMP.
            32. Leaves not entire. Margins toothed.
                38. Leaves rough with long petioles. Separate male and female flowers. Fruit covered with burs. COCKLEBUR.
                38. Leaves smooth, somewhat white on underside when young. LAMB-SQUARTER.

24. Flowers are red, pink, blue, lavender, or purple.
    39. Plants spiny or thorny.
        40. Leaves sessile (not stalked). Prickly margins. Flowers in dense, globular, prickly heads.
            41. Plants biennial. Seed stalk arising from a basal rosette of spiny leaves.
                42. Leaves glabrous. Heads bend over (nod). MUSK THISTLE.
                42. Leaves covered with fine hairs giving a gray appearance. Heads do not nod. SCOTCH THISTLE.
            41. Plants perennial. No basal rosette of spiny leaves.
                43. Plant spreads by rhizomes. Stems very branched. Leaves lobed, spiny, and hairy on underside. CANADA THISTLE.
                43. Large taproot, no rhizomes. Leaves lobed, spiny, and hairy on underside. PLUMELESS THISTLE.
        40. Leaves petioled (stalked). Leaves with spines on midrib. HORSE NETTLE.

39. Plants not spiny or thorny.
    44. Leaves opposite. Stem with milky sap. Flowers in dense cluster. Fruit a large pod. COMMON MILKWEED.
    44. Leaves alternate.
        45. Plants vining, twining, or climbing.
            46. Leaves heart-shaped. Flowers about 2" across. TALL MORNING GLORY.
            46. Leaves arrowhead-shaped.
                47. Leaves with narrow basal lobes. Flowers about 1" diameter. FIELD BINDWEED.
                47. Leaves with squared off basal lobes. Flowers 1½ to 2" diameter. HEDGE BINDWEED.
        45. Plant not vining, stems erect.
            48. Stems encircled by short membranous sheath at base of leaf. Annual plant. PENNSYLVANIA SMARTWEED.
            48. Stems with membranous sheaths. Perennial plant spreading by rhizomes. RUSSIAN KNAPWEED.

Table 12-1 lists information on weeds. A review of the terminology used in Lab Exercise 2 to describe leaves, and in Lab Exercise 3 to describe flowers will help you learn the identification of weeds that will be presented by the instructor. All weeds listed spread by seeds. Some perennials also spread by rhizomes as noted in the table under "Growth Habit". All leaves are alternate unless otherwise noted.

Table 12-1. Information on, and identification of, weeds that are common in cropland and range.

Name	Scientific Name	Growth Habit	Life Span	Plant and Inflorescence or Flowers	Seed
Quackgrass	*Agropyron repens*	Sod-former. Rhizomes.	Perennial	Spreads by seeds and rhizomes. Stems erect 2 to 3 feet tall. Pubescent basal leaves, glabrous above, auricles usually prominent. Scattered short pubescence on leaf blades. Inflorescence a slender spike.	⅜ to ½ inch long. Slender. Rachilla short, somewhat wedge-shaped. Several florets often cling together through threshing process.
Wild oats	*Avena fatua*		Summer annual	Stems erect, smooth, 1 to 4 feet tall. Leaves 3 to 8 inches long. Panicle inflorescence similar to cultivated oats. Long, dark awns.	Hairy, varied in color. Encased in hulls like cultivated oats.
Downy brome	*Bromus tectorum*		Winter annual	Stems erect, forming bunch 1 to 2 feet tall. Sheath covered with dense, short hairs. Open panicle. Florets spreading, only slightly overlapping.	Slender, bears an awn. Seed arched, bent inside out.
Hairy chess	*Bromus commutatus*		Winter annual	Stems erect, forming bunch 1 to 2 feet tall. Blade and sheath densely pubescent. Open panicle. Florets much overlapped, clinging closely.	Boat-shaped. Same size as downy brome.
Japanese brome	*Bromus japonicus*		Summer or winter annual	Stems erect, hairy. Leaf blades and sheaths have soft hairs. Open panicle. Spikelets ½ inch long on long, drooping pedicels.	Stiff awn ¼ to ½ inch long. Bends outward at maturity.
Sandbur	*Chenchrus pauciflorus*		Summer annual	Stems erect, sometime spreading. Rooting at nodes if in contact with soil. Leaves smooth, twisted. Spike inflorescence with sharp, spiny burs.	Burs, each with 1 to 3 seeds.
Large crabgrass	*Digitaria sanguinalis.*		Summer annual	Prostrate stems 4 inches to several feet long. Leaf sheaths and blades usually pubescent, light green, turning purple with first frost of autumn. Auricles absent. Whorl of three to six racemes near the top of the stems.	Small, oblong, pointed. Brown to purplish-colored.
Barnyardgrass	*Echinochloa crusgalli*		Summer annual	Stems thick, coarse, smooth, mostly erect, branching at base. Leaves smooth, light green. Panicle bearing several compact side branches, green or purplish-colored. Florets covered with short stiff bristles.	Tan to brown. Oval with longitudinal ridges on convex surface.
Witchgrass	*Panicum capillare*		Summer annual	Stems spreading and branched. Leaves covered with dense soft hairs. Panicle much branched, becoming open at maturity.	Small. Greenish or grayish. Shiny, smooth, readily separated from lemma and palea when ripe.

Name	Scientific Name	Growth Habit	Life Span	Plant and Inflorescence or Flowers	Seed
Yellow foxtail	*Setaria glauca*		Summer annual	Stems erect. Leaves flat, often having a spiral twist, with many long hairs on the upper surface of the blade near the base. Inflorescence cylindrical.	Oval, with pointed ends, smooth. Mostly yellowish, but some dark brown. Slightly larger than green foxtail
Green foxtail	*Setaria viridis*		Summer annual	Stems erect. Auricles absent. Blade glabrous. Inflorescence cylindrical. Bristles purple to yellow.	Usually thresh free of bristles. Oval, with pointed ends, smooth. Green to yellow.
Giant foxtail	*Setaria faberii*		Summer annual	Similar to yellow foxtail only larger. Short hairs on upper leaf surface, leaves flat.	Mostly green-colored. Intermediate in size between green and yellow foxtail.
Johnsongrass	*Sorghum halepense*	Warm-season. Rhizomes	Perennial	Erect, 5 to 9 feet tall. Panicle inflorescence with whorls of spreading branches. Related to sorghum and sudangrass.	Mostly enclosed in the glumes. Glumes tan to purple. Attached pedicels are knobbed and of unequal length.
Yellow nutsedge	*Cyperus esculentus*	Rhizomes with tubers	Perennial	Triangular stem. Fibrous root system. Leaves narrow, grass-like, emerging from base of stem. Panicle inflorescence. Yellow flowers.	3-sided, 1/16 inch long, blunted end, yellowish-brown
Prostrate knotweed	*Polygonum aviculare*		Summer annual	Stems bluish-green, forming a mat from taproot. Papery stipule at each node. Leaves bluish-green, oblong, smooth. Flowers very small, yellow or white in clusters at leaf axils.	Small, triangular, slender, reddish-brown.
Wild buckwheat	*Polygonum convolvulus*		Summer annual	Stems twining or creeping, branched at base. Leaves alternate, simple entire, somewhat cordate, tapering to sharp tip. Flowers inconspicuous, extending from leaf axil.	Dull, black, triangular. Sometimes part of brown calyx is present.
Pennsylvania smartweed	*Polygonum pennsylvanicum*		Summer annual	Stems 2 to 5 feet long, often smooth, swollen at nodes. Leaves simple, lanceolate, small sheath (ocher) at nodes. Pink flower clusters.	Glossy, black, flattened.
Red sorrel	*Rumex acetosella*	Rhizomes	Perennial	Short upright stem. Leaves alternate, simple, hastate. Flowers small, bright red.	Triangular; reddish-brown. Smaller than curly dock. Often covered with dull, reddish-brown floral parts.
Curly dock	*Rumex crispus*	Seeds only	Perennial	First-year plant forms dense rosette of leaves, later sends up erect stems 3 feet or more high, branched at top. Leaves alternate, simple lanceolate, wavy along margins. Stipules fused to form a sheath around each node. Flowers in whorls along ascending branches in top of plant.	Triangular, with sharp edges. Glossy reddish-brown. Smaller than wild buckwheat.

Name	Scientific Name	Growth Habit	Life Span	Plant and Inflorescence or Flowers	Seed
Common lambsquarter	*Chenopodium album*		Summer annual	Stems erect, branched, angular or ridged, glabrous. Leaves alternate, simple, somewhat triangular in shape, margin with few broad teeth, frequently with purple spots on underside and at base of petiole. Flowers small, greenish.	Sepals give it a five-pointed appearance. Seed inside is circular, black, and sometimes glossy.
Kochia	*Kochia scoparia*		Summer annual	Stems erect, much branched, may become red in autumn. Leaves alternate, simple, nearly linear, pubescent. Inconspicuous flowers in leaf axils on upper part of plant.	Flattened, egg-shaped with winglike hulls, unless threshed off.
Russian thistle	*Salsola kali*		Summer annual	Stems branched, often reddish-striped. Leaves alternate, simple, awl-shaped, stiff. Flowers small, green in the axils of the leaves.	Calyx nearly covers fruit, giving it a bell or umbrella shape.
Water hemp	*Acnida altissima*		Summer annual	Stem smooth, erect, 1 to 5 feet tall. Leaves narrowly ovate to lanceolate, long-petioled. Flowers small, greenish. Dioecious plant. Male plants more prostrate.	Enclosed in a bladder-like hull. One seed, darkened, oval.
Rough pigweed	*Amaranthus retroflexus*		Summer annual	Stems erect, branched, somewhat hairy, and may have reddish markings above a reddish root. Leaves alternate, simple, ovate. Flowers in dense terminal panicle.	Ovate, very small glossy. Black.
Wild mustard	*Brassica kaber*		Winter annual	Stem erect with spreading branches, sparsely pubescent. Leaves alternate, pubescent, somewhat ovate with irregular lobing, usually deeply lobed at base. Flowers yellow. Seed pods long and slender.	Small, globular, resembling bird shot. Color varies from black to reddish-brown.
Shepherd's purse	*Capsella bursa-pastoris*		Summer or winter annual	Stems erect, covered with gray hair. Leaves in rosette at base are coarsely lobed. Leaves on stem are sessile, clasping stem with pointed lobes, coarsely serrate. Inflorescence an elongated raceme. Seed pod triangular, two-parted, about ¼ inch long.	Small, yellowish, and shiny.
Hoary cress	*Cardaria draba*	Rhizomes	Perennial	Stems erect, pubescent, branched. Leaves alternate, simple, dentate margins. Flowers white, borne in finely branched clusters. Seeds borne in two-celled heart-shaped pods.	Slightly flattened, egg-shaped. Reddish-brown, one side may be grooved.
Virginia pepperweed	*Lepidium virginicum*		Summer or winter annual	Rosette growth early, later stems erect, branched. Leaves alternate, lanceolate, pinnately lobed. Flowers white, many small two-seeded pods in inflorescence.	Small, flattened, egg-shaped. Orange to reddish-brown.
Field pennycress	*Thlaspi arvense*		Summer or winter annual	Stems erect, branched at top, glabrous. Leaves alternate, simple, dentate. Few branches bearing white flowers and seed pods extended upward. Seeds borne in heart-shaped pods containing 10 to 12 seeds.	Flattened, disc-shaped, with 10 to 14 curved ridges on either side, giving a fingerprint appearance. Dark brown to black.

Name	Scientific Name	Growth Habit	Life Span	Plant and Inflorescence or Flowers	Seed
Puncturevine	*Tribulus terrestris*		Summer annual	Stems hairy, prostrate, branching from the base and extending in every direction to form a dense mat 6 to 8 feet across. Pinnately compound, bright green leaves covered with silky hair. Small yellow flowers in axils of leaves.	Pod a cluster of five burs, each with two sharp, long, rough spines which break apart at maturity.
Leafy spurge	*Euphorbia esula*	Rhizomes	Perennial	Stems erect, branched at top, glabrous, with milky sap. Leaves alternate, simple, linear with smooth margins. Greenish, paired, heart-shaped sepals in inflorescence. Seeds borne in three-lobed capsule.	Oval, smooth. Light gray to dark brown.
Velvetleaf	*Abutilon theophrasti*		Summer annual	Stems erect 3 to 8 feet tall. Leaves large, heart-shaped, covered with soft velvety hairs. Flowers yellow, borne in leaf axils, about 1 inch across.	Dull gray, somewhat heart-shaped, approximately ⅛ inch long.
Common milkweed	*Asclepias syriaca*	Rhizomes	Perennial	Stems grow 2 to 5 feet tall and exude a milky sap when broken. Leaves large, opposite, oblong with prominent veins.	Borne in pods 3 to 5 inches long. Seed brown, flat, oval, winged.
Hemp dogbane	*Apocynum cannabinum*	Rhizomes	Perennial	Stems erect, 1 to 2 feet tall and exude a milky sap when broken. Leaves erect, opposite, narrow, smooth. Flowers greenish-white with petals only slightly longer than sepals. Seed pod long and slender.	Thin and flat, with a tuft of soft silky hairs at one end.
Sunflower	*Helianthus annuus*		Summer annual	Stout, erect stems, 2 to 5 feet tall, many branches. Leaves rough, hairy, with serrated margins. Head inflorescence with yellow ray flowers and dark brown disk flowers, 1 to 5 inches in diameter.	1/8 to ½ inch long, oval to wedge-shaped and flattened. White to dark brown with lighter stripes or spots.
Field bindweed	*Convolvulus arvensis*	Rhizomes	Perennial	Stems slender, prostrate or twining, glabrous. Leaves with almost parallel sides and narrow basal lobes (blunt-pointed). Flowers bell-shaped, white to pink or purple, 1 inch or less in diameter, 2 small bracts on flower stalk.	Three-angled, flat on two sides. Seed coat roughened, dark gray to reddish-brown.
Hedge bindweed	*Convolvulus sepium*	Rhizomes	Perennial	Stems twining. Leaves alternate, simple, entire, hastate. Basal lobes squared off to give two points on each side. Flowers 1½ to 2 inches across on long peduncle with large bracts at base, white to rose in color.	Plump, larger than field bindweed. One side nearly flat. Black to dark brown.
Field dodder	*Cuscuta compestris*		Parasitic annual	Stems threadlike, yellow, with small suckers which attach to the host plant. Leaves only minute scales. Small white flowers in dense clusters.	About the size of red clover, one side rounded, other flattened. Yellow to brown and roughened with minute pits.

Name	Scientific Name	Growth Habit	Life Span	Plant and Inflorescence or Flowers	Seed
Ground cherry	*Physalis heterophylla*	Rhizomes	Perennial	Stems hairy, erect, forming a bush 1 to 3 feet tall. Leaves 2 to 3 inches long, hairy, crenate margins. Flowers bell-shaped, 5-lobed, greenish-yellow, dropping. Fruit round, berry-like, yellow, enclosed in papery calyx.	Small, yellow, flattened, oval, 1/16 inch diameter.
Black nightshade	*Solanum nigrum*		Summer annual	Stem smooth, erect, very branched, 1 to 3 feet tall. Leaves ovate, wavy margins. Flowers white, 5-lobed, 1/4 inch diameter in small clusters. Fruit berry-like, black at maturity	Flattened, 1/16 inch diameter, yellow to dark brown.
Horsenettle	*Solanum carolinense*	Rhizomes	Perennial	Stems erect, with yellowish spines. Leaves alternate, simple, ovate, and irregularly lobed, yellow spines on veins, midrib, and petiole. Flower violet or white. Fruit resembles small yellow tomato, smooth globular, full of many seeds.	Oval, flattened. Yellow to light brown, like small tomato seeds.
Buffalobur	*Solanum rostratum*		Summer annual	Stem erect, branched, hairy, densely covered with long, stiff, yellow spines. Leaves petioled, densely hairy with round lobes and spiny. Flowers yellow.	Enclosed in spiny bur. Seeds round, flattened. Dull, brown-black.
Common ragweed	*Ambrosia artemisiifolia*		Summer annual	Stems rough, hairy, erect, branched. Leaves nearly smooth, deeply cut into a number of lobes. Flowers imperfect. Male flowers in small inverted clusters at tips of branches, female flowers fewer, borne at the bases of leaves in forks of upper branches.	Enclosed in woody hull. Top-shaped, pointed, with several longitudinal ridges ending in short, spiny projections. Light brown.
Western ragweed	*Ambrosia psilostachya*	Rhizomes	Perennial	Stems hairy, bushy, many branched, less than 2 feet tall. Leaves rough, very deeply lobed, sometimes compound. Flowers imperfect, similar to common ragweed.	Similar to common ragweed, only more rounded and only slightly ridged.
Giant ragweed	*Ambrosia trifida*		Summer annual	Stems coarse, rough, may be 12 to 18 feet tall. Leaves opposite, large, slightly hairy, 3 to 5 lobes. Flowers imperfect, similar to common ragweed.	Similar to common ragweed, only larger with less prominent ridges.
Plumeless thistle	*Carduus acanthoides*	Seed only	Perennial	Large taproot. Stem erect, branched. Leaves much like musk thistle except underside hairy, narrower, more deeply lobed, more finely divided. Flowers reddish-purple, ½ to 1 inch in diameter.	Similar to musk thistle, but only ⅓ the size.

Name	Scientific Name	Growth Habit	Life Span	Plant and Inflorescence or Flowers	Seed
Musk thistle	*Carduus nutans*		Biennial	Stems erect 2 to 6 feet high. Leaves glabrous, deeply lobed, 3 to 6 inches in length, and very spiny. Solitary head inflorescence, nodding at the end of the stem or branches, 1½ to 2½ inches across, purple in color and fragrant.	About ⅛ inch long, oblong, one edge curved while other is almost straight. Pointed at tip with the base rounded toward a small, depressed scar. Surface longitudinally grooved and shiny. Yellowish-brown in color.
Russian knapweed	*Centaurea repens*	Rhizomes	Perennial	Stem erect and branching. Leaves alternate, simple. Upper leaves linear with entire margin, basal leaves linear with dentate margin. Flowers thistle-like, yellow to purple, ¼ inch in diameter.	Flattened, oval, ⅛ inch long. Ivory to greenish color.
Canada thistle	*Cirsium arvense*	Rhizomes	Perennial	Dioecious. Stems erect, branched, nearly glabrous. Leaves alternate, oblong, very irregularly lobed, hairy beneath, spiny-toothed margins. Flowers purple to white, lacking spines, ½ inch or less.	Straight to banana-shaped, smooth. Light to dark brown. No lines on seed.
Scotch thistle	*Onopordium acanthium*		Biennial	Stems erect 1 to 8 feet high, covered with spines. Leaves covered with dense, fine hair which gives grayish appearance. Leaves oblong, lobed, very spiny, often 12 inches or more long, and extend into stem. Purplish flowers 1 to 2 inches in diameter, solitary, but do not nod.	About 3/16 inch long. Wrinkled surface gives ridged appearance. Gray-brown.
Morning glory	*Ipomoea purpurea*		Summer annual	Stems slender, hairy, and viney, 4 to 10 feet long. Leaves 2 to 4 inches broad, deeply cordate, pointed at tip. Flowers white to pink to purple, bell-shaped, usually larger than those of field bindweed.	Three-angled, flat on two sides. Dull brownish black. Similar in appearance, but larger than field bindweed.
Skeleton leaf bursage	*Franseria discolor*	Rhizomes	Perennial	Lobed leaves are glabrous and green above, but woolly white beneath, deeply lobed.	Enclosed in small burs, with several stout distinctly hooked spines.
Cocklebur	*Xanthium pennsylvanicum*		Summer annual	Stem rough, hairy, spotted, 2 to 4 feet tall. Leaves triangular, toothed or lobed, rough with long petioles. Plant monoecious. Male and female flowers together in clusters in upper leaf axils. Fruit a hard, prickly, bur.	½ inch long, dark brown, slender.

STUDY GUIDE

Complete the following table to help you learn the material in this exercise:

	Scientific name	Life Cycle	Propagation	Classification
Quackgrass				
Wild oats				
Downy brome				
Hairy chess				
Japanese brome				
Sandbur				
Large crabgrass				
Barnyardgrass				
Witchgrass				
Yellow foxtail				
Green foxtail				
Giant foxtail				
Johnsongrass				
Yellow nutsedge				
Prostrate knotweed				
Wild buckwheat				
Penn. smartweed				
Red sorrel				
Curly dock				
Common lambsquarter				
Kochia				
Russian thistle				
Water hemp				
Rough pigweed				
Wild mustard				
Shepherd's purse				
Hoary cress				

	Scientific name	Life Cycle	Propagation	Classification
Virginia pepperweed	____________	____________	____________	____________
Field pennycress	____________	____________	____________	____________
Puncturevine	____________	____________	____________	____________
Leafy spurge	____________	____________	____________	____________
Velvetleaf	____________	____________	____________	____________
Common Milkweed	____________	____________	____________	____________
Hemp dogbane	____________	____________	____________	____________
Sunflower	____________	____________	____________	____________
Field bindweed	____________	____________	____________	____________
Hedge bindweed	____________	____________	____________	____________
Field dodder	____________	____________	____________	____________
Ground cherry	____________	____________	____________	____________
Black nightshade	____________	____________	____________	____________
Horsenettle	____________	____________	____________	____________
Buffalobur	____________	____________	____________	____________
Common ragweed	____________	____________	____________	____________
Western ragweed	____________	____________	____________	____________
Giant ragweed	____________	____________	____________	____________
Plumeless thistle	____________	____________	____________	____________
Musk thistle	____________	____________	____________	____________
Russian knapweed	____________	____________	____________	____________
Canada thistle	____________	____________	____________	____________
Scotch thistle	____________	____________	____________	____________
Morning glory	____________	____________	____________	____________
Skeleton leaf bursage	____________	____________	____________	____________
Cocklebur	____________	____________	____________	____________

Index

C

G

H

M

N

O

P

Q

R

S

T

U

V

W

X

Y